EXPERIMENT, THEORY, PRACTICE

BOSTON STUDIES IN THE PHILOSOPHY OF SCIENCE

EDITED BY ROBERT S. COHEN AND MARX W. WARTOFSKY

VOLUME 46

PETER LEONIDOVICH KAPITZA

P. L. KAPITZA
Institute for Physical Problems, Academy of Sciences, Moscow, U.S.S.R.

EXPERIMENT, THEORY, PRACTICE

Articles and Addresses

D. REIDEL PUBLISHING COMPANY
DORDRECHT : HOLLAND / BOSTON : U.S.A.
LONDON : ENGLAND

Library of Congress Cataloging in Publication Data

CIP

Kapiťsa, Petr Leonidovich, 1894–
Experiment, theory, practice.

(Boston studies in the philosophy of science ; v. 46)
Translation based on the 2d ed. author's
Ėksperiment, teoriia, praktika, published in 1977.
Includes bibliographies and index.
1. Physics–Addresses, essays, lectures. 2. Physicists–Biography–Addresses, essays, lectures. 3. Science–Russia–Addresses, essays, lectures. I. Kapiťsa, Petr Leonidovich, 1894– Ėksperiment, teoriia, praktika. II. Title. III. Series.
Q174.B67 vol. 46 [QC71] 501s [530] 79-27617
ISBN 90-277-1061-9
ISBN 90-277-1062-7 pbk.

Published by D. Reidel Publishing Company,
P.O. Box 17, 3300 AA Dordrecht, Holland

Sold and distributed in the U.S.A. and Canada
by Kluwer Boston Inc., Lincoln Building,
160 Old Derby Street, Hingham, MA 02043, U.S.A.

In all other countries, sold and distributed
by Kluwer Academic Publishers Group,
P. O. Box 322, 3300 AH Dordrecht, Holland

D. Reidel Publishing Company is a member of the Kluwer Group

This English translation is based on the second Russian revised and updated edition of 1977, edited by A. S. Borovik-Romanov and P. E. Rubinin (Nauka, Moscow). To the complete contents of that edition, several texts have been added by the author for this volume (Papers 9, 40, 41). This English volume has been edited by Robert S. Cohen.

Printed in The Netherlands

EDITORIAL PREFACE

In this splendid collection of the articles and addresses of P.L. Kapitza, the author remarks on the insight of the 18th century Ukrainian philosopher Skovoroda who wrote: "We must be grateful to God that He created the world in such a way that everything simple is true, and everything complicated is untrue." At another place, Kapitza meditates on the roles played by instinct, imagination, audacity, experiment, and hard work in the development of science, and for a moment seems to despair at understanding the dogged arguments of great scientists: "Einstein loved to refer to God when there was no more sensible argument!"

With Academician Kapitza, there are reasoned arguments, plausible alternatives, humor and humane discipline, energy and patience, a skill for the practical, and transcendent clarity about what is at issue in theoretical practice as in engineering necessities. Kapitza has been physicist, engineer, research manager, teacher, humanist, and this book demonstrates that he is a wise interpreter of historical, philosophical, and social realities. He is also, in C. P. Snow's words, strong, brave, and good (*Variety of Men,* N.Y. 1966, p. 19).

In this preface, we shall point to themes from Kapitza's interpretations of science and life.

On scientific work. Good work is never done with someone else's hands. The separation of theory from experience, from experimental work, and from practice, above all harms theory itself. Can there be too much theory? None of the work of theorists acquires value until after experimental verification; and yet theory can lose this insight, can begin to work on its own, and the value placed on it depends then upon methodological and esthetic considerations.

Understanding the phenomenon proceeds from the phenomenon itself; the determining role of the experiment in the study of phenomena; the need for unity of theory and experiment as a prerequisite for the harmonious development of science; a negative attitude toward any dogmas in science, and the need for new ideas based upon experiment; recognition of the inexhaustibility of matter.

The manifestations of human creative genius are always beautiful!

On education in science. Science is like music for non-experts: enjoying it, living it, being interested in it.

Most university professors give top marks to those students who *know* most rather than to those who *understand* best.

Who will teach the teacher, if not his own student? Only they, our students, can save us from a precocious stagnation of the brain.

Never try to prevent scientific arguments. . . to the contrary, people should even be goaded a bit into a healthy intense discussion.

We cannot plan without seeking and locating talented people; we cannot make a department for writing cantatas unless we have a Handel!

Teach! . . . in school, in industry, to engineers, everywhere, for adopting new scientific techniques is not simply an administrative act. Rather this process needs to be approached as one would approach a 'pedagogical poem'. [How reminiscent of the great Marxist teacher, A. S. Makarenko, who stated in his own pedagogical poem, *The Road to Life,* that his "fundamental principle has always been: *the utmost possible demands on a person, but at the same time the utmost possible respect for him.*"]

How to educate? The most satisfied are those who are creative – scientists, artists, theater directors, with no division between work and leisure. But creativity is broad and "it emerges in any form of human activity where man has to decide for himself what to do, rather than follow instructions." Make the free time of ordinary persons creative!

On the history of science. Rutherford, who predicted the main trends of nuclear physics, worked not through intuition but by deep thinking and thorough experimentation.

The tragedy of Lomonosov's isolation was redoubled by the lack of a scientific community at home; only a mature scientific community can evaluate the cognitive power of a scientific advance independently of its immediate practical significance.

The creative work of a scientist cannot develop successfully outside a collective.

How could Benjamin Franklin, who was never earlier engaged in physics, lead the development of an entire branch of science in only a few years of work, in a small American town far from the centers of international science, and being already a mature man, standing by himself?

Not great erudition, but imaginativeness, perspicacity, audacity. . . at the initial stages, the precision and punctuality normal for professional scientists may rather hamper the advancement of daring assumptions.

The tragedy of isolation: Cavendish was the first to discover Coulomb's law, far more accurately than Coulomb was to do, and Ohm's law long before Ohm. [An individual isolation like that of the genius Leonardo, not the isolating social circumstances of the great Lomonosov.]

People are divided into three categories: some stand in front, and spend their powers to further science, culture and humanity – these are the progressive people. Others, the majority, go along with progress, at the side, they do not interfere and do not help; and then finally there are people who stand behind and hold back, conservators, cowardly and without imagination. For those who go ahead it is hardest of all. . . one such was Paul Langevin.

'New phenomena': such as cannot possibly be predicted on the basis of theoretical concepts already in existence: Galvani, Oersted, Hertz, Becquerel, Thomson, Michelson and Morley, Hess, Meitner and Hahn. . . we must not believe that no new discoveries will be made in the future.

The great role of 'methodological inventions'; one of the greatest being electronic cybernetic machines that bring problems within control that were beyond accessibility.

Evolution, guided by the 'wisdom of nature' is based on 'trial and error'; man started to transform his environment by the same method but this process of nature's conquest by man was consolidated by his starting to generalize the experience of successful tests, storing and transmitting them to other people. . . the mechanism for social inheritance.

Trial and error are still the basis of the study of nature, but the process which limits the number of trials and errors necessary for the solution of a given problem can be regarded as the beginning of a scientific approach.

Separation of theory from practice causes the time lag in the introduction of a scientific discovery into practical life: the modern laser is based on the phenomenon of stimulated emission, the basic theory of which was provided by Einstein in 1916. Scientific experimental technique was sufficiently advanced at that time for the laser to have been built then.

Open problems in science. All life processes are electromagnetic in nature and involve the passage of electric current along nerve fibres. But we know next to nothing about the physical nature of these processes. . . I shall remain a sceptic until the polymer fibres – nerves for the transmission of information and muscles capable of contraction – are made artificially in the laboratory. Only then will these processes really be understood.

. . . the energy problem, which has become the central problem of our epoch, is naturally the number one problem of physics.

The main task facing the physicist is the deeper experimental study of the hydrodynamics of a hot plasma as is needed for reaching a thermonuclear reaction at high pressure and strong magnetic fields.

The future of science? From chance to plan in the course of the last 50 years, from development by private initiative to an integral part, a necessary component, of the social system.

The most important scientific problems of the near future: conquest of outer space; nuclear waste disposal; cheap electric energy; controlled thermonuclear reactions. . . creation of polymers with specified properties. Our age will not only be the age of the utilization of atomic energy, but also the age when man learns to produce polymers. . . the main 'building material' in living nature.

Only one feeling - the sense of small, i.e. the determination and detection of small quantities of organic matter, is more perfect in the animal than by existing instruments. To 'surpass a dog's smell' is one of the problems of physicists in the future.

Is present-day knowledge of inanimate nature adequate for describing all the phenomena characteristic of the living world? Most phenomena are described by existing laws, but all the same I dare say that one of the main properties of living Nature - to reproduce itself - may be the manifestation of various forces in Nature which as yet are still unknown and unexplained by the known laws of interaction between elementary particles.

Global scientific problems of the immediate future: the technical-economic aspect, linked with depletion of natural resources; the ecological aspect, namely biological equilibrium between man and living nature in an environment contaminated on a global scale; a social-political aspect, in recognition of the need to solve these problems on a global scale.

Nature must be cured of diseases just as people are.

Modern technology is capable not only of detroying life on a global scale, but also of stimulating it.

We, in the U.S.S.R., can exploit the Lake Baikal resources without disturbing the natural equilibrium - something the capitalist countries have been unable to achieve. Here, too, experience shows the superiority of socialism over capitalism in solving such ecological problems.

In ancient times global problems were described in artistic and religious books . . . the Last Judgement, the Universal Flood.

Problems in the social sciences. The life of the socialist state is being built and developed on a scientific basis. [In *Pravda*, a few days before the German invasion of the U.S.S.R.]

The causes of war are amenable to scientific analysis. . . they can be identified.

. . . the most progressive social order is the one which allows a most balanced development of the spiritual qualities of the individual as a necessary foundation for a happy life.

. . . just as successful development of biological sciences must be based on physics and chemistry, so also the science of the laws underlying the organization of society must be based on the science of man's higher nervous activity. . . . Only on this scientific basis can the right forms of organization be sought for man's toil and leisure.

And - this is the main thing - only on a scientific basis can a healthy and effective structure of society be created.

Why even in our time, the time of the scientific-technological revolution, are the social sciences evolving so feebly? I dare say that the more natural explanation is, as in all such cases, the unfavorable conditions existing right now. . . somewhat similar to those which applied to the natural sciences in the Middle Ages, when the Church took upon itself a monopoly of scholastic-dogmatic interpretation of all natural phenomena.

Marx founded the basic science of the laws of development and construction of human society - historical materialism. To this is related the science of man's higher nervous activity, founded by Freud and Pavlov, the first to place the study of thought processes on an experimental basis. By them were found the laws of man's perception of the external medium, the emergence of conditioned reflexes, and the influence of the subconscious on man's activity.

The future of humanity will be determined by the natural selection of social structures themselves, and this selection will, in the end, determine the fate of mankind. Probably for the time being it is not possible for man to influence this selection, and we must here still rely on being governed by the 'wisdom of nature'.

Modern society strives to organize people's lives in such a way that they will have an *impression* of freedom despite the lack of it . . . the *Brave New World* of Aldous Huxley.

There is a certain similarity between the impact of science and the impact of art on the organization of social life. Scientific laws are a generalization of experience, fixed for all time and internationally accepted. Similarly, art is a generalization of processes which take place in the lives of people and of society. A truly great work of art . . . becomes international and everlasting, an impact at least as great as the impact of science. Cocteau: "Poetry is indispensible, although why – I know not."

Which state structure is most advanced in the evolutionary development of mankind?. . . Such an evaluation should be based on the quality of a country's spiritual culture.

Science and society. Science, too, needs propaganda and public relations work if it is to take root among the people.

Pure science must provide the knowledge necessary for transforming nature into a tool for the progress of human culture.

Practical results? The man who picks the apple is given all the credit, whereas those who have planted the tree and cultivated it are largely ignored.

We have not yet managed to make proper use of the enormous freedom for scientific thought in our country.

Errors do not constitute a pseudo-science. To the contrary, pseudo-science is non-recognition of errors, a major impediment to the healthy growth of science.

. . . creativity, the most valuable element in science; planning must accordingly stimulate freedom of creation rather than restrict it. Planning and creativity are not necessarily contradictory.

Soviet and capitalist science differ; this difference does not apply to the contents of research, since the discovery of a natural phenomenon or its correct theoretical explanation are unrelated to the place or time of their occurrence. The main distinction of our (Soviet) science lies in its purposefulness and in the evaluation of its achievements.

. . . in the year that Rutherford died there disappeared forever the happy days of free scientific work which gave us such delight in our youth. Science has lost her freedom. Science has become a productive force. . . . rich but enslaved, part of her veiled in secrecy.

On dialectics. The use of dialectics in natural science requires an exceptionally deep knowledge of experimental facts and their theoretical generalization. It is, as it were, a Stradivarius violin, the most perfect of violins, but to play on it, one needs to be a musician and know music. Without this it will play out of tune, the same as an ordinary violin.

. . . everything born has to die – such is the dialectics of nature. Death spares no-one – infusoria, men, institutions, states. . . eventually our entire planetary system.

. . . as for dialectical thinking, the teacher can stimulate it by citing examples in which discrepancies between theoretical concepts and experimental data have led to major discoveries in physics.

On philosophy. We scientists tend to underestimate these philosophical questions of our time.

. . . our philosophers must participate in the (world ideological) struggle on an equal footing, just as our sportsmen do . . . in this struggle our ideologists will lose the privilege they enjoy in our country, where they have no opposite views to contend with. . . they will have to show their own worth.

. . . religion is capable of a blithe disregard for the laws of causality and consequently provides answers to problems that can have no scientific solution. . . it is possible for many religions to exist, whereas science - like the multiplication table - is unique.

. . . a person's spiritual happiness is associated with the feeling of freedom.

This volume depends upon the careful work of the editors of the several Russian editions, A. S. Borovik-Romanov and P. E. Rubinin, to whom we are most grateful. Translations have been checked and revised for this edition, but to a substantial extent they derive from earlier English publications as shown in the Acknowledgments. In particular, attention should be drawn to Professor D. ter Haar's edition in three volumes of the *Collected Papers of P. L. Kapitza* (Pergamon Press, London and New York, 1964, 1965, 1967).

We are pleased to acknowledge the fine editorial craftsmanship of Dr D. J. Larner, as well as the immensely skilled editorial assistance of Carolyn R. Fawcett. Ms. Fawcett also prepared the index. We are grateful to Dr Rubinin for providing the materials for the Bibliography of the published writings of Academician Kapitza. Readers will note that there are English translations of many of the published journal papers in the regular translation journals from the Russian that have been published by the American Institute of Physics for several decades. [In the present volume, notes in square brackets have been provided by the editor.]

Center for Philosophy and History of Science ROBERT S. COHEN
Boston University MARX W. WARTOFSKY
February 1980

TABLE OF CONTENTS

PART FOUR

PART FIVE

PART SIX

PART SEVEN

ACKNOWLEDGMENTS

The editor and publisher of the English translation of *Experiment, Theory, Practice* acknowledge with thanks the permission granted by the following institutions to use translations, in whole or in part, and other materials from their publications:

American Institute of Physics, New York, N.Y. for articles that first appeared in *Soviet Physics-Uspekhi*;

Paper 8, 'Energy and Physics' (from *Sov. Phys.-Usp.* **12**, 169–173 (1976))

Paper 19, 'The Centenary of the *Journal of Experimental and Theoretical Physics* and the Role of Journals in the Development of Science' (from *Sov. Phys.-Usp.* **16**, 928–933 (1974))

Paper 32, 'Lomonosov and World Science' (from *Sov. Phys.-Usp.* **8**, 720–728 (1966) with a new Foreword prepared by the author for the Russian edition of this volume)

Paper 42, 'Global Problems and Energy' (from *Sov. Phys-Usp.* **20**, 547–553 (1977))

American Academy of Arts and Sciences, and *Daedalus*, Journal of the Academy, Boston, Massachusetts:

Paper 20, 'Basic Factors in the Organization of Science, and How They are Handled in the U.S.S.R.', first published in the issue of Spring 1973, *The Search for Knowledge*.

The Nobel Foundation, Stockholm:

Paper 9, 'Plasma and the Controlled Thermonuclear Reaction', which also appears in *Les prix Nobel en 1978* as well as in *Nobel Lectures* (Elsevier, Amsterdam and New York, 1979) and *Science* **205**, 959–964 (7 September 1979).

The Royal Society of London:

Paper 30, 'Recollections of Lord Rutherford', (*Proc. Roy. Soc. A* **294**, 123–137 (1966))

Paper 37, 'Lev Davydovitch Landau', (*Biog. Memoirs of Fellows of the Royal Society* **15**, 141–146 (1969))

Paper 43, 'Scientific and Social Approaches for the Solution of Global Problems' (*Proc. Roy. Soc. A* **357**, 1–14 (1977))

The Oxford University Junior Scientific Club:

Paper 1, 'The Production of and Experiments in Strong Magnetic Fields', *Transactions of the Oxford University Junior Scientific Club*, 5th series, no. 4, 129–133 (1931)

Pergamon Press, Oxford and New York, for articles that appear in the *Collected Papers of P. L. Kapitza*:

Paper 16, 'Experiment, Theory, Practice' (adapted from *Coll. Papers* **3** 185–189, as revised by the author for the Russian edition of this volume)

Paper 17, 'Effectiveness of Scientific Work' (from *Coll. Papers* **3**, 215–217)

Paper 18, 'Applying the Achievements of Science and Engineering', (adapted from *Coll. Papers* **3**, 218–224)

Paper 33, 'The Scientific Activity of Benjamin Franklin' (adapted from *Coll. Papers* **3**, 141–152)

Paper 34, 'The Physicist and Public Figure, Paul Langevin' (from *Coll. Papers* **3**, 208–214)

Paper 40, 'The Future of Science' (from *Coll. Papers* **3**, 190–206)

FOREWORD

Peter Leonidovich Kapitza is a scientist with a wide range of interests. A major experimental physicist, he has made substantial contributions to the physics of magnetic phenomena, the physics and technology of low temperatures, quantum physics of the condensed state, electronics and plasma physics.

Peter Leonidovich was born on July 9, 1894 in the family of a military engineer. He began his scientific work under the professorship of A. F. Ioffe at the Electromechanics Faculty of the St. Petersburg Polytechnic Institute, from which he graduated in 1918.

In his first original scientific work Peter Leonidovich devised a new method for the preparation of Wollaston fibres – fine platinum or gold wires (less than 1 micron thick) made by drawing in a silver envelope which is subsequently dissolved. P. L. Kapitza applied an electrolytic process for dissolving the silver, thus reducing the risk of rupture of the fine wires. In his next work P. L. Kapitza proposed an original model of an X-ray spectrometer in which the intensity of X-rays reflected by the crystal was made several times greater as a result of their focusing by the curved cylindrical surface of the crystal.

The third published work of P. L. Kapitza was made together with N. N. Semenov. It proposed a method for the determination of the magnetic moment of the atom on the basis of the interaction between an atomic beam and a heterogeneous magnetic field. Later this procedure was carried out in the well-known experiments of Stern and Gerlach.

In 1921, P. L. Kapitza was sent for research work in Britain, where he spent a long time at the Cavendish Laboratory of Cambridge University; the director of the laboratory was Ernest Rutherford. In 1923, Peter Leonidovich for the first time placed a Wilson cloud chamber in a strong magnetic field and observed a bending of the trajectory of alpha particles. In the course of this research, he realized the need for the creation of hyperstrong magnetic fields. He showed that electromagnets with iron cores are useless for this purpose and coils traversed by a very strong current must be used instead. The main difficulty here stems from the overheating of these coils. P. L. Kapitza proposed an original method for overcoming this obstacle – namely, by the creation of brief magnetic fields following the passage of a very strong current through the coil; during such brief intervals the coil does not overheat.

After trying a variety of current sources he selected a motor-generator of a special design. In this generator the energy required for creating the magnetic field was accumulated in the form of the kinetic energy of the rotor. In his device Peter Leonidovich was able to obtain a magnetic field strength of 320 kilo-oersted with a pulse duration of the order of 10 milliseconds. This principle of the creation of pulse fields has gained a wide acceptance in many laboratories. With the progress of technology it became possible to store energy in condensers; in terms of the magnitude of the magnetic energy obtained in the coil, however, Kapitza's results remain unsurpassed. Kapitza also devised original procedures for the measurement of various physical parameters in pulse fields.

One of the most important results of Kapitza's research into the physical properties of matter in strong magnetic fields was his discovery of a linear relationship between the magnetic field and the electric resistance of a number of metals subjected to very strong magnetic fields. This law, discovered by Kapitza in 1928, found its theoretical explanation some 30 years later following the discovery of the intricate topological structure of Fermi surfaces in metals.

P. L. Kapitza also studied the magnetostriction of para- and diamagnetic substances in strong magnetic fields and discovered the abnormally high magnetostriction of bismuth monocrystals. He observed the very high anisotropy of this magnetostriction: following the imposition of a magnetic field along the trigonal axis, the bismuth expanded in the direction of the field and contracted when the field was applied perpendicularly to the axis. Kapitza also studied the Zeeman effect in strong magnetic fields and observed the Paschen-Back effect.

Later P. L. Kapitza embarked upon the study of low temperature physics. Here too he began with a critical appraisal of the available methods for obtaining low temperatures and designed a new original installation for the liquefaction of helium. In this device P. L. Kapitza managed to eliminate the need for a preliminary cooling of the helium with liquid hydrogen. Instead, the helium in the new device was cooled by performing work in a specially designed compressed-gas motor. A distinctive feature of this piston-type motor was that the gaseous helium also served as lubricant.

Practically all modern helium liquefiers are based on the design proposed by P. L. Kapitza. He was appointed director of the newly built Mond Laboratory of the Royal Society in Cambridge, which was intended for research into strong magnetic fields and low temperatures.

In 1934, P. L. Kapitza returned to Moscow and founded the Institute of

Physical Problems, the equipment of which was purchased from the Mond Laboratory by the Soviet government with the assistance of Ernest Rutherford. There he continued his work on strong magnetic fields and various physical and technological aspects of low temperatures.

In the field of low temperature technology, P. L. Kapitza proposed a new method for air liquefaction with a low pressure cycle using a special turbo-expansion device with a high efficiency.

The highly efficient radial compressed gas turboengine developed by Kapitza, with an output of 80–85%, still serves as a world model for modern large-scale oxygen production plants by air fractionation using low pressure only.

Large low-pressure air fractionation plants ranging in capacity from 10,000 to 65,000 m^3 of oxygen per hour are in operation and under construction in the U.S.S.R. In the Western industrialized countries, low-pressure air fractionation turbines based on Kapitza's design produced about 53 billion m^3 of oxygen in 1970. Nearly one half of the oxygen obtained is used in ferrous and nonferrous metallurgy. Other large-scale applications of oxygen include the chemical industry and rocket technology.

It is pertinent to note in this connection that Kapitza's research in extreme fields and liquefiers exemplify a rare combination of a major scientist and a major engineer in one person.

Kapitza was among the first who applied large technical installations in the laboratory and at the same time made direct practical use of some of the latest achievements of physics. He was a true pioneer of what has become today a characteristic trait of the recent scientific-technological revolution.

In the field of low-temperature physics P. L. Kapitza began a series of elegant experiments on the properties of liquid helium. As a result of these studies he discovered the superfluidity of helium in 1937. He found that the viscosity of liquid helium at temperatures below 2.19 K during its passage through narrow slits is so far below that of any low-viscosity liquid known that it can be regarded as equal to zero. Consequently he named this state of helium superfluid. In the course of his research into the aberrant properties of liquid helium Kapitza designed a series of extremely sensitive and instructive experiments which demonstrate the highly unusual properties of liquid helium at temperatures below 2.19 K.

Kapitza's work on the properties of liquid helium constitutes a brilliant example of the approach applied by an experimental physicist toward the solution of a complex problem. In reading his articles one experiences an aesthetic satisfaction, being able to follow in Kapitza's steps, one experiment after another, culminating in the major discovery of the existence in helium of two liquids having totally different properties and capable of moving against one another.

This discovery opened a new trend in physical research – namely, the quantum physics of the condensed state. For its explanation, new quantum concepts had to be introduced – the so-called elementary excitations or quasiparticles.

In his studies of heat transfer in liquid helium Kapitza established an important fact – namely, that in the course of heat transfer from a solid body to liquid helium a temperature jump develops in the boundary zone, the magnitude of which rises sharply with the decline of temperature – the so-called Kapitza jump. This finding also demonstrated the need for a quantum approach to the understanding of the apparently classical [heat] transfer phenomenon.

In 1946, the method for the production of oxygen proposed by Kapitza was unjustly condemned, and he was dismissed from his post of director of the Institute for Physical Problems; Kapitza was even forbidden to work in the Institute which he himself had created. During these difficult years Kapitza showed great courage; in his own dacha, he set up a small domestic laboratory where he undertook intensive research. At first he undertook a series of elegant studies in mechanics and hydrodynamics. In the late 1940s, however, his interest turned to a totally different realm of physics – namely, the prospects for the creation of powerful continuous-action UHF oscillators. Kapitza was able to solve the intricate mathematical problem of electron movement in UHF oscillators of the magnetron type.

On the basis of these calculations he built UHF oscillators of a new type – the planotron and the nigotron. The output of the nigotron attains the record value of 175 kW in a continuous operation. While studying these powerful generators Kapitza discovered an unexpected phenomenon: when a helium-filled flask was placed in the beam of electromagnetic waves emitted by the generator, the helium developed a very bright discharge and the quartz walls of the flask melted. Kapitza assumed on these grounds that plasma can be heated to a very high temperature by the use of powerful UHF electromagnetic waves.

In 1955 Kapitza was reinstated as director of the Institute for Physical Problems. The unjust allegations against him were officially dismissed and he is still engaged in extensive research in the fields of strong current electronics and plasma physics at the institute.

Kapitza designed an installation for the production of a stationary high-frequency discharge. For this purpose he combined a nigotron with a chamber serving as resonator for UHF oscillations. Having filled this chamber with gases (helium, hydrogen, deuterium) at a pressure of 1–2 atmospheres Kapitza found that the gas develops a constricted discharge in the center of the chamber, where the UHF oscillations have a peak intensity.

Using various methods for the determination of plasma Kapitza was able to show that the plasma electrons of this discharge have a temperature of about one million degrees. These studies, which Kapitza pursued intensively, provided a fresh approach to the creation of a thermonuclear reactor with a complete solution of all calculations involved.

The first part of this book comprises Kapitza's articles, lectures and reports on extreme magnetic fields, a new original helium liquefaction plant, a description of the famous experiments which led to the discovery of superfluid helium, and problems associated with the production and utilization of oxygen. The same part contains also Kapitza's article entitled 'On the Nature of Ball Lightning', a work inspired by the above-mentioned experiment on discharges in the radiation field of powerful generators. Also included here are the introduction to Kapitza's book *High Power Microwave Electronics*, the introductory lecture at the opening of the 1973 international conference on magnetism in Moscow, and the report 'Energy and Physics', read at the scientific session dedicated to the 250th anniversary of the U.S.S.R. Academy of Sciences.

Apart from his remarkable scientific achievements, Peter Leonidovich Kapitza is known as a major organizer of science. In his capacity of director of the Institute for Physical Problems, member of the Presidium of the U.S.S.R. Academy of Sciences, and editor-in-chief of the country's leading physics journal (*ZhETF*) he devoted much energy to specific aspects of research organization. As in his research work proper, here too he is a prominent innovator, opposed to bureaucratic management methods, searching for a progressive approach to the organization of such a delicate mechanism as a team of creative researchers. The second and third parts of the book contain Kapitza's articles and addresses on these topics.

The second part includes reports on the setting up of the Institute for Physical Problems at the U.S.S.R. Academy of Sciences and the organization of research there. Readers of the older generation may remember that much of what Kapitza spoke in the late 1930s and early 1940s sounded rather strange at the time but today has become standard procedure in the majority of institutes.

The third part deals with Kapitza's views of the more general aspects of the organization and planning of science and its links with industry.

Peter Leonidovich Kapitza devoted great attention to the selection and education of youth with a talent for creative research. He was one of the founders of the Moscow Physical-Technical Institute and is chairman of the coordinating council of the Institute.

Kapitza always personally conducts the sessions of the State Examination Board for the final examinations of those students of the Moscow Physical-Technical Institute who prepare their theses at the Institute for Physical Problems. He always takes part in the entrance examinations to postgraduate courses and each time compiles the problems for the examinations. Typically, these problems have no standard solution. They always deal with a specific experiment or phenomenon in physics. To solve them, the student must find out by himself the effects and interactions that are essential to the problem and those which can be disregarded.

Some of these problems can be found in the fourth part of the book, which also contains Kapitza's views on the creative education of the youth.

The fifth part is devoted to Kapitza's teacher and friend Ernest Rutherford. It comprises articles on the scientific work of the great British physicist, a report read before the London Royal Society ('Recollections of Lord Rutherford'), and various other materials.

The sixth part represents a collection of Kapitza's articles devoted to various prominent scientists. Many of these articles are based on his addresses at jubilee sessions in memory of these scientists. True to his thorough approach, in each public address Kapitza contributed much to the history of science by making a profound analysis of the research work of many scientists, striving all the time to reveal the objective factors and individual traits responsible for their achievements. Kapitza depicts vivid portraits of persons who were close to him – E. Rutherford, I. P. Pavlov, P. Langevin, L. D. Landau.

Kapitza is not only a major scientist and a prominent research administrator; he is also a great public figure. He is deeply involved in all aspects of the development of human society. He is an active member of the Soviet National Committee of the Pugwash Movement of scientists for peace and disarmament. Kapitza has spoken often on the future development of human society, especially in connection with such topics as the struggle for disarmament, the problem of environmental pollution, the ecological crisis. Some of these addresses are collected in part seven of the book.

Kapitza works daily in his laboratory despite the heavy burden of administrative and public commitments. Many young scientists can envy his inexhaustible energy and his personal involvement in work.

A. S. BOROVIK-ROMANOV

PART ONE

1. THE PRODUCTION OF AND EXPERIMENTS IN STRONG MAGNETIC FIELDS

In the light of modern knowledge we take it that the structural character of the atom is of an essentially dynamic nature: that is, the atom consists of a system in which moving charged bodies move round a central nucleus, the properties of the atom being wholly dependent on the number and orbits of the electrons. We therefore see that these properties, for instance the magnetic moments, coherence forces, spectra, etc., may be changed if we can find a means of disturbing the motion of the orbital electrons. The most efficient way of doing this is by subjecting the atom to an outside magnetic force.

The internal magnetic field produced in the atom by the motion of the electrons in their orbits is of very great magnitude, but should it be possible to obtain an external field of the same magnitude it is obvious that the motion of the electrons would be altered very considerably as the coupling energy between them will be of the same order as the perturbation produced by the field; we should then expect to get some very significant results. However, when we come to consider the order of the fields inside the atom, we find that even for the most loosely bound electrons it would be in the neighbourhood of 1,000,000 gauss. As such a field would be about 30 times larger than is usually available in laboratory work the object of my research has been to develop a method for obtaining fields of this order.

The usual way of generating a strong magnetic field is by means of an electromagnet, but the field is strictly limited owing to the fact that the iron gets saturated. An increase in strength can only be brought about by an enormous increase in the weight of the magnet and of the current used. The largest electromagnet ever built is that of Professor Cotton; the diameter of the iron cores is about 1 metre, between the pole pieces there is room for a man to stand, and an immense current is required. The magnetic field only increases very slowly with increasing size of the electromagnet, and even Professor Cotton's magnet will not produce a field much stronger than 60,000 gauss in a space sufficiently large for experiments.

Text of a lecture of 27 February, 1931, from *Transactions of the Oxford University Junior Scientific Club*, 5th series, no. 4 (October 1931), 129-133.

A more successful method was found to be with a coil. Very large currents are required as the magnitude of the field in a coil is proportional to the exciting current. The obvious way, then, of creating large fields is to increase the current, but here we meet with difficulties; not only do we have to provide a source of very high current, but the magnitude of the current is strictly limited by the heating which it produces in the coil. One method for reducing this heat effect would be to carry away the heat as it was generated, or alternatively and more directly, by cooling the coil to very low temperatures. The resistance would thereby be reduced very considerably, and even in certain metals to zero when the metal becomes a supraconductor. The difficulty in this case is that the magnetic field produced by the coil will destroy the supraconducting state and also very rapidly increase the resistance to a value very close to that at room temperature. Neither of these methods seem very feasible, and even if carried out most efficiently would not allow a greater field than 50,000–60,000 gauss to be produced. If we assume we can make an efficient coil with an inside diameter of 1 cm, it can be shown that in order to produce a field of 1,000,000 gauss a power of 50,000 kw. is required, and in one second the coil will be heated to 10,000°C., and it is obvious that we could not deal with such a large temperature rise.

The main idea of our method of attacking the problem has been to make the time of duration of the field very short, when the coil has practically no time to overheat. The time actually chosen was 0.01 sec.

This condition, of course, created a new set of difficulties, firstly a very large current was required, and secondly all the measurements had to be made in a very short space of time.

Our first experiments were done using accumulator batteries having a very small capacity and a small internal resistance. In this way we were able to produce fields of 100,000 gauss by charging the accumulators for a few minutes and then discharging them in 0.01 sec., but further increase was impossible as it was found difficult to break currents of several thousand amperes sufficiently suddenly.

In our later experiments when larger powers were required we used a single phase A.C. turbo-generator. It is well-known that such a machine will give very large impulses of current when short circuited, and this is carefully avoided in usual practice as it might cause a serious accident. Our machine was purposely designed on the opposite lines so that large impulses were obtainable on short circuit. Considerable revision in design and careful calculation were necessary as the electro-dynamical forces might easily result in breakdown of the windings. The machine was of the size normally delivering 2000 kw. at continuous rating, but on short circuit on the test bench it gave us 220,000 kw. When short circuited on a coil of equal impedance only half of the power will be available; half of it will be lost in the machine and the other half will go to the coil. In this way the required 50,000 kw. are obtainable.

Fig. 1.

Normally the current in such a coil never remains constant, but with a certain design of the armature it was possible to obtain a current wave with a flat top, which gave a constant magnetic field for several thousandths of a second.

The greatest difficulty encountered was that the coils tended to burst owing to the electrodynamical forces trying to enlarge the diameter. We worked out a method of reinforcing the coils with steel bands, and designed a coil of such a shape that the electrodynamical forces together with the reaction forces of the reinforcement would be reduced to a uniform (hydrostatic) pressure on the copper. The force on the outside reinforcement of our present coil reaches a value of 140 tons.

The next problem was the design of a special switch to break the current synchronously with the current wave. As the current was only of 0.01 sec. duration, the time available for making and breaking was only a few ten-thousandths of a second, during which time a copper plate had to separate several millimetres from a brush. The acceleration required to move the copper plate 1 kg. in weight this distance is about 1000 times that of the gravity field, and the force required is over a ton. An exceptionally strong and carefully designed cam shaft mechanism is used for this apparatus.

The controls are so arranged that by means of various timing devices the pressing of a single button carries out the experiments automatically, and oscillograph records are obtained giving the currents in the coil thus measuring the magnetic fields.

We then had to overcome the difficulty of the shock caused by the sudden stopping of the generator. When the machine is short circuited the angular velocity of the armature, which weighs 2½ tons, is reduced by 10 per cent, in 0.01 sec., and a strong couple results which tends to turn the whole machine on its foundation. To avoid the results of this shock on our measurements, the coil is placed 20 metres from the generator so that measurements are over before the earthquake reaches the coil.

The short time of experiment led to certain difficulties in methods of observing and measuring, but on the whole the loss in time was compensated by gain in the magnitude of the phenomena observed due to the very strong fields; it also has the great advantage that the effect of the temperature change on different phenomena is practically eliminated, as during 0.01 sec. the temperature remains more or less constant.

We have so far studied the effects of strong magnetic fields on various phenomena: for instance, in the study of the Zeeman Effect we found that the splitting of the lines was so large that an ordinary prism spectrograph could be used having a large luminosity, and the time of exposure could be reduced to 0.01 sec. without interfering seriously with the accuracy of the results.

The study of the change of resistance of various metals in strong magnetic fields proved to be of great interest; in some cases there was an increase in resistance of about 20 to 30 per cent, whereas in ordinary fields there is only an increase of a fraction of a per cent. Moreover, we found that with strong fields a linear law of increase of resistance with increase of field obtains, while with ordinary fields the increase of resistance is proportional to the square of the field.

We have also measured the susceptibility of certain metals in strong fields. For this purpose we designed and constructed a special balance having a frequency of about 2000–3000. As in our case the magnetic forces were magnified about 100 times, this balance was sufficiently sensitive to measure the susceptibilities of most substances.

Another line of research has been the study of magnetostriction. In ordinary fields this phenomenon is only known in ferromagnetic substances, but in strong fields we found it specially marked in various others such as bismuth, tin and graphite which have a crystal structure of low symmetry. In bismuth crystals in strong magnetic fields the crystal increases in length in the direction of the trigonal axis, but decreases perpendicularly to it.

It will be seen that a wide range of research is opened up in the study of various phenomena in strong magnetic fields lasting for a very short time, but that special technique and apparatus are required.

2. A NEW METHOD FOR THE LIQUEFACTION OF HELIUM

Liquefied gases serve as coolants for the attainment of low temperatures close to absolute zero. Liquid air boils at 81 K (–192°C), liquid hydrogen at 20 K. Among all known gases, however, helium has the lowest boiling temperature. Helium liquefies at a temperature as low as 4.2 K. By letting it boil under a reduced pressure, however, a temperature of 0.8 K was obtained. Temperatures even closer to absolute zero were reached by using the magnetic properties of strongly magnetic substances cooled down to the temperature of liquid helium; the lowest temperature attained was slightly below 0.1 K.

At the low temperatures obtained by means of liquid helium, the thermal motion of atoms and molecules in solid substances nearly ceases, causing a marked change in their physical properties and the appearance of many strange physical phenomena such as the superconductivity discovered by Kamerlingh Onnes.

In some metals such as lead, mercury, tin, etc., the resistance to electrical current vanishes abruptly at the temperature of liquid helium. All that could be established so far is that the resistance of superconducting lead is one hundred million times less than that of the best grade of copper at room temperature. In effect, the resistance in the superconducting state is so low that a current initiated in a closed circle can circulate for days on end without any appreciable attenuation.

The greatest obstacle facing studies in this field concerns the preparation of liquid helium. This is a time-consuming process which requires complex equipment and trained personnel.

A brief look at the methods for the liquefaction of gases is sufficient for understanding the problems associated with the liquefaction of helium.

Gases are liquefied by cooling them down in the process of adiabatic expansion and using the principle of heat regeneration. Before being let into the expansion engine, the compressed gas is passed through a regenerating heat exchanger. After the expansion, the cooled gas again enters the heat exchanger, where it cools down the gas entering the expansion engine. The

---

An article in the journal *Sotsialisticheskaya Rekonstruktsiya i Nauka* [Socialist Reconstruction and Science], 1934.

temperature of the gas entering and leaving the expansion engine will drop to the liquefaction level, whereupon part of the gas will leave the expansion engine in the liquid state.

Technical difficulties associated with the construction of the expansion engine have so far prevented the application of this gas liquefaction process to the production of liquid helium.

The piston movement inside the cylinder of the expansion engine must be as frictionless as possible; moreover, the piston must hermetically seal the cylinder. Both requirements can be met by lubricating the gap between piston and cylinder. At the temperature of liquid helium and hydrogen, however, absolutely all substances are solid and totally unsuitable for lubricating purposes; moreover, most of them become fragile like glass. For this reason the production of liquid helium and hydrogen has been based on a different principle – namely, the Joule-Thomson phenomenon. Using this procedure, Dewar liquefied hydrogen for the first time in 1898 in Britain, and Kamerlingh Onnes, in Holland, liquefied helium in 1908.

The Joule-Thomson effect can be outlined as follows. Under certain conditions a compressed gas cools down, even without performing any external work, in the course of its expansion through a valve into a vessel of a lower pressure. The cooling effect results from the internal work performed in overcoming the attractive forces between the gas molecules. This phenomenon is used in combination with the regenerational heat exchange described above.

The main shortcoming of this method lies in its extremely low output owing to the small magnitude of the Joule-Thomson effect. Thus, air expanded from 30 atmospheres to 1 atmosphere cools down from room temperature to 17°C, whereas an adiabatic expansion in the same range lowers the temperature by 165°.

In the liquefaction of helium by this method matters are even worse than with compressed air: because of the negligible magnitude of the interaction forces between helium atoms, the Joule-Thomson effect is not only very small but it can be of use only after the helium has been cooled down to a very low temperature. For this reason the helium, once compressed, has first to be cooled to the temperature of liquid hydrogen boiling under a reduced pressure (14 K). The yield of this method is only 1% of that obtained by adiabatic expansion. Moreover, this method necessitates the use of large quantities of liquid hydrogen and liquid air: over 100 kg of liquid air have to be prepared first, then liquid hydrogen on the next day, all this in order to obtain a few liters of liquid helium on the third day.

After a year of intensive work at our laboratory, we managed to build an expansion engine free from any lubrication problems. In fact we abandoned the idea of using a lubricant and even a tight piston. Our piston moves quite quite freely inside the cylinder, thus allowing the compressed gas to flow through the gap between cylinder and piston. However, the machine is so designed that the expansion of the gas lasts only a few hundredths of a second, so that the amount of gas escaping through the gap during this brief interval does not exceed 2–3%.

Our expansion engine is rather like an explosive mechanism. The piston fires away at great speed and then slowly returns to its original position. A crank mechanism is inapplicable in our case; we replaced it with a hydraulic device.

Having overcome all technical obstacles (selection of nonfragile materials, valve design) we managed to build a machine having an efficiency of 0.7 (see Figure). Here the piston and cylinder are not subject to any wear. The

small fraction of the gaseous helium escaping through the gap apparently serves as a gaseous lubricant. Thus, in our present liquefier the preliminary cooling down of the gaseous helium is brought about with liquid air only. Though not necessary in principle, this approach speeds up the work to a considerable extent and helps reduce the size of the heat exchangers. The process consumes a very small quantity of liquid air – namely, two liters per liter of liquid helium obtained, and not more than 10–15 kg of liquid air per working day.

Our adiabatic liquefier is at least 10 times as efficient as the previous installations. Its main advantage, however, is the saving of time. Within two hours of operation the installation yields liquid helium in a quantity sufficient for starting the experiments. Because of this simplified procedure, liquid helium can be made available to a large number of laboratories.

3. PROBLEMS OF LIQUID HELIUM

I experience some difficulty in outlining my work in the field of liquid helium. This audience is, naturally, accustomed to the analytical thinking so essential in any field of research; yet I suspect that the problems of physics are rather remote to many of you.

As in any scientific work, research in physics can be divided in three parts: first, the objective sought; second, the means for achieving this objective; and third, the results obtained and their significance.

Turning to the second part, which deals with the methods of work, it is of major interest to the physicist as it often bears the seeds of success or failure. However, I find it very difficult to discuss such matters as work procedures, experimental technique, methods and accuracy of measurements, with people who have never worked in a laboratory, to say nothing of this particular field. A person who loves and understands music but is not a professional musician finds it just as difficult to evaluate the technical difficulties in the performance of a musical piece. Yet this does not prevent him from enjoying the music, loving it, and being interested in it. I believe this remark applies to all branches of creative work.

For this reason I shall dwell here mainly on the objective sought in the design of each experiment, and the results obtained. As for the experimental techniques used, they will be only briefly mentioned.

The purpose of any research is dictated by the status of science in the specific field. This status of science and the problems arising from it must be well known not only to the research worker himself but also to his listeners. And here lies my great difficulty.

In my opinion, physics is one of the weakest points in the scientific knowledge of the broadly educated man. All of us are well acquainted with history, having read the works of such illustrious historians as Klyuchevskii, Tarle and others; philosophy and economics are closely connected with our social structure, and we have to deal daily with them. We are also familiar with

---

A Report at the General Assembly of the U.S.S.R. Academy of Sciences, 28 December, 1940.

Translated from *Sovetskaya Nauka* [Soviet Science] **1**, 33 (1941).

Darwinism and other basic concepts of natural science, which we tend to understand readily. Technical matters also enjoy a considerable popularity as technology provides the foundation of industry, which is the focus of public interest, to say nothing of the radio, telephone, automobile and other devices which have become part of our daily life. Matters are by far less satisfactory with the basic concepts of physics, and even more so with mathematics.

If we ask any educated person about the quantum theory and even about more specific questions, for example, about the photoeffect, spectra, etc., or in the field of mathematics, about group theory, laws of probability, I think only one out of ten would give an answer reflecting some knowledge of these matters.

My position is made even more difficult by the fact that today the theme of my lecture is still distant from daily life and remains little known. Now, generally there are two kinds of endeavors in science, the distinction between which may be illustrated by the following analogy. In exploring our natural resources we can either deepen the exploitation of already known geologic deposits, or devote our energy to a search for new, previously unknown ones. Both fields of work are obviously of great importance to us; yet we assess them in different ways. Once the practical value of a given ore is well known, its exploitation becomes a matter closely linked with everyday life, and it is not difficult to evaluate the importance of any further surveys. On the other hand, one finds it difficult to evaluate the discovery of new mineral deposits that have no immediate application at present, and indeed the full significance of such findings can only emerge some time later.

This is true of most fields of research. From the realm of physics let us consider, for example, such a major contribution as Faraday's discovery of induction. Today we know that this discovery has made possible the construction of the innumerable electric motors and generators which form a dense network all over the globe and are so essential for any technological process. Yet many decades elapsed between the discovery of induction and its application in daily life; Faraday and most of his contemporaries did not live to see the colossal practical value of this scientific achievement.

Many similar examples can be cited. Hertz, the discoverer of radio waves, rejected even the possibility of their application in wireless telegraphy; his stance was quite correct from his own point of view, as at the time nothing was known of the existence of an upper atmospheric layer which reflects the radio waves and makes them circle the globe, thus making long-distance radio links possible. Röntgen, upon discovering the X-rays which also bear his name, did not realize that they would turn out to be almost the only known therapeutic means to treat cancer. Obviously one cannot expect an immediate and definitive evaluation of every new and surprising discovery. Today we can only stress the unexpectedness of the phenomenon, i.e. how much it contradicts the established views of the nature of things.

In physics, as in other sciences, there are many fields more or less thoroughly dominated by theories, hypotheses and assumptions. As science develops, however, the correctly established facts become and remain firmly entrenched, whereas theories are in a constant process of change, expansion, improvement, updating and adjustment. This evolution brings us irrevocably closer to a true picture of the surrounding nature, the understanding of which is essential for us to achieve its domination and control. The greatest stimuli to the evolution of a given theory come in the wake of the discovery of unexpected experimental facts which contradict established concepts. If the emerging contradiction can be made acute enough, the theory must change and advance.

Thus, the search for such contradictions and their interpretation is a major moving force in the development of physics, and of every other science. It provides us with a foundation for an objective evaluation of a scientific achievement which as yet has no immediate practical application. In other words, the value of a newly discovered natural phenomenon is measured by its capacity to demand changes in the prevailing theories and concepts.

A correct understanding of the significance of a given research is of particular significance first and foremost to the scientist involved, as it steers his endeavors. We believe that these considerations should guide the experimenter in planning his work with the objective in mind.

In order to explain the purpose and significance of our studies of the properties of liquid helium, I shall begin with a general outline of the theoretical concepts involved.

During the last 50 years, the progress of experimental physics was dominated by two theoretical views. The first of them was the atomic concept of matter. The development of this concept, especially in combination with the laws of thermodynamics, has produced a series of brilliant generalizations, the most important of which has been, of course, the kinetic theory of matter. However, this fruitful evolution led at the beginning of the century to a most curious blind alley. In effect, it followed from the theoretical generalizations that an equilibrium between matter and radiation was impossible since it turned out that the entire energy of the thermal motion of the atom would have to be continuously converted into radiation energy.

This conclusion, named the Rayleigh-Jeans paradox, is well known to physicists. Permit me to dwell on it for some time as it is highly instructive.

Attitudes to the theory varied sharply from one physicist to another. Many physicists, dominated by their inner conservatism, tend to consider any thoroughly understood and well-assimilated theory as something absolute and immutable. Such an attitude to theory, let me note, is much more prevalent on the continent than in Britain. Most of the leading British scientists regard experiment as the main factor, relegating theory to the status of an auxiliary tool. More than 140 years ago, Davy said: "One good experiment is worth more than the ingenuity of a brain like Newton's" (1799). This sentence is frequently cited to the present day, notably by such contemporary scientists as J.J. Thomson and Rutherford. Naturally, it should be interpreted as hyperbole, as a protest against the deification of theory. It is noteworthy that this same paradox is known in Germany as the 'Jeans-Rayleigh catastrophe', apparently reflecting the fatal danger with this remarkable scientific contradiction constitutes for theory.

We know all too well that this 'catastrophe' has been extremely fruitful for science. Quantum theory developed from it. Next to the atomic concept, it represents the second largest theoretical achievement of modern physics. One can only wish we had more and more such 'catastrophes', considering the outstanding benefit derived from this one. History shows that such 'catastrophes', whatever their magnitude, actually stimulate the progress of science.

Many of you probably know that Planck was the first to find a way out of this dead end. His solution was simple and many interpreted it as purely formal at first. By making some transformations in the classic radiation formula and introducing a new constant, Planck was able to show that the disequilibrium between matter and radiation could be eliminated. However, a true understanding of the profound universal significance of Planck's constant came some time later. We owe this achievement of physics to Einstein, who was the first to realize the fundamental importance of Planck's discovery and gave it a general physical interpretation now called Einstein's law. I believe that in terms of the practical implications for the development of science, this remarkable work of Einstein is even more important than his famous relativity theory.

Quantum theory began its fruitful development after these works. The basic ideas of quantum theory can be outlined as follows. Natural processes are not continuous as was previously thought; in reality there is a consecutive change of elementary states which are the only stable states of matter taking part in natural processes.

Now we understand the natural processes to be of a discontinuous character, in much the same way as the atomic distribution of mass inside matter. And perhaps this does not now seem to us as unexpected as in the beginning.

Matter and energy are in effect joined by an equality sign – not only in theory but in recent years also experimentally; in other words, they are interconvertible. Now if matter occurs in nature strictly in the form of discrete masses, we should expect energy phenomena to have the same discrete character. This is by no means a proof, but merely an indication that such a relationship would be quite natural.

In the early stages of its development, quantum theory was extremely beneficial to the study of the atom, largely because of Bohr's ideas. Now we are well acquainted with the structure and properties of the atom. The processes of radiation from the electron shells of the atom are now quite precisely known.

The progress of atomic physics has in fact led to the great advances of quantum theory and the remarkable generalizations made by Schrödinger, Heisenberg and Dirac.

It is noteworthy that the mathematical apparatus required by the quantum conception of natural processes has been considerably simplified. Indeed, the application of the tools of celestial mechanics to the study of the atom as a system composed of numerous electrons moving around a nucleus would create far greater mathematical difficulties than those encountered now.

Despite all these successes, however, it would be wrong to think that the quantum theory is now complete and will not develop further. In effect, major advances in this field are possible and even necessary, and even fundamental changes of our basic concepts should not come as a surprise.

In our search for new contradictions in nature we should turn our attention to those areas of physics where these quantum concepts face their most crucial test. From this viewpoint I believe that our efforts should be devoted to those areas of physics where the quantum phenomena are least known and understood. We owe the creation of quantum theory to experimental evidence collected from studies of the atom, a source largely exhausted nowadays. The following two branches of physics seem most promising in this respect.

First, the atomic nucleus. The nucleus contains elementary particles so close to one another that the laws established for particles as far apart as those in the atomic shell might turn out to be inapplicable here. One could assume on these grounds that nuclear physics would require a thorough revision of quantum theory.

Second, condensed-state physics. The basic, and quite general, ideas of quantum theory must naturally apply to phenomena occurring in our environment, where atoms and molecules, by interacting with one another, form gases, liquids and solid bodies. But there are cases, notably in our study of matter at room temperature, when the quantum nature of processes cannot manifest itself. Indeed, the thermal motion of atoms figuratively speaking, irons out the features of the quantum nature of processes, thus making them imperceptible. Matters would be much the same if we attempt to study the laws governing the impacts of billiard balls on a pool table placed on a ship at sea. Obviously this can only be done in a calm sea. The same principle applies to the study of the quantum nature of phenomena in the course of processes taking place in the condensed state. Their manifestation reaches a peak when the thermal motion of atoms remains at a sufficiently low level. Hence the interest among physicists in the study of phenomena at extremely low temperatures, notably the range of phenomena at the temperature of liquid helium, which I shall describe now.

To make my point in a more concrete way, allow me to introduce the following simple experiment (see Figure 1) which, for all the oversimplification it entails, will bring the idea of thermal motion somewhat closer to those among you who are not familiar with this field. The projector shows the frame *1*, which contains tens of bicycle ball-bearings between two parallel glass sheets. The balls appear in the drawing as solid black circles. The frame is suspended on the springs *2* to another frame *3* in such way that it can oscillate in its own plane.

The frame can be made to perform horizontal oscillations by means of the connecting rod *4* and the crank *5*, attached to the axis of the small electric motor *6*. Under such circumstances the balls will begin to move between the glass sheets. Their motion resembles that of thermally excited atoms. The quicker the oscillations, the greater the velocity of the balls, as is the case when the temperature rises.

Let me now draw your attention to a certain detail of the frame design. The lower part of the frame contains a series of partitions *7*, which form six hollow pits out in the general space of the frame. When the frame oscillations are small, the balls within the respective pits move inside them only and the motion in one pit does not affect that in the next. This condition is shown in the top drawing. On the other hand, a rapidly oscillating frame duplicating the condition observed at high temperatures, yields the picture shown in the bottom drawing. Here the balls traverse the entire space of the frame, apparently unaffected by the partitions.

Let us assume now that we face the opposite task – namely, to determine, judging from the balls' motion, whether any such pits exist, the walls of these pits being made of a transparent, invisible material. Obviously this could only be done by studying the ball movements at slow oscillations of the frame, that is at a low temperature.

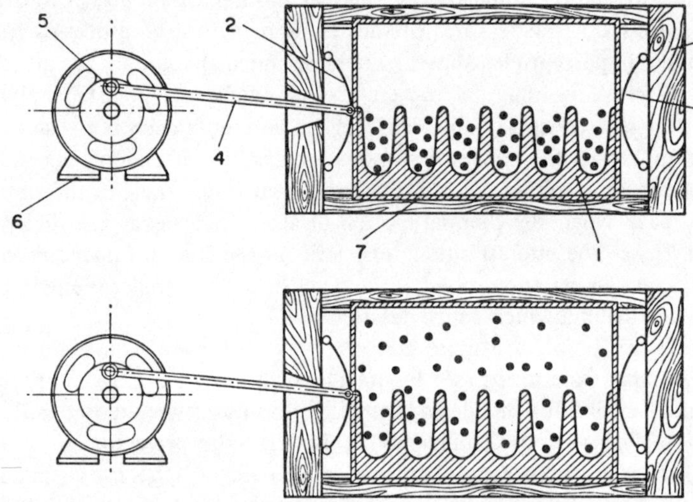

Fig. 1.

The same is true of studies of quantum properties in the condensed state. The limitations which the partitions of the frame in our model impose on the movement of the balls are somewhat similar to those which the quantum nature of phenomena dictates to processes involving atoms in the condensed state. At sufficiently low temperatures, the quantum nature of atomic interactions can reveal a series of physical phenomena not observed at higher temperatures. The search for such phenomena explains the extraordinary interest in the study of the properties of matter at temperatures close to the absolute zero.

The anticipation of discovery of new properties of matter at low temperatures has been amply rewarded. One of the earliest findings was the discovery of the anomalous behavior of the heat capacity of bodies and gases at low temperatures, a behavior explicable in terms of the quantum theory, as shown by Debye and Einstein.

According to the quantum theory the heat capacity of bodies approaches zero in the vicinity of absolute zero. Indeed, in the temperature range explored by us – between 0 K and 4 K – we found that the heat capacity of bodies is tens of thousands of times less than at room temperature. It is noteworthy that this particular property of matter makes it possible to perform the cooling operations necessary for experimentation at low temperatures. Liquid helium has a very low heat of evaporation, less than one thousandth of that of an equal volume of water. Calculations show that with such a negligible heat of evaporation, it is practically impossible to cool a body if its heat capacity remains at its room temperature level.

Research into physical phenomena at extremely low temperatures was initiated by Kamerlingh Onnes, who managed to liquefy helium in 1908 after great difficulties.

Helium was extremely difficult to liquefy. This is so because its atoms are highly symmetrical and are attracted to one another by very weak forces. At normal pressure its liquefaction point is only 4.2° above absolute zero. If helium is subjected to evaporation under a reduced pressure, its temperature can be lowered to just 0.8° above the absolute zero. A further lowering of the temperature has been made possible by the recently developed method of demagnetization of paramagnetic salts. This brought us as close as a few thousands of a degree from absolute zero. Perhaps one day we will get even closer to absolute zero, though actually reaching it is impossible in principle.

Of all the extremely interesting physical phenomena observed at low temperatures, superconductivity is perhaps the most remarkable. At some very low temperature, certain metals totally lose their electric resistance. Under such circumstances a current once established in the metal will circulate indefinitely as long as the metal is kept at the low temperature required.

This frictionless movement of electricity in a conductor contradicts our accepted concepts of the motion of electrons (the carriers of electricity in metals) across a crystalline lattice, as in normal conditions such motion necessarily involves a loss of energy.

A theory for this phenomenon has not yet been offered, despite a number of highly interesting attempts. Physicists agree that this phenomenon reflects the quantum nature of processes at low temperatures, but its explanation eludes us[1].

Of the other phenomena observed at low temperatures only, and apparently also due to their own quantum nature, the properties of helium

---

[1] The quantum theory of superconductivity was developed as late as 1957 by J. Bardeen, L. Cooper and J. Schrieffer.

itself are of particular interest, and they will be discussed here in greater detail.

Liquid helium has a very low specific weight: roughly 1/7 of that of water. It is extremely transparent and therefore hardly visible at all in comparison with water.

Research into the properties of liquid helium has led to the discovery of a whole series of interesting phenomena. Kamerlingh Onnes himself found at the time that helium exists in two states or phases. The first phase is the normal one, named helium I. It exists down to a temperature of 2.19 K, below which it turns into another fluid phase named helium II, and persists as a liquid down to the lowest temperatures attained so far.

Upon close examination, helium I represents a boiling liquid, since even incident light makes it boil. To protect the helium from the ambient heat, vessels containing it are surrounded by two series of vacuum envelopes filled with liquid air. Without such precautions the liquid helium would evaporate within ten minutes or so.

The second phase of helium differs markedly from the first. Helium II does not boil, and it appears that its free surface forms a totally stationary plane. Helium II possesses a number of unique physical properties. Perhaps the most remarkable of them is its extremely large thermal conductivity, discovered by Keesom and his daughter. This property was observed in the course of measurements of the heat conductivity of helium II in fine capillary tubes. At room temperature, metals have the greatest heat conductivity, especially silver and copper. Many of you have probably experienced how a copper rod heated at one end can burn the hand holding it at the other end. In effect, the heat conductivity of helium in capillaries was found to be more than a million times greater than that of copper. Naturally enough, Keesom referred to this property as thermal superconductivity, in analogy with the electrical superconductivity of metals.

Experiments made in Canada showed that the viscosity of liquid helium II is extremely low, in fact several times less than in helium I.

Viscosity is a property of liquids which measures their fluidity. If different liquids are passed through the same tube under the same pressure, it will be found that some of them flow more readily than others. A liquid that flows more readily has a greater fluidity and a lower viscosity. In other words, viscosity can be regarded as a measure of the internal friction in the process of flow. It can be shown experimentally, for example, that viscosity is low in water, greater in oil and even more so in a resin. By a sufficiently sensitive experiment, even gases can be shown to possess a viscosity, though of low magnitude. The viscosity of liquid helium is roughly 1/1,000 that of water, and it even decreases somewhat during the transition from helium I to helium II.

We were deeply fascinated by this phenomenon. It can easily be shown that these properties of liquid helium contradict our conventional ideas of the mechanism of viscosity and thermal conductivity.

What is, in effect, the mechanism of thermal conductivity? We regard heat as the motion of atoms within matter. When one part of a body is hotter than the rest, its atoms perform more energetic oscillations than elsewhere. Interaction forces transmit this more energetic movement of atoms from the hotter to the cooler part. It appears that the discrepancy in the energy of oscillation tends to vanish as the heat spreads evenly throughout the body. Thus, the thermal conductivity can be regarded as the capacity of atoms to transmit their oscillations to one another; the more pronounced this property of transmission, the greater the heat conductivity of a given substance.

Now let us try to explain the mechanism of viscosity. When a liquid flows along a tube, the layer adjacent to the wall of the tube remains immobile; the next layer moves at some velocity, then a third layer flows at a somewhat greater velocity, etc. Between these layers there is a sliding motion accompanied by friction. This friction results from the fact that the moving atoms of a given layer lag behind those of the next layer, so that the interaction forces involved impede the flow. Hence a loss of energy, which is the actual cause of the viscosity of a liquid. It follows from this picture of events that viscosity is proportional to the extent to which the motion of the atoms of a given layer affects that of the atoms of the next layer; in other words, viscosity is proportional to the rate of heat transfer in the body.

We would expect on these grounds that an increase of the thermal conductivity of a substance necessarily entails a greater viscosity, yet in helium these two parameters are inversely related. Now the question arises, why is this colossal rise of thermal conductivity in helium II accompanied by a decrease of viscosity?

To settle this contradiction, we assumed that the enormous thermal conductivity reported by Keesom is only apparent. In fact there are two known mechanisms of heat transfer: one of them is the heat transfer from one atom to another, of the type just described and observed in solid bodies; the other mechanism involves convection. Holding a hand over a heat source such as a radiator we immediately feel the heat, carried by a current of hot air. This heat transfer together with a moving current of matter is known as convection. A hand placed below the radiator would feel no heat whatsoever since the flow of hot air is directed upward, and normal thermal conductivity in air is very low. In such a poor heat-conducting medium as air, convection is normally the only form of heat transfer. In view of the high fluidity of helium, heat transfer by convection is naturally assumed to be of considerable magnitude; moreover, this mechanism of heat transfer might easily explain the high heat conductivity observed by Keesom.

Calculations soon showed, however, that to be compatible with an intensive heat transfer by convection currents the viscosity of helium II would have to be far below the level measured by the Canadian workers.

Let us note at this juncture that a low viscosity is not easily measured. Theory shows that the actual value of viscosity can be masked by the presence of a turbulent movement in the liquid, that is, in addition to the calm flow there is a superimposed turbulence which so distorts the results of measurement that the apparent observed value may turn out to be many times greater than the actual one.

I shall not dwell in greater detail on this strictly experimental topic, in order to avoid a minute technical description of measurement techniques[2]. Let me only point out that measurements of the viscosity of helium were repeated with the above consideration in mind. We were able to build a viscosimeter (a device for the determination of viscosity) with a slit only half a micron wide (one micron being one thousandth of a millimeter), through which the helium was made to flow. The experiment was so designed as to avoid the adverse effect of turbulence to a considerable extent. Under such circumstances it became evident that the observed viscosity of helium II was at most one thousandth of the value previously found.

We also managed to show that the value of viscosity obtained by us actually represented its possible upper limit, as in fact the actual value could have been anywhere below this limit. In other words, even our narrow slit did not fully eliminate the deleterious effect of turbulence. That work, which we published three years ago, aroused considerable discussion and criticism.

We immediately began to look for possible errors in the method used. Here it would be instructive to tell of a particular objection raised against us.

This criticism was based on another highly interesting property of helium II, namely, its ability to flow in a film over the walls of vessels. If the level of

[2] Full experimental details can be found in the following publications: *DAN SSSR* **18** 21 (1938); *ZhETF* **11** No. 1, 1 (1941); **11** No. 6, 581 (1941); [See *Collected Papers of P. L. Kapitza*, Vol. II, No. 34, p. 505, No. 38, p. 581 and No. 39, p. 625, for English translations of these papers.]

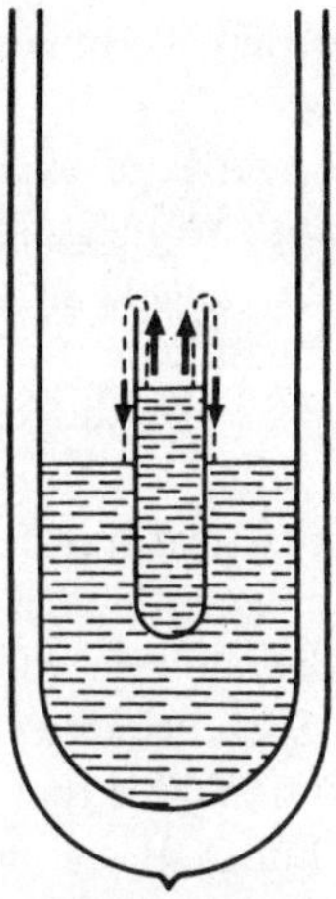

Fig. 2.

helium II in the test tube shown in Figure 2 is above that of the surrounding helium, the levels become equalized rather quickly. Studies of this phenomenon showed that helium readily creeps over the surface in the form of a thin film. The critics of our experiments maintained that in my viscosimeter I did not measure the outward flow of helium through the slit but rather that the flow took place by the above-mentioned creeping movement, and consequently my viscosity data were too low. However, in these experiments I did consider the possibility of error as a result of this phenomenon. It is noteworthy, however, that this criticism, which originated from scientists in the U.S.A. and Canada, disregarded the fact that helium can creep in a thin film the thickness of which, as measured by Kikoin and Lazarev, is less than one hundredth of a micron, and only when its viscosity is one million times less than the limit established by us. Thus it turned out that the criticism of the high fluidity of helium was based on a phenomenon the explanation of which required an even greater fluidity.

We suggested that helium II is an ideally fluid liquid and proposed the name *superfluidity* for this property, in analogy to superconductivity.

It might be assumed in view of the above considerations that the abnormal thermal conductivity of helium II could be explained in terms of its superfluidity. However, a closer quantitative analysis of the experimental evidence revealed fresh difficulties, which I shall describe in greater detail.

Calculations showed that to explain the thermal conductivity values observed by Keesom the convection velocity must be assumed to be about

50 m/s. This is a considerable velocity indeed, and therefore I decided to measure it more accurately.

For this purpose we set up a series of experiments in which a procedure was worked out for a more sensitive measurement of greater thermal conductivity values of helium II than available in Keesom's work. We managed to improve the technique for the measurement of temperature differences down to an accuracy of a few millionths of a degree. The procedure involved will not be described here as it would divert us from our subject.

Thus we were able to observe a heat transfer at least 20 times greater than that observed by Keesom. Consequently, the convection velocity necessary for explaining such thermal conductivity would have to be of the order of 1,000 m/s rather than 50 m/s. Such velocities obviously do not exist in convection currents. It is indeed unthinkable that helium can move in a capillary at a speed exceeding that of a bullet. It can be shown that sources of energy for such powerful convection currents simply do not exist.

As it turned out, our findings led to even more fundamental difficulties when it became clear that convection had to be ruled out as a possible mechanism of heat transfer. Returning to the previously mentioned conventional mechanism of thermal conductivity involving the transmission of thermal motion from one atom to the next, it can easily be shown that here too we face a basic contradiction.

Let us reconsider this mechanism of thermal conductivity. If a layer of atoms is suddenly heated, the atoms involved will oscillate more intensively than those of the neighboring layers. These oscillations will be transmitted from one layer to the next in the form of a heat wave spreading from the heated part throughout the body. It has been shown that the velocity of such a heat wave cannot exceed that of elastic oscillations in the body, i.e. the velocity of sound. The velocity of sound in helium II was found to be 230 m/s, while the velocity obtained in our thermal measurements is several times greater, as it turned out, a discrepancy which contradicts the conditions of the assumed mechanism of heat transfer.

We spent about a year in attempts to solve these contradictions.

How could we proceed with our search for the true mechanism of this heat transfer in the absence of any guiding thought? Indeed, our findings basically contradicted all the established theoretical concepts.

We had to grope along, testing various physical factors which might affect the thermal conductivity. Thus we examined the heat transfer of helium II as a function of pressure, gravity, time, etc. Only negative results were obtained; the thermal conductivity remaining unchanged at its high level.

Finally, an accidental observation gave our work an impetus in a totally new direction. We found that pressure pulsations transmitted quite accidentally from the helium pipeline of the laboratory into the helium in the capillary caused marked changes in the thermal conductivity. Though minute in magnitude, the pulsations reduced the heat conductivity of helium II by a factor of tens of times. The question arose, how can these slight pressure pulsations exert such a marked effect on the thermal conductivity of helium?

The most natural explanation could be summed up as follows. Liquid helium II is known to be, relatively speaking, a readily compressible fluid, roughly 100 times more so than water. Because of this property the pressure pulses, by compressing the fluid, generated currents of helium in the capillary used for studying its thermal conductivity. We assumed indeed that these currents affect the thermal conductivity. To test this assumption, we set up experiments for measuring the thermal conductivity of helium flowing in a capillary. It was found that the heat conductivity of helium II flowing in a capillary is 100 to 1,000 times lower. These tests also revealed that the helium flows readily along the capillary as long as heat flows through the helium. Thus a relationship between the flow and thermal conductivity of liquid helium was established, providing the key to our further research.

Indeed, if the currents of helium affect its thermal conductivity, it can be assumed that the heat transfer proper can generate currents. But how can helium currents be observed experimentally in a thin capillary only 0.5 mm in diameter? A difficult task undoubtedly, but perhaps these currents can break through outward at the free end of the capillary and be observed there.

For this purpose, we built the device shown schematically in Figure 3.

The glass capillary tube *1* is placed horizontally. One of its ends is bent upward and sealed with the glass bulb *2* which contains a heating element. Opposite the free end of the capillary tube, the lightweight beam *3* bears the vane *4*. The beam is suspended on a fine quartz filament through the long glass bar *5*. The beam and the bulb connected with the capillary are placed in a Dewar vessel considerably below the level of helium II. Liquid escaping from the bulb as a result of the heating inside strikes against the vane, which is deflected as a result.

The force of the pressure exerted by the helium can be measured from the twisting of the quartz filament, which in turn is determined from the displacement of the light spot reflected by the mirror *6* attached to the glass bar *5*.

It became clear that even the slightest heating of the helium inside the bulb causes an exit of liquid which exerts an easily detectable pressure on the vane.

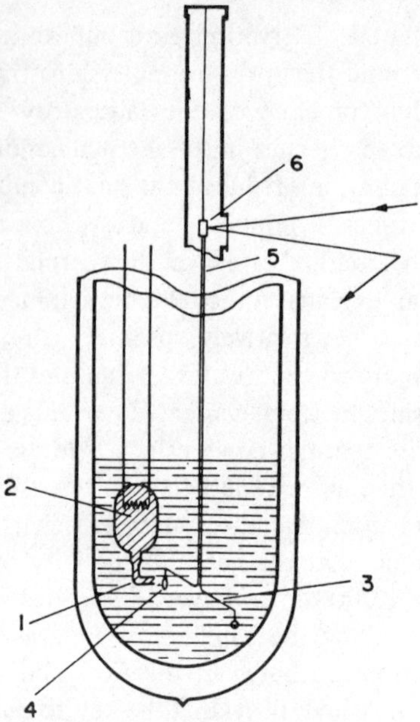

Fig. 3.

As for the nature of the observed helium current escaping through the capillary opening, it can be determined by the following simple experiment. The vane *4* was made of small diameter, only slightly larger than the capillary opening itself. Tests showed that the pressure exerted on the vane is essentially the same regardless of whether the vane is located just beside the capillary opening or as far as 15 capillary diameters away from it. When the vane was placed at a large distance, however, it had to be located exactly in line with the continuation of the capillary axis; the pressure exerted by the helium ceased when the vane was moved even slightly off the line. Obviously the helium leaves the capillary in the form of a well-directed jet. We shall not enter here into a detailed quantitative analysis of this phenomenon and its dependence on the load; suffice it to note that by measuring the pressure of the helium jet on the vane it was possible to determine the velocity of helium outflow (5–6 cm/s).

If helium flows continuously out of the capillary in the form of a jet, how is the bulb replenished with helium? Obviously helium must refill the bulb somehow, judging from the fact that the amount of liquid inside the bulb does not diminish. Any replenishment can only take place through the same capillary, and if the vane is placed close enough to the aperture, then the inflowing helium must also exert a pressure on the vane. The experiment showed, however, that the pressure exerted by the jet on the vane is unrelated to the distance between the vane and the aperture; in other words, when the vane was placed as close as possible to the capillary aperture, the jet pushed it with just the same force as when the vane was far away from it. This means that helium re-enters the capillary in a very thin film around the aperture.

To analyze the mechanism of this re-entry of helium in greater detail we designed experiments based on the following considerations. The outflowing helium jet must exert a reactive force on the capillary tube, i.e. one opposite to the direction of the jet. It can be shown that this force equals the force of the pressure which the jet exerts on the vane. The liquid helium entering the capillary can likewise exert some force on it. By detecting and measuring this

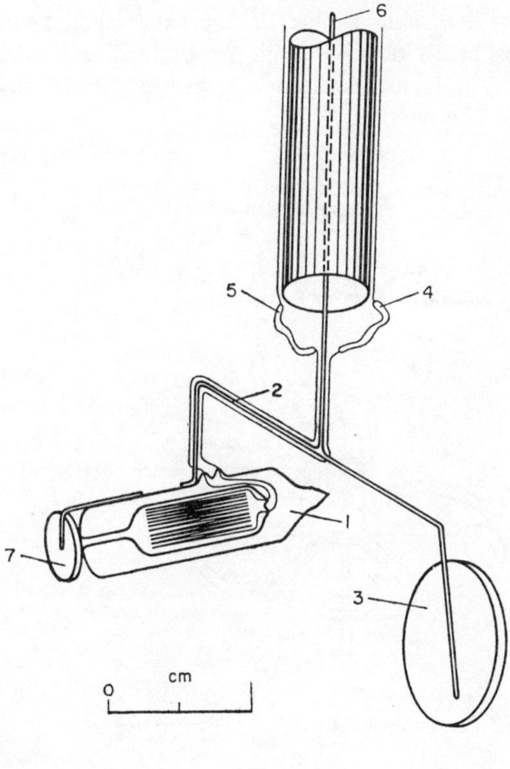

Fig. 4.

force we can more accurately determine the manner in which helium enters the bulb via the capillary. To determine this force we built the device shown in Figure 4.

We made a very small bulb with a capillary *1*. The bulb, which weighed just 0.25 g, was suspended by means of the beam *2* on the glass rod *6*, instead of the vane which occupied the corresponding place in Figure 3. Balance was achieved with the heavier, disklike counterweight *3*. The helium inside the bulb was heated by current supplied via conductors in the form of the fine silver bands *4* and *5*.

Experiments with this suspended bulb showed that the jet does exert a reactive force equal in magnitude and nature to the pressure exerted by the jet on the disk. To show this more accurately we attached the disk *7* to the bulb itself; then the pressure of the helium on the disk must balance the reactive force exerted by the jet on the bulb, and the residual force must be attributed to the inflowing helium. In the experiment, however, such a residual force was hardly detected at all. We assumed on these grounds that helium flows in by some mechanism which does not involve any pressure on

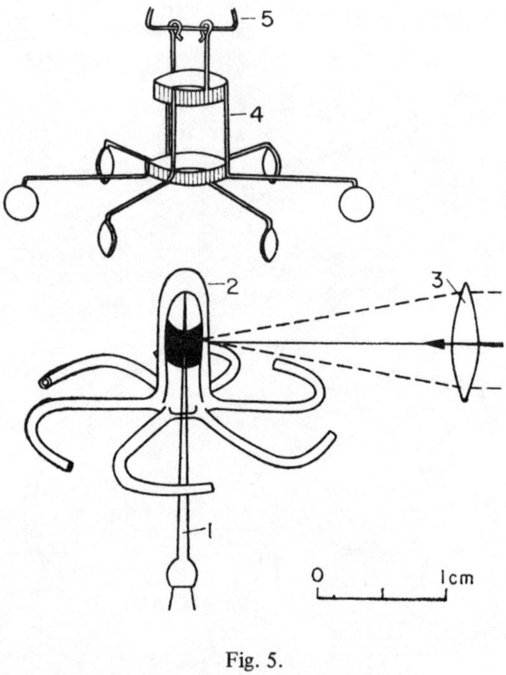

Fig. 5.

the bulb. The only possible mechanism of helium entry without any reactive force is a creeping movement in a thin film on the surface.

This phenomenon is so interesting and clear-cut that it can be demonstrated to you in a simple device built specially for this purpose. The device is shown in Figure 5. We named it 'the little spider'. It consists of two glass caps *2* which are sealed below and attached to 6 capillaries shaped as spider legs. The glass cap rests on the very sharply pointed needle *1* and therefore can rotate freely. Naturally, the entire 'spider' is completely immersed in liquid helium II. Now if the helium contained in the vessel *2* is heated by means of the light beam *3*, it will begin to escape out of each leg in the form of a jet the reactive force of which will make the spider turn, as you can see.

Now if by immersing the loop *5* one places the spider inside the two

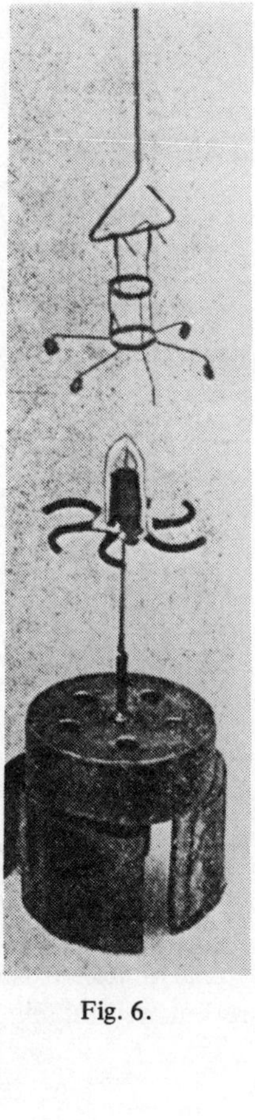

Fig. 6.

extremely light rings *4* with fine rods bearing 6 terminal disks so fitted that each disk is located 1–2 mm away from each capillary opening, the spider will no longer be able to rotate upon being heated with a light beam. A photograph of this device appears as Figure 6.

These experiments reveal the mechanism of the movement of liquid helium in the capillary as a result of a heat current. A scheme of this movement is given in Figure 7. Here the helium is heated inside the bulb *1*, and the

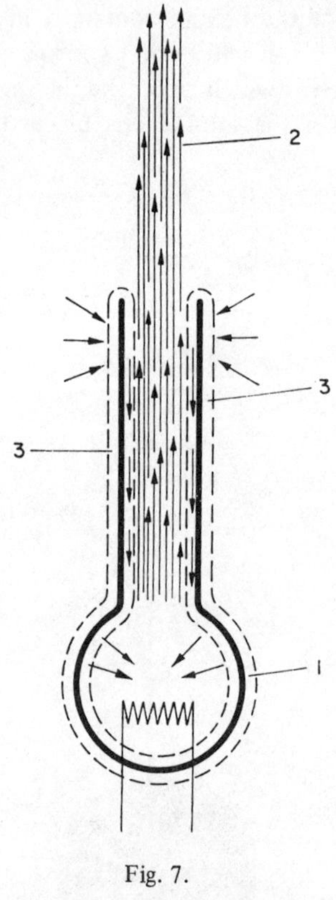

Fig. 7.

resulting current in the capillary escapes in the form of the jet *2*. Helium enters by moving in the opposite direction along the capillary walls as a thin superficial film *3* which, upon entering the bulb, leaves the surface to become free helium again. This creeping motion of helium on the surface, closely similar to that discussed previously, actually explains the equalization of helium levels in the two vessels shown in Figure 2. In both cases such movement is possible only if the flowing helium II behaves as a nonviscous fluid.

Now, having established the mechanism of helium movement in the capillary tube – one generated by the heat current and detected by strictly experimental means – we could proceed to draw conclusions from our study aimed at explaining the cause of the colossal thermal conductivity observed in helium II.

We have reasons to assume that the helium which moves in a thin film on the surface is in a physical state different from that of the helium flowing in the opposite direction in the capillary core. We assume that its energy state is somewhat different because of the effect of molecular forces from the capillary walls. Speaking in thermodynamical terms, its heat function differs from that of free helium.

As it turned out, this assumption can fully explain the great thermal conductivity of helium in the capillary. Actually the experiment revealed the following picture: the helium flowing into the bulb on the inner surface of the capillary absorbs heat upon leaving the surface and reverts to a free state. This very process creates the impression of a colossal thermal conductivity. We shall illustrate this mechanism with the following example.

If we want to produce a cooling effect, and wish to use for this purpose either a jet of cold water at 0°C or ice at the same temperature, we see that owing to the latent heat of melting the cooling effect will be more pronounced in the second case. We regard the cooling produced in the heated bulb as analogous to that caused by melting ice. Helium, having entered by traveling close to the surface, passes into a different energy state as it leaves the walls, thus absorbing heat generated by the heating element. On the basis of this mechanism it can be shown that the heat transfer becomes fully explainable in quantitative and qualitative terms, and that a thermal superconductivity does not exist in helium II[3].

Further, we tested our explanation of heat transfer in helium II inside a capillary by measuring its thermal conductivity in free space rather than inside a capillary. Thereby we measured the thermal conductivity under

---

[3] The concept of a flow of helium in a given physical state inside a liquid, opposite a thin film of liquid in a different physical state, provided a basis for L.D. Landau's quantum theory of superfluidity. According to this theory, liquid helium represents a mixture of two fluids (two components) in different quantum states. Quantum theory made it clear that the countercurrents observed in the experiment represent a head-on movement of these two components toward one another throughout the volume of the liquid.

conditions which rule out any possibility of heat transfer by film movement. These experiments were made, for example, with a glass tube containing a heating element and a thermometer. The tube was freely suspended in helium II on extremely fine conductors. Helium II films could move from colder to warmer areas, but only on these conductors, which have a very small area, so that only a negligible fraction of the heat could be transported by any film moving on them. Clearly the bulk of the heat had to traverse the mass of liquid helium. Measurements of the heat conductivity of the free helium in such experiments produce normal values, as could be expected from our mechanism, that is, not more than in helium I, or roughly 1/100,000th of the corresponding value in copper.[4] Here, I believe, I can conclude the description of our experiments with liquid helium.

Let me point out once again that the description given so far is strictly schematic and designed solely to illustrate the course of our thoughts. But in my opinion even this rather general description provides a picture of the evolution of research in this highly interesting field of modern physics. We saw that the observed discrepancy between the simultaneous existence of a great thermal conductivity and a low viscosity in liquid helium led to experiments which revealed that this viscosity is not just low but practically imperceptible. In view of this finding we assumed that helium II is a 'superfluid' whereas its thermal conductivity is of the convection type. This assumption again placed us in a blind alley as it could not explain the high thermal conductivity of helium II. To get out of this impasse we had to find a movement of helium in the capillary tube. We managed to observe such movement in a series of experiments which I have described in greater detail. In order to explain the great thermal conductivity of helium on the basis of the observed pattern of movement, we suggested that there is a difference between the heat functions of helium in thin films and in the free state. This hypothesis proved fruitful; it led to the prediction that the thermal conductivity of helium would be strictly normal in the absence of surface phenomena.

However, it would be incorrect to conclude that all problems of liquid helium are solved and that the matter is exhausted. Further research constantly reveals many contradictions and unclarities in this field, and a great deal of

---

4 Our further studies of heat transfer from a heated body inside free helium II showed that the entire thermal resistance in this case is concentrated in an extremely thin boundary layer. Inside this layer there is a temperature jump the magnitude of which is inversely related to temperature, or more accurately, inversely proportional to the third power of temperature. This phenomenon strongly impedes physical research at temperatures far below 1 K.

interesting work still lies ahead. A detailed discussion of these topics would bring us too far; therefore let me point out only a few aspects.

Boundary layers, which perform such a major role in the behavior of liquid helium II, create many problems in research. Little is known, for example, of the mechanism of helium flow in thin layers on the surface, or about the possible velocities of this movement. On the basis of general theoretical considerations, the surface layer of helium II participating in the countercurrent should be very thin. Then it turns out, however, that the velocity of flow would be very high, of the order of 200 m/s. Closer analysis shows that in principle no physical law precludes the existence of such high speeds in a thin film. Yet to accept their existence as a fact one must actually confirm such velocities experimentally.

Another question to be answered is whether the 'superfluidity' of helium II is confined to its surface layers or spread throughout the mass of helium. Analysis of the available experimental evidence gives no clear-cut answer to this question; to the contrary, it leads to a series of remarkable discrepancies which have to be examined experimentally.

A multitude of unsolved problems remains to be coped with.

At this stage it would be worthwhile to consider the possible significance of the existing data for the progress of modern theoretical physics. As we pointed out previously, the superfluidity of helium II is a phenomenon closely similar to superconductivity. In both cases, at temperatures close to absolute zero where manifestations of the quantum character of phenomena can be expected, both electricity and matter itself begin to flow without any losses. It would be surprising indeed if both of these phenomena are not covered by a single theory, by a still obscure aspect of quantum processes in the condensed state.

In superconductivity the electrons, as carriers of electricity, flow without any friction across the crystal lattice. Superfluidity, on the other hand, applies to atoms capable of moving in an organized manner with regard to one another, likewise without any friction. Theorists search for quantum relations which could explain such frictionless movement, and conceivably they have greater chances of solving the problem by studying the interactions between identical atoms of liquid helium II than those of electrons with the atoms which constitute the crystalline lattice of the metal.

Here we could end this description of our works, were it not for an extraordinary idea suggested to me for a practical application of the great fluidity of liquid helium. I would like to tell you about it not because I am convinced of its feasibility but in order to demonstrate that any phenomenon discovered in nature inevitably opens new prospects which eventually become applicable in one way or another. Such applications may be quite unexpected indeed, in fields which the discoverer is unacquainted with and could hardly take into consideration. The far-reaching idea of the application of liquid helium to airplane design came from Professor L.G. Loitsyanskii. Though still remote from reality and liable to arouse many objections, his idea deserves indeed to be told here.

The idea concerns the testing of the wings and fuselage of airplanes for streamlining. Now engineers have to use very bulky and expensive aerodynamic tunnels where airplanes are tested in full scale. It is known that reduced airplane models are inadmissible for this purpose since the similarity theory which underlies any experimentation with models is totally inapplicable here. Any size reduction in an aerodynamic tunnel must be matched by a corresponding reduction of the kinematic viscosity of the surrounding medium. This kinematic viscosity is the quotient obtained from the viscosity divided by the density of the medium. To reduce it, attempts were made to increase the air pressure in the tunnels. This approach raised the air density while leaving the viscosity unchanged, thus decreasing the kinematic viscosity. To achieve all this, however, was very expensive and difficult.

It is noteworthy that in nearly all media the kinematic viscosity is either greater than, or roughly equal to, the kinematic viscosity of air. Liquid helium is the only exception to this rule. The idea of Professor Loitsyanskii consists in using a current of liquid helium in studies of the aerodynamic properties of airplane models. This idea is theoretically correct and experimentally far-reaching indeed; one day it may find its application. Meanwhile, however, much remains to be done on the properties of liquid helium itself as well as on the techniques for implementing this experiment.

What matters to us now is the unexpected prospect of finding new applications of a recently discovered phenomenon.

I am convinced that many such applications, just as unexpected and richly rewarding, will be found in future. Yet we must now wait until the research scientist himself comes up with the idea. Such proposals must be the fruit of common interest and cooperation of creative people from various fields of life. Stimulating interest in science and making it more and more accessible to an ever expanding circle of Soviet workers is a necessary condition for a rapid application of recent scientific achievements. Science too needs public relations work if it is to take roots among the people. In the U.S.S.R., the Academy of Sciences is best equipped for this task.

4. OXYGEN

I suppose you realize the difficulties I face in choosing the theme of my present report. You study humanities, economics, history. However, they do not teach you technology or its foundation – namely, physics. Perhaps you have even forgotten most of the physics you studied in secondary school. For this reason I shall speak today not on a purely scientific subject but on more general matters – that is, about oxygen, and I shall tell you of the emergence of an interesting scientific problem in research laboratories, its gradual transition from small-scale laboratory experiments to technology and industry, and finally, how this matter began to exert a major influence on the national economy. Using this example I intend to show how science influences the progress of culture in our country.

We can also cite another, rather well-known example of vast practical consequences arising from an apparently trifling observation. Just over a century has passed from the day on which the Italian physician Galvani, while experimenting with a frog, suspended its leg on the metal railing of his balcony, and found that mere contact with two different metals caused spasmodic movements of the leg muscles in the absence of any external electrification.

Such muscular contractions were already known at the time, but only under the influence of electric discharges. Volta, who was more perceptive than Galvani, was the first to realize that the Italian physician had discovered a new source of electricity, namely, an electrochemical element. Having correctly interpreted Galvani's observations, in 1799 Volta created his famous 'voltaic pile' which consisted of a stack of copper and zinc disks separated by disks of cloth moistened with an acid. This device provided a considerable electromotive force. Thus a source of direct electric current was obtained, and a study of the physical properties of electricity became possible.

Many scientists working in the 19th century alone managed to establish all the fundamental electromagnetic phenomena caused by electric currents, and worked out a theory to explain them. Perhaps you know that these

---

Lecture read at the Central School of Party Organizers, Central Committee of the Communist Party of the U.S.S.R., 1944.

discoveries paved the way for the creation of powerful sources of electrical energy and its wide application in technology and in the national economy.

We all realize now that our life is unthinkable without electric current. In historical terms, modern electrotechnology is rather young; its growth covers essentially a single century, and our grandfathers saw its birth.

On the other hand, the story of applications of oxygen, which will be told here, being a matter of our own time, is still unfinished. Here too one can see quite clearly how a scientific discovery, once born in the laboratory, soon begins to exert an ever increasing influence on technology and the national economy. I chose this particular problem as the theme of the present lecture because I took an active part in the work on the industrial applications of oxygen.

Oxygen gas as a component of air was discovered long ago. In the second half of the 18th century three scientists – Scheele in Sweden, Priestley in Britain and Lavoisier in France – established almost simultaneously that oxygen supports combustion and named it at first 'fiery' or 'vital' air. Later Lavoisier named it 'oxygen' in recognition of the fact that it produces acids with many combustible substances (with phosphorus, phosphoric acid; with carbon, carbonic acid, etc.). The importance of oxygen to human life soon emerged in detail, and it was found that oxygen exerts a beneficial effect on sick people suffering from respiratory difficulties.

In the days of Lavoisier, chemists used to obtain oxygen from manganese peroxide, which occurs in nature. Later this element was obtained from a number of acids and salts containing a large proportion of it. Priestley and Scheele obtained oxygen for medical purposes from potassium chlorate, named Berthollet salt after the French chemist who discovered it.

Potassium chlorate, which in its physical properties resembles ordinary table salt and has the form of colorless transparent flakes, fuses upon heating and decomposes in the process, yielding oxygen. One hundred grams of Berthollet's salt yield about 29 liters of oxygen. Oxygen obtained in this manner was stored in pillows and was used to maintain the breathing of gravely ill patients. That was long ago. Obtained by such a difficult, low-yield procedure, oxygen found no wide application.

With the progress of physics it became clear that a different approach must be sought for the production of oxygen.

Air is obviously the most abundant source of oxygen. For a long time, however, this source remained unavailable to mankind until scientists found a way to increase its natural concentration of 21%. This happened some 70 years ago, when the Swiss physicist Pictet and the French physicist Cailletet almost simultaneously managed to liquefy air. At that time many gases had already been liquefied but there were other gases named 'permanent' by Faraday because of his erroneous conviction that they cannot be liquefied in any way. Air was assigned to the latter category because it could not be liquefied at the time, not even upon high compression.

Soon it became clear that the liquefaction of a gas requires not only a sufficiently high pressure but also a sufficiently low temperature known as the critical temperature. For air this temperature, as we know today, is $-141°C$ at a pressure of 32 atmospheres. Many attempts were made to find a method for achieving such low temperatures, and Pictet was the one to succeed. In the 1870s, Pictet obtained liquid air and filled a small vessel with it. This was regarded as a major scientific discovery at the time; among other honors, Pictet received a medal and was elected honorary member of various research societies.

The main difficulty in these works was the fact that liquid air could not be kept in such state for any length of time; it quickly evaporated upon being poured into a vessel. No one dared think that a liquid which could not be stored in any way could have any industrial application. Liquid air remained in the status of a curiosity item which could be produced in only one or two well-equipped laboratories in the world.

This state of affairs persisted for some twenty years until the British scientist Dewar concluded on the basis of theoretical considerations that vacuum can be used for thermal insulation.

Heat is transmitted by the impact of moving molecules. Heat transmission becomes difficult if only a small number of molecules are available. Dewar showed that if a vessel with double walls is made and the space between them is evacuated, ice placed in the vessel will remain for a long time without melting because of the negligible inflow of heat through the walls. Such vessels bear Dewar's name. Now they are widely used for the storage of food at a high or low temperature. Their popular name is 'thermos flask'. The discovery of the principle underlying the thermos was a major scientific event at the time. Dewar vessels made it possible to store liquid air and thus to examine its physical properties in greater detail.

Approximately in the 1890s, Bailey and Linde, in Britain and Germany, respectively, while studying liquid air as a mixture of two liquid gases, nitrogen and oxygen, found simultaneously that the partial evaporation of liquid air leads to its oxygen enrichment because the boiling temperature of liquid nitrogen is slightly below that of liquid oxygen. Closer examination revealed that under atmospheric pressure the difference between the boiling points of these two liquids is considerable, about 13°C. Linde was the first to realize that this phenomenon opens new prospects for a cheap production of oxygen and therefore may have great practical importance. These events took place some 25 years after liquid air was obtained for the first time.

Since antiquity, man has used fractional distillation of mixtures of various liquids on the basis of differences between the boiling points of the components. One example is the process for the production of alcohol in the distillation of home brew in the countryside. A similar distillation process is widely used in the petroleum industry.

Linde's idea was to subject liquid air to fractional distillation in order to separate the more readily boiling nitrogen from the oxygen. Even his early experiments easily yielded rather pure oxygen. This new method, far cheaper than the old chemical procedure used for filling oxygen bags for sick people, made it possible to produce oxygen in bulk, in quantities of the order of hundreds and even thousands of cubic meters. Thus oxygen became available as a stimulant not only of human respiration but also of large-scale processes such as combustion.

The idea emerged that if a combustible gas such as acetylene is made to burn in pure oxygen, – that is, in the absence of nitrogen which does not participate in the combustion and represents a harmful admixture which takes away part of the generated heat, then a much hotter flame can be obtained. Experiments showed that such a hot flame can locally melt any metal and can therefore be used to fuse two pieces of metal together without the aid of any solder, for example, to weld iron with iron. This development marked the beginning of autogenous welding.

A method for the autogenous cutting of metals was devised soon afterwards. You probably know that autogenous methods of metal working have now reached vast applications, among other areas, in the airplane industry and in shipbuilding. This was made possible by the discovery of methods for a cheap and large-scale production of oxygen directly from air.

Procedures for the fractionation of liquid air, originally proposed by Linde, have been in process of constant development for over fifty years. Oxygen production is steadily increasing. Current annual consumption of oxygen in the U.S.A. is 250 million cubic meters, an impressive growth from the early days of the present century, when the production of several hundred liters from Berthollet's salt was regarded as an achievement.

Now the following question arises rather naturally. We have managed quite successfully to intensify combustion processes, having obtained a very hot flame by feeding pure oxygen to the autogenous burners. But in view of the numerous oxidation processes which take place in nature, would it not be desirable to intensify them too?

For all their diversity, nearly all energy transformations in nature are associated in one way or another with processes of oxidation. Respiration is in effect a form of oxidation. At the expense of energy derived from this process we can work, move, and maintain our body temperature. The same principle applies to all living creatures, down to the majority of primitive bacteria. In addition, 90% of our technology is based on the utilization of oxygen. The combustion of gasoline vapor in the cylinders of internal combustion engines; the burning of coal in the furnaces of heating and power plants or in blast furnaces; the combustion of iron pyrites, and many other major technological processes, are based on oxidation. Imagine now that having speeded up respiration and having intensified the flame of the welding apparatus we proceed with a similar stimulation of all technological processes involving the use of oxygen. What benefit will this approach bring to the national economy?

To answer this question one must make an economic calculation for which it is necessary to know, firstly, the specific advantages of such oxygen intensification in each case and secondly, whether the expected advantages would justify the expenses incurred in the production of oxygen.

To begin with, the cheapest possible method for the production of oxygen must be found. Science is in a position to answer this question. Since the basic cost of oxygen is determined by the energy expended we must find the minimum amount of power, in terms of kilowatts for example, that must be used in order to obtain 1 m^3 of oxygen per hour from air. As it turns out, this magnitude is rather small, 0.08 kWh. It can be shown that this is the minimal power consumption for the production of oxygen from air. But what is the actual expenditure? In the present plants, the power outlay is 15 times greater than the theoretically minimal value. This is so because the existing methods for the production of oxygen are still far from perfection. Can they be improved in any way? Indeed they can.

Here I cannot enter into details about the reasons for the existing losses in the production of oxygen. Let me only point out that once the engineers have found the magnitude and cause of a loss they usually find a way to cope with it. Having determined the production costs of oxygen, we can judge in economic terms the desirability of its application in specific fields of technology today. On the basis of these data we can forecast the implications of the application of oxygen in different fields of our industry.

In view of the importance of this problem to the development of Soviet industry, a special agency (Glavkislorod, of which I am the director) has been established at the U.S.S.R. Council of People's Commisars. This agency comprises a Technical Council of prominent specialists from those manufacturing branches where the application of oxygen seems most promising. Academician I.P. Bardin is in charge of the agency department which deals with oxygen applications in metallurgy. You all know that he is an extremely knowledgeable and experienced engineer. An article written by him on the applications of oxygen in metallurgy was published in the 'Bulletin' of the agency. Since this field has first priority in the future applications of oxygen I shall discuss it in greater detail, citing some data from Bardin's article.

For example, what are the advantages from the introduction of oxygen blowing to blast furnaces? Two blast furnaces were already working with oxygen at the time, one of them in the Chernoreche, the other near Dnepropetrovsk. The latter is a large blast furnace already in operation for 5 or 6 months. Regrettably, experiments there, having reached their most interesting point, had to be discontinued because of the evacuation during the war. But the results obtained so far are interesting enough. Bardin concluded without any hesitation that the addition of sufficient amounts of oxygen to the air blast (a pure oxygen blast has not yet become feasible) will increase the output of iron 3.5–4 times. This is so because oxygen speeds up the reduction of ore in the blast furnace.

It became evident from these experiments that enriching the air by 1% oxygen improves the output of the blast furnace by 10%. Oxygen can also be profitably used in the subsequent stages during which the iron is transformed into steel in converters or open-hearth plants. Apart from speeding up the process to a considerable extent, this approach even yields a better grade of steel in the absence of nitrogen. Great savings can thus be achieved.

Let us assume, says Bardin, that metallurgy in the U.S.S.R. will attain the U.S.A. level, namely, the smelting of 90–100 million tons of steel annually. If we manage to achieve this, the transition to oxygen processes will save 10 billion rubles in investment alone. The saving in the cost of iron will be 16–17%. Naturally, we take into account that the amount of processed ore will increase in proportion to the output; in other words, the process itself is merely intensified without being changed.

Another fact of possible interest to you as economists emerges in this connection. Production costs are not the only factor in any intensification of output, as labor costs must also be considered.

I shall present the following simple example. Let us assume that one ton of a given product has to be manufactured. For this purpose two workers are hired for 300 rubles each. Thus a ton of the product costs 600 rubles. Now suppose that we have mechanized and intensified the manufacturing process. Under such circumstances we no longer need the same two workers for producing the same quantity of product. We shall be satisfied with a single worker who must, however, be more qualified. He will work the same hours as any of the former two men. However, he will have to be paid 700 rubles – that is, more than the former two workers put together. This will increase the product cost by 100 rubles despite the use of fewer man-hours. Is this approach economically sound or not?

If the matter is treated on a national scale the answer is affirmative. Any worker made redundant by this process can begin new studies. The training of manpower costs less than the benefit which the resulting more highly qualified labor brings to the national economy. Indeed expenditure for training is a negligible fraction of the cost of products of human labor. Therefore, any discussion of the economic soundness of the intensification of manufacturing processes should be conducted not only in money terms but also in man-hours, taking into account such factors as labor productivity and labor savings achieved.

Moreover, the transition of workers to more highly qualified jobs entails a higher professional level and a better living standard for them. On these grounds, the application of oxygen for the intensification of various manufacturing processes should be regarded as a positive development even if some preliminary calculations reveal an apparent loss in money terms, in view of the fact that the least qualified workers are thus released from their jobs. According to Bardin's calculations the use of oxygen in industry will reduce manpower by 40%.

I gave this example from the iron and steel industry because this branch is most thoroughly known and we possess reliable experimental data to support our conclusions. Calculations show that from this viewpoint the application of oxygen in many other fields of the national economy is highly desirable.

Oxygen is also applied in the nitrogen-fertilizer industry, in the production of cellulose, for the extraction of gold from ores, in the manufacture of inexpensive explosives (the so-called oxyliquit), etc. Further details on these matters can be found in the 'Bulletin' published by Glavkislorod. But even the present outline illustrates adequately the possible advantages of the application of oxygen for the intensification of industrial processes.

I as an engineer and physicist, together with my collaborators at the Institute for Physical Problems, have worked on the development of more efficient methods for the production of oxygen. As pointed out, the existing installations for the production of oxygen waste energy far in excess of the theoretical limit. Hence we have the task of rationalizing the process of oxygen extraction from air and producing a cheaper oxygen by cutting the energy expenditure.

Other problems remain, however. Apart from the cost element we need a much greater oxygen output to cope with demand. In the present case this is not so easy as it becomes clear that quantity soon turns into quality. Indeed, the first blast furnace using oxygen will consume as much of this element as the entire autogenous industry of the U.S.S.R. combined.

In striving to cover this rising demand for oxygen by conventional methods, engineers inevitably face a basic difficulty. Technology involves a peculiarity more acutely felt than is realized by engineers, although the problem can be rather clearly defined theoretically.

I shall illustrate this difficulty with the following example. In enlarging the size of any piston engine in order to increase its output it turns out that beyond a given size the engine weight per unit output will increase rather than diminish. Thus, if a steam engine of 100 horsepower (here we take totally arbitrary figures for characterizing the proportional values) weighs one ton, an engine of 1,000 horsepower would weigh more than 10 tons. Beyond a certain size any further enlargement of piston engines reduces their output per unit weight. For this reason the only practical way of building a more powerful piston engine is by increasing the number of cylinders rather than enlarging them.

Modern airplane engines provide a suitable example. Because of weight considerations, power gains are achieved by increasing the number of cylinders rather than their dimensions. Such engines now have up to 24 cylinders, and the latest American models as many as 48[1]. Engine weight is the main obstacle in increasing the size of airplanes.

Any policy of large-scale oxygen production by the use of piston expanders and compressors for cooling purposes would soon reach the limit of acceptable size for oxygen plants, and any further expansion of oxygen output would require increasing the numbers of piston engines rather than their size.

A similar situation exists in thermoelectric power stations. Were we to equip our present powerful thermoelectric power plants with Watt-type

---

[1] This lecture was delivered before the introduction of turbopropeller and turbojet airplane engines. With piston engines any further increase of airplane size became impossible.

piston machines such as those still occasionally found on obsolete Volga ships or at small power stations, these machines would have to be of a size beyond the limits of feasibility. It is well known that the power growth problem was solved with the introduction of the steam turbine invented by Laval and Parsons, the output of which per unit weight is many times greater than that of the piston-type steam engine. This is why the large modern electric power stations operate with turbines only.

Air was initially liquefied by a method based on the so-called Joule-Thomson effect, which can be defined as follows: the greater the pressure of a compressed gas, the deeper its cooling when allowed to expand freely. This pressure was usually about 200 atmospheres. Later Claude in France and Heiland in Germany achieved the cooling effect by letting the compressed air expand in a special piston engine named 'expander' [or expansion engine] and operating in much the same way as the steam engine. It is widely known that the steam engine works by the expansion of hot steam which, having performed the work, leaves the engine in a much cooler state. The similarity lies in the fact that upon expanding, the compressed air likewise performs work and cools down. This is the principle of the piston refrigeration engine, (expander). Upon entering the cylinder the compressed air expands, performs work and cools down. Calculations show that for a large-scale production of liquid air for a subsequent fractionation of oxygen, a large output can only be obtained by introducing turbines and discarding all piston-type compressors and expanders.

The possibility of using refrigeration turbines was pointed out long ago by the well-known British physicist Rayleigh. As long as 40 years ago he proposed the use of turbines for the liquefaction of air. His reasons, however, were somewhat different and not particularly related to the necessity for a large-scale liquefaction of air. Rayleigh was concerned mainly with lubrication problems in piston-type expanders. All liquid lubricants freeze at low temperatures. The turbine, on the other hand, requires no lubrication at all. Many attempts have been made since toward the use of turbines as expanders, largely without success.

Let me tell you now about our own work in this field, as we were the first to obtain liquid air by using a turbine as an expander, and with satisfactory technical characteristics at that. Our basic idea is simple enough to have us wondering why no one ever thought of using it.

The general course of thought can be outlined as follows, simply speaking. To achieve refrigeration we built piston-type expanders and in order to improve their efficiency we resorted to high pressures, just as in energy production there was a trend toward piston engines working under a high steam pressure. Later the piston engines of power plants were replaced with turbines in order to obtain an even greater output. Obviously a similar step had to be done for a large-scale production of liquid air. Pursuing this analogy, engineers began to use conventional types of steam turbines as expanders in refrigeration technology. As it turned out, these machines did produce cold, but at a low efficiency.

This example teaches us once again that analogies should be approached with caution. Spellbound by the analogy of thermal processes in refrigeration and steam engines, engineers overlooked a factor of crucial importance. They failed to realize that air, owing to its great compressibility at low temperatures, becomes so dense that in physical properties it resembles water rather than vapor. This means that refrigeration turbines must be designed after the principle of water turbines rather than after their steam counterparts, namely, by using some modification of the well-known Jonval design of reactive turbines. When I drew the attention of the designers of our oxygen plants to the fact that they were using the wrong type of turbine they disregarded my remark. The answer they gave me was roughly as follows: steam turbines are being used abroad; your suggestion runs contrary to the policy of those firms; it is merely the abstract theory of a scientist.

Then we decided to design and build in our Institute a refrigeration turbine similar to a water turbine and test experimentally whether it will have the same high efficiency which characterizes water turbines. This work took 2–3 years and proved successful. Now our turbine enjoys general recognition in the U.S.S.R. and abroad, and it even received a government award.

This example clearly illustrates the tendency of people to neglect the perfectly obvious, that air in the process of cooling acquires a new property normally associated with liquids, although it remains in the gaseous state. Hypnotized by the conventional approach to the problems, designers do not readily accept anything new even if it provides a solution to the problem.

Once this peculiarity of air at low temperatures was taken into account in turbine design, new prospects emerged for a large-scale production of oxygen. To publish numerical data is not wise in war time. Let me tell you, however, that there is a factory where our turbines have been used successfully for several thousand hours so far. In Moscow, one third of the oxygen output is produced in this manner. (One of the earliest experimental models of the rotor of a radial turboexpander is shown in the figure.)

This trend in the production of oxygen is steadily expanding. Life, however, always places obstacles in the way of any new development. Though essentially

simple, the idea faces a series of new technical difficulties in the process of its implementation. In the construction of a high-velocity turbine operating in the dense medium of a cold gas, for example, the rotor becomes unstable. A new type of stabilizer had to be designed to cope with this problem.

Psychological obstacles also have to be overcome. People cling to old concepts; this applies also to Soviet industry, where conservatism prevails. Gradually overcoming these common facts of life, we are beginning now to have an edge over the West in the fields of extreme cold technology and applications of oxygen. And an interesting fact becomes evident in connection with the way in which such problems are being solved in the U.S.S.R. national economy.

Complex, large-scale innovations are, as it turns out, somewhat easier to implement in the U.S.S.R. than in the West. Indeed, the achievement of a minor invention in our country is often difficult, but industry readily adopts major innovations representing vast technological advances. To explain the causes of this phenomenon I shall give the following example from everyday life.

Let us assume that the implementation of a major discovery requires the participation of two or three branches of industry. For example, the combustion of waste gases from open-hearth plants with the use of oxygen is energetically profitable. In the U.S.S.R., three government agencies are interested in this process, namely, those in charge of electric power stations, where the waste gases can be used for thermoelectric power stations; iron and steel industry, where the productivity of open-hearth furnaces can be increased with the introduction of oxygen in smelting; and the oxygen industry, which must supply the gas to the open-hearth furnaces. All three government agencies are organs of an integrated socialist economy, whereas in capitalist countries these branches of industry usually belong to firms financially unrelated to one another. Such innovation may prove highly advantageous to one of these firms, the second may derive a small profit whereas the third may even suffer losses.

Although the national economy as a whole derives a benefit in the final analysis, combining the interests of three private firms is a legally and financially difficult task. In the U.S.S.R., on the other hand, one government decree can enforce the matter on all three agencies, thus paving the way for a successful implementation of a complex technological development.

Many similar examples can be given. We have another great advantage of which we still make little use in the implementation of discoveries in our complex economy. We can set up technical experiments on a very large scale without fearing any risks. Indeed one should not avoid risks when dealing with innovations. A discovery can never be implemented with a 100% certainty of success, as unforeseen difficulties always appear. If on a national scale there are prospects for savings amounting to billions, the risk of losing a few millions is obviously a well-calculated one. In the capitalist countries even a major company cannot afford to risk such considerable sums of money and would not dare commit itself to such a major experiment. In the U.S.S.R., on the other hand, the difficulties arise mainly from conservatism and routine approaches on the part of certain bureaucratic officials.

In presenting this report, I took into account that all of you as responsible Party activists and leading public figures should be in the front line of the struggle against the conservatism and inertness of U.S.S.R. economic officials. I hope that each of you, upon returning to your respective places all over the country, will remember my words and contribute toward the progress of our Socialist land.

5. ON THE NATURE OF BALL LIGHTNING

The nature of ball lightning remains obscure. This is so because ball lightning represents a rather rare phenomenon which no one has so far managed to reproduce quite convincingly in laboratory conditions. There are many hypotheses about the nature of ball lightning [1,2], but our present remarks have apparently not been published so far. They deserve attention mainly because their verification leads to a well-defined avenue of experimental research.

In our opinion the existing hypotheses of the nature of ball lightning are unacceptable because they contradict the law of the conservation of energy. Indeed, the luminescence of ball lightning is usually attributed to energy released in some molecular or chemical transformation; in other words, it is assumed that the source of energy for the luminescence of ball lightning lies within the lightning itself. This explanation faces the following basic difficulty.

It follows from the fundamental concepts of modern physics that the potential energy of a gas in any chemical or active state is less than the energy required for the dissociation and ionization of molecules. This makes it possible to determine quantitatively the upper limit of energy that can be stored in a gaseous sphere filled with air and corresponding in size to the ball lightning.

On the other hand, the intensity of radiation from the surface of ball lightning can be determined quantitatively. Such rough calculations show that the upper limit of the duration of luminescent light emission is far below what is actually observed in ball lightnings. This conclusion has been confirmed experimentally from the published data on the duration of light emission from the cloud formed as a result of a nuclear explosion [3]. Immediately after the explosion, this cloud undoubtedly represents a fully ionized gaseous mass, which means that it could be regarded as containing a maximum store of potential energy. One could assume on these grounds that the cloud should retain its luminosity longer than the most long-lasting ball lightning of a similar size. Yet this is not the case.

Since the energy stored in the cloud is proportional to the volume (d^3)

Dokl. Akad. Nauk SSSR **101**, 245 (1955).

whereas the emission is proportional to the area ($\sim d^2$) it follows that the duration of energy discharge from the ball will be proportional to d, its linear dimension. A nuclear explosion cloud of diameter d = 150 m will be totally discharged within less than 10 s [3], so that a ball 10 cm in diameter will be exhausted within less than 0.01 s. In fact, however, it is evident from the literature that ball lightning of such size usually persists for several seconds and often as long as one minute [1,2].

Thus, if energy sources unknown to us do not exist in nature, it follows from the law of conservation of energy that energy is continuously being supplied during the luminescence of the ball lightning, and we must look for the source of this energy outside the boundaries of the ball lightning. Since ball lightning usually appears to be 'suspended' in the air without any direct contact with a conductor, the most natural and apparently the only available mechanism of energy supply is the absorption of intense external radio waves by the ball lightning.

Let us adopt this assumption as a working hypothesis and find out whether it agrees with the most typical phenomena of ball lightning [1,2,4].

Comparison of the behavior of ball lightning with that of the luminous cloud formed as a result of a nuclear explosion reveals a major difference. The nuclear cloud, once formed, grows steadily and then silently fades away. On the other hand, ball lightning remains constant in size throughout its luminescence period, and ofter breaks down with an explosion. The nuclear explosion cloud, being filled with hot gases of low density, is lighter than air and therefore moves only upward. As for ball lightning, sometimes it remains immobile and in other cases it moves, but this movement has no preferential direction with regard to the Earth and does not depend on the direction of wind. Now we shall show that this typical difference can be fully explained by our hypothesis.

It is known that an effective absorption of electromagnetic waves by an ionized gaseous cloud, called 'plasma', can only take place in case of resonance, that is, when the characteristic period of electromagnetic oscillations of the plasma coincides with the period of the absorbed radiation. At the ionization intensities which cause the bright luminosity of ball lightning, the resonance conditions are entirely determined by the external dimensions of the ball.

Assuming that the absorbed frequency corresponds to the characteristic oscillations of the sphere, it follows that the length λ of the absorbed wave must be roughly four times greater than the diameter of the ball lightning (more precisely, $\lambda = 3.65d$). Now if the gas is only slightly ionized in this volume, then the period of plasma oscillations depends primarily on the degree of ionization, in which case the corresponding resonance wavelength will always exceed the wavelength which is determined by the size of the ionized volume, and which equals $3.65d$, as noted above.

The absorption mechanism in the formation of ball lightning can be outlined as follows. At first there is a plasma volume of small size in relation to $(\pi/6)d^3$. If its ionization is low, however, a resonance with a wavelength $\lambda = 3.65d$ will nevertheless be possible and effective absorption of radio waves will take place. This in turn will increase the ionization and the sphere will begin to grow until it reaches the diameter d. Then the resonance nature of the absorption process will depend solely on the form of the ball lightning, so that the latter will become stable in size.

Indeed, let us assume that the intensity of absorbed waves is increasing. Then the temperature of the ionized gas will rise somewhat and the sphere will expand. This enlargement, however, will bring it out of resonance; consequently, the absorption of electromagnetic waves will decrease, the sphere will cool down and will return to a size corresponding to resonance. This explains why the observed diameter of the ball lightning remains constant during the emission of light.

The observed ball lightnings range in size from 1 to 27 cm [4]. According to our hypothesis these magnitudes, multiplied by four, yield the wave band which is responsible for the formation of the ball lightning in nature. Thus, the commonest ball lightning diameters, 10–20 cm [1] correspond to a wavelength of 35–70 cm.

The most favorable areas for the formation of ball lightnings will obviously be those where radio waves reach peak intensity. Such locations will correspond to electrical potential antinodes obtained in a variety of possible interference phenomena. Owing to the higher strength of the electric field at the antinodes, the position of the latter will indicate the possible sites of ball lightning. According to this mechanism, ball lightning will move so as to follow the displacement of antinodes, regardless of the direction of wind or convection air currents [1,2].

As a possible example of such fixed location of ball lightning we shall consider a case in which radio waves fall on a conductive earth surface and are reflected from it. Interference phenomena will create standing waves, and at distances equal to the wavelength λ multiplied by 0.25, 0.75, 1.25, 1.75, etc. spatially fixed anti-nodes will appear in which the strength of the electric field is twice the magnitude of the incident wave. The high potential prevailing near these areas will create favorable conditions for the establishment of an initial breakdown as well as for the further development and maintenance of ionization in the cloud forming the ball lightning. Thus, an absorption of electromagnetic oscillations by the ionized gas can only take place in certain areas parallel to the earth terrain. These factors determine the spatial position of the ball lightning.

This mechanism explains why ball lightning usually develops at a relatively short distance from the earth and often moves in a horizontal plane. At any rate, the minimal distance between the center of the ball lightning and the conductive surface will be equal to 1/4 wavelength, so that the gap between the reflecting surface and the edge of the ball margin will be roughly equal to the ball radius.

In cases of intense oscillations, a number of discrete ball lightnings may appear in a series of antinodes at a distance of one-half wavelength from one another. Such strings of ball lightnings do occur; they are known as 'bead lightnings' [or coupled lightnings] and have been photographed [2].

Our hypothesis can also explain why ball lightning sometimes breaks down with a thunder which causes no destruction [1,2]. When the power supply is suddenly cut off, a small ball lightning cools down so quickly that it turns into a sphere of rarefied air the rapid filling of which generates a comparatively small shock wave. On the other hand, a slow de-excitation is associated with a calm, noiseless process of extinction.

Our hypothesis provides a satisfactory explanation for what is perhaps the most mysterious property of ball lightning – its capacity to penetrate into buildings through windows, slits, and frequently even through chimneys. Having penetrated into the room, the luminous sphere hovers for a few seconds or runs along conductors [1,2,4]. So many reports of such behavior have been described that their veracity is beyond doubt.

A highly interesting case was reported in which a ball lightning entered into an airplane traversing a thundercloud at an altitude of 2,800 m [5]. According to our hypothesis all these phenomena result from the fact that ball lightnings penetrate into closed rooms by following the path of short-wave electromagnetic oscillations which spread through openings, chimneys or conduits as along a wave guide. Usually the chimney size corresponds to the critical section of a wave guide where waves of a length of up to 30–40 cm can propagate freely; this also fits the size of ball lightnings observed inside buildings [1].

Thus, the concept of ball lightning as a phenomenon generated by short-wave electromagnetic oscillations sheds light not only on various other commonly observed but mysterious properties of ball lightning, such as its constant size, sluggish motion, the formation of chains and the explosion or blast wave which accompanies its extinction, but also its capacity to penetrate into buildings.

A pertinent question arises: can St. Elmo's fire, the long-known glowing, brushlike luminescence, be attributed to electromagnetic oscillations of a more limited power? The prevailing explanation [6] of this phenomenon speaks of a leakage of charges at prominent spiky points as a result of a direct voltage arising in the presence of a large potential difference between the earth and the cloud. This explanation sounded quite natural as long as the phenomenon was observed on the ground, where a closed path of the direct current can be demonstrated. Meanwhile, however, St. Elmo's fire was observed for rather long intervals on the fuselages of flying airplanes [7]. It appears, therefore, that here too our hypothesis could assist in solving this difficulty.

Although our hypothesis quite successfully solves a number of major problems and helps us toward a better understanding of ball lightning, we must admit that much remains to be clarified as it is necessary to demonstrate that electromagnetic waves feeding ball lightning actually exist in nature. The first question to be answered in this connection is why electromagnetic oscillations of wavelengths necessary for the generation of the ball lightning have not been detected during a storm.

Until such waves are actually found during a thunderstorm, we may venture the following assumption. In view of the rarity of ball lightning it is conceivable that the corresponding radio waves also occur rather infrequently, and chances are even slimmer that such waves would be detected by receivers operating in the shortwave range of 35–70 cm, which is not used very much. The next step toward a verification of our assumptions would therefore be the elaboration of a suitable experimental method of observation in an attempt to detect radio emission of the corresponding shortwave band during a thunderstorm.

As for the source of these radio waves, observations of ball lightnings have revealed two facts which might provide some insight. First, it is known that ball lightnings commonly appear toward the end of a storm; and second, an ordinary lightning immediately precedes the ball lightning.

It is evident from the first fact that ionized air promotes the generation of radio waves, whereas the second finding indicates that these waves are created by a thunder discharge. This leads to the natural conclusion that the radio waves originate from an oscillatory process which takes place in an ionized atmosphere either at the cloud or at the ground. In the latter case, if the waves originate from the ground, the area enveloped by intensive radio emission will be limited and adjacent to the site of the ball lightning. The intensity of radio waves may fall sharply with the distance from this site, which means that a sensitive apparatus is necessary for observations carried out at some distance away. On the other hand, if the radio waves are emitted by the thundercloud itself, they would occupy considerable areas and could easily be detected even by a receiver of low sensitivity.

Finally, a second possible course of research for testing our hypothesis is the creation in laboratory conditions of a discharge resembling a ball lightning. For this purpose one must possess a powerful generator of constant-intensity radio waves in the decimeter band, and an ability to focus them into a small volume. In the presence of an electric field of suitable voltage, the necessary conditions will be established for breakdown without electrodes, which, through an ionization resonance absorption by the plasma, will develop into a luminous ball approximately one quarter wavelength in diameter.

REFERENCES

[1] Brand, W., *Der Kugelblitz*, Hamburg, 1923.
[2] Stekol'nikov, I. S., 'Fizika molnii i grozozashchita', [The Physics of Lightning Protection from Thunderstorms]. *Izdatel'stvo AN SSSR*, Moscow, 1943, p. 145.
[3] *The Effects of Atomic Weapons*, London, 1950, § 2.15.
[4] Rossmann, F., 'Ueber den Kugelblitz', *Wetter und Klima*, March–April 1949, p. 75.
[5] Durward, J., *Nature* **169**, 563 (1952).
[6] Benndorf, G., *Atmosfernoe elektrichestvo* [*Atmospheric Electricity* (Translated from the German)]. Moscow, GITTL, 1934, p. 51.
[7] Schonland, B. F. J., *The Flight of Thunderbolts*, Oxford, 1950, p. 47.

6. HIGH POWER ELECTRONICS

Electronic processes are used on an ever expanding scale in modern electrical engineering. Yet there is one field where electronics still finds little application. Indeed, electronic processes are widely used in measuring instruments (cathode ray oscillographs, photocells, high-frequency measuring devices, amplifiers, etc.), in cybernetics (automation, computers, stabilizers, etc.), communications (radio, television, radar, etc.). However, the application of ultrahigh frequency [UHF] electronics for the solution of problems in the field of energetics is still in its initial stage.

I have proposed the name 'high power electronics' for this particular field of application. This name is, of course, arbitrary, as the boundary where 'high power' begins cannot be determined. For this reason I believe that the concept of high power electronics should refer to that field of electronics where UHF electronics serves to obtain a direct energetic effect, i.e. for the generation of electromagnetic oscillations which are transformed not only into electromagnetic waves but also into heat, energy of accelerated corpuscular beams, and other forms of energy.

In my opinion, the introduction of UHF electronics to the field of power energetics is one of the most promising trends in the development of modern electronics. The basic advantages of UHF energetics are quite evident, namely, the possibility of concentrating a large amount of electromagnetic energy in a small volume, and the extraordinary flexibility which characterizes the transformation of UHF energy into other forms of energy (concentrated supply of heat, acceleration of elementary particles, the creation, heating and maintenance of a plasma, etc.). The insufficient use of electronics in the field of power engineering can be attributed to the lack of efficient and reliable methods for the generation of UHF energy and for its transformation into other forms of energy.

The development of high power electronics requires that these problems be solved on a broad scientific basis.

Our work began with a theoretical study of the process of generation of

---

From the book *High Power Microwave Electronics* [In Russian: *Elektronika Bol'shikh Moshchnostei*, Moscow, 1962]

powerful UHF oscillations. We started from the assumption that powerful oscillations can effectively be generated only by electronic processes which take place in permanent (crossed) magnetic and electric fields, and we devised a method for the theoretical analysis of these processes. This method is sufficiently general and complete; it provided us with a clear-cut quantitative theory of processes taking place in magnetron-type generators.

This method and the main results obtained were described in our large work entitled *High Power Electronics,* which was completed in April 1952 and gave the whole trend its name. Our research in high power electronics began to expand in the wake of this work.

The initial part of the work (including its experimental and theoretical aspects) was carried out by myself in close cooperation with S.I. Filimonov and S.P. Kapitza. V.A. Fok showed constant interest in theoretical matters and gave much valuable advice. I am grateful to my friends and collaborators for their participation in my research despite the adverse circumstances under which it took place in 1946–1952.

The Tasks of High Power Electronics

Electronics encompasses a broad circle of physical phenomena involving the passage of electric current through a gas. Of particular interest are two important properties of electric current passing through a gas, properties which reveal totally new prospects for electrical engineering and are therefore widely used in practice.

The first property of electric processes associated with the passage of current through a gas is their extremely low inertia, which makes them easily controllable. The physical reason for this property became clear with the discovery of electrons, namely the charge is carried by electrons, the mass of which is several thousand times less than that of ions. The motion of electrons can be controlled quickly and effectively during the passage of electric current through a gas, in contrast to the corresponding process in metals.

The second major physical property of electric current traversing a gas is that the movement of electrons in a sufficiently rarefied gas takes place with very little 'friction' and consequently with negligible losses, so that the electrons carrying the electric current can be made to move at very high velocities. In metals, on the other hand, large currents with low ohmic losses involve a vast number of slowly moving electrons. In a gas, currents with similar losses can be achieved by means of a small number of fast electrons because losses become negligible at sufficiently low pressures.

The freedom of movement of electrons in a gas and their low inertial property have long been used in electron tubes, which have contributed much to the progress of modern radio engineering, especially in the UHF field. Yet electronic processes have not yet attained the characteristics required for their application in energetics. The prospects open for physical processes associated with the passage of electric currents through gases are made use of in energetics only for the solution of minor problems, if at all*.

Is this so because energetics does not need any quick processes? Absolutely not. High power electronics can indeed solve a whole series of otherwise insoluble problems in electrical engineering. Let us note here only a few of these problems: power transmission along wave guides over large distances with little losses; the generation of intense, well-directed beams of electromagnetic waves and corpuscular beams; direct uses of atomic energy; an effective method for the separation of isotopes. Even this partial list shows the great promise in the development of high power electronics.

Are there any basic factors which prevent the development of high power electronics? I believe that this question should be answered in the affirmative. Such obstacles do exist. Though insignificant at first glance they actually preclude any progress at this time. Only by overcoming these obstacles will it be possible to create high power electronic processes.

Electrons moving in vacuum (in the absence of ions) form a little negatively charged cloud. The identity of charges causes a mutual repulsion which impairs the regularity of movement. If the cloud has a low density, the repulsion caused by the space charges will have a negligible disturbing effect on the motion; as power grows, however, the density of the cloud increases and so do the repulsive forces. These forces can attain a magnitude at which any further increase of power can totally disturb the course of the electron movement. In conventional electronic devices such as electron tubes this phenomenon appears at a comparatively low power. To increase the admissible power by making the apparatus larger turns out to be senseless as it can be shown that the linear dimensions must increase with the square of the processed power, so that at a sufficiently large power all dimensions become impractically great.

The disturbing effect of space charges is the main limiting factor in the application of electronic processes to the field of power engineering. How can we cope with the limiting effect of space charges? This can be done in two ways, which often prove quite effective.

---

* Written in 1952.

The first of these methods is widely used; it consists in a compensation of space charges by positive gas ions. It is well known that if an electronic process takes place in a medium other than a high vacuum, the negatively charged electron cloud is pierced by positive ions which, because of their great inertia, do not participate in the dynamics of the process but are able, because of their own charges, to neutralize the mutual repulsion of the electrons. This approach yields electronic processes involving a considerable amount of power. Among its applications are mercury rectifiers, thyratrons, and other gas-filled devices.

This approach, however, suffers from two essential and unavoidable drawbacks which largely invalidate the major advantages of electronic processes. The first of them arises from additional losses caused by the collision of the rapidly moving electrons with gas molecules present in the working space. The second and most important drawback is that the presence of extraneous ions drastically reduces the prospects for an effective electrical control of the motion of electrons.

The other means for the control of space charges is more effective. It consists in a compensation of space charges by forces caused by the motion of electrons in a permanent magnetic field. To illustrate the mechanism of this process, let us consider one of the simplest and best known cases in which a magnetic field is used to compensate for the harmful effect of space charges when the electron cloud moves in parallel to the lines of force of the magnetic field. Under the influence of the space charge the electrons acquire transverse velocities which are perpendicular to the magnetic field; the Lorentz force arising in the process bends the trajectory of the electrons in a plane perpendicular to the basic motion and brings the electrons back into the cloud. As a result the moving cloud does not spread out and retains a constant cross section.

The focusing effect of a permanent magnetic field is well known and widely used in various applications for compensating for the repulsive action of space charges. This remarkable property of magnetic fields has its manifestations also in other, more complex causes, where it often remains unnoticed despite the fact that the underlying physical mechanism is analogous to that outlined above. The magnetron is a vivid example of a device where the focusing effect of a magnetic field on the electronic process is implemented in a concealed form.

It is known that the magnetron generates ultra-high-frequency oscillations which are created by the uniform movement of an electron cloud possessing a periodic distribution of charge densities. A sharply defined cloud with a high charge density can be formed only as a result of the fact that the repulsive forces between the electrons are offset by the permanent magnetic field which participates in the process. This is brought about by a process known as phase focusing. Hence the enormous output of modern magnetrons under pulse operation. This output reaches hundreds of kilowatts per square centimeter of cathode working area, around which the electron cloud moves. Such output is generated by pulses lasting just a few microseconds. This fact, however, does not affect the basic aspect of the problem since the time necessary for the establishment of electronic processes accounts for a negligible fraction of the pulse period.

It is evident from the actual attainment of such outputs that the limitations which space charges impose upon electronic processes involving high power can be eliminated if the motion of electrons takes place in a permanent magnetic field.

One is naturally inclined to ask why electronic processes in a magnetic field have not yet been used for the development of a high-power UHF electrical engineering. In my opinion this state of affairs can be attributed to three factors.

First, the great promise which electronics holds for the development of high-power electrical engineering has not yet been sufficiently realized.

Second, little work has been done so far on the theoretical foundations of problems which can be solved by high-power electronics.

Third, the physical nature of phenomena occurring in the various electronic devices is not well known theoretically, and various difficulties associated with the quantitative aspects of these phenomena and instruments are yet to be solved.

Our work was intended to provide a solution for all these problems in theoretical as well as experimental terms.

In my opinion the main task was to work out a feasible, instructive approach to a theoretical analysis of the electronic processes which take place in a permanent magnetic field. The inadequacy of the existing theoretical level can be illustrated by the calculation procedure usually applied in the design of magnetrons: after long years of what is essentially an empirical work, a large number of diverse magnetrons are made, and those with the best performances are selected among them; then on the basis of mathematical formulae derived from a similarity theory these magnetrons can be recalculated for other working conditions, whereupon their basic parameters remain unchanged. The magnetrons selected by this empirical procedure are used in satisfactorily operating devices the efficiency of which attains 60–70%.

This approach enables the design engineers to answer the present needs of radio engineering, but it contributes next to nothing toward the understanding and exploitation of all the possibilities hidden in the electronics of the magnetron.

In our studies, we strove to reveal the mechanism of electronic processes which take place in the presence of a magnetic field, and work out a calculation procedure applicable to them. This task consists in the solution of equations of motion which, though well established, are rather complex. So far these equations have only been numerically integrated, an approach which does not clearly reveal the physical nature of the phenomena involved. The solution of these equations is based on the periodicity of the processes concerned, which in turn results from the permanent magnetic field and the high-frequency oscillations. If this periodicity is mathematically eliminated by averaging, one obtains a simple and sufficiently precise solution which sheds light on the physical nature of the phenomena and leads to mathematical expressions suitable for practical calculations. The electron trajectories, previously determined by numerical integration, can thus be obtained in a clear form in elementary functions.

We were able to show the application of this method to various specific problems of electronics and to undertake the experimental confirmation of the obtained results. We made a thorough analysis of the processes which take place in the planotron and the magnetron. The planotron resembles the magnetron; the difference between them is that the magnetron has an enclosed design whereas the planotron can be visualized as a magnetron which has been opened up and spread out in a plane, which explains its name. In addition, we applied our method to the conventional multichambered magnetrons and compared the theoretical results obtained with the published experimental data.

We examined the resonance effect of the high-frequency field on the circular (Larmor) motion of charged particles, which brought us to a theoretical analysis of the isotope separation process based on this effect. The calculation was made by averaging; it led to a quantitative description of the process, which may also have other applications.

Finally, we made a general review of other, still highly problematic, tasks which could be solved by the planotron in high-power electronics. We showed how the task of energy transfer over long distances can be solved. Briefly, theoretical considerations reveal the reversibility of electronic processes in planotrons and magnetrons, which therefore cannot only convert direct current into high-frequency oscillations as is now the case but also vice versa, namely, high-frequency oscillations into direct current. This opens prospects for the transmission of electric energy along wave guides in the form of high-frequency waves.

All these considerations are still of a rather problematic nature; they merely indicate that the development of high-power electronics can be highly beneficial to science and technology alike.

Before entering the service of energetics, electrical engineering in the past century was largely confined to electrical communications (telegraphy, signaling, etc.). Perhaps history will repeat itself here: now electronics serves mainly for radio communication, but its future lies in the solution of major problems of energetics.

7. ON SOME STAGES OF RESEARCH IN THE FIELD OF MAGNETISM

I was asked to give the introductory address at the opening of the conference. Though I find little joy in the fact, it seems I am one of the oldest members of this conference, and therefore I accepted the honor of conducting its first session.

My thesis at the Polytechnic Institute in 1918, in A.F. Ioffe's laboratory, was in the field of magnetism, and all these years I have remained faithful to this realm of physics. Perhaps you will be interested to hear of those stages in the development of magnetism which have produced the deepest impression in me over the many years of my involvement in this field.

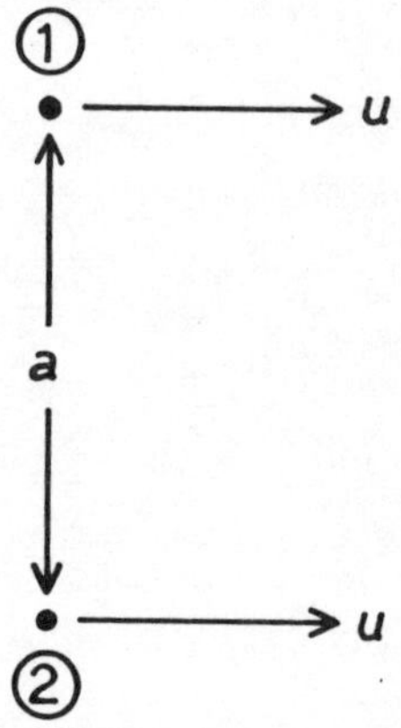

Fig. 1.

I shall begin with a very simple phenomenon. Point *1* (see Figure 1) contains the charge e, whereupon at point *2* located at a distance a from it there appears an electric field the intensity of which is

$$E = e/a^2. \quad (1)$$

An address at the opening of the World Conference on Magnetism, 1973.

Now if the charge e' is located at point *2*, a mechanical force of an electrostatic nature will develop between the two charges:

$$F_e = ee'/a^2. \tag{2}$$

If the charge e moves at the velocity u, a point located at distance a will also become the site of a magnetic field of intensity

$$H = (u/c)(e/a^2) \tag{3}$$

where c is the velocity of light. If the charge e' at point *2* moves in parallel with the charge e at the same velocity u, it is affected by an electromagnetic force. It is well known that this force is

$$F_m = -(u^2/c^2)(ee'/a^2). \tag{4}$$

Thus the entire force between such charges moving in parallel will be

$$F = F_e + F_m = (ee'/a^2)(1 - u^2/c^2). \tag{5}$$

Clearly the main force results from the electrostatic interaction and we subtract from it the force of the magnetic interaction (i.e. the term with the coefficient u^2/c^2) arising from the motion of the charges e and e' in the magnetic field.

Let us assume now that the observer also moves at the same velocity as the charges and parallel to them. Here we encounter a paradox since such an observer, in contrast to the one at rest, will characterize this same force of interaction between particles as an electrostatic one. In other words, a magnetic field will not exist for him.

This simple case reveals quite clearly the nature of the electromagnetic field; it is one of my favorite examples in my lectures to students. Today this paradox is easily explained by the theory of relativity. I mention it here because Maxwell deals with it in a somewhat more general form in the last chapter of his *Treatise on Electricity and Magnetism.* But in those days it was not at all easy to interpret this phenomenon.

I cite this example in order to draw your attention to the fact that the first edition of Maxwell's treatise was published exactly 100 years ago. (The photograph shows the title page of this remarkable book, which I bought as long ago as 1921 for a few shillings from a second-hand bookseller in Cambridge.) I think it is our duty to mention this memorable date at a conference on magnetism. Just as Newton's *Principia* constitute a theoretical generalization of the experimental works of Galileo and established the

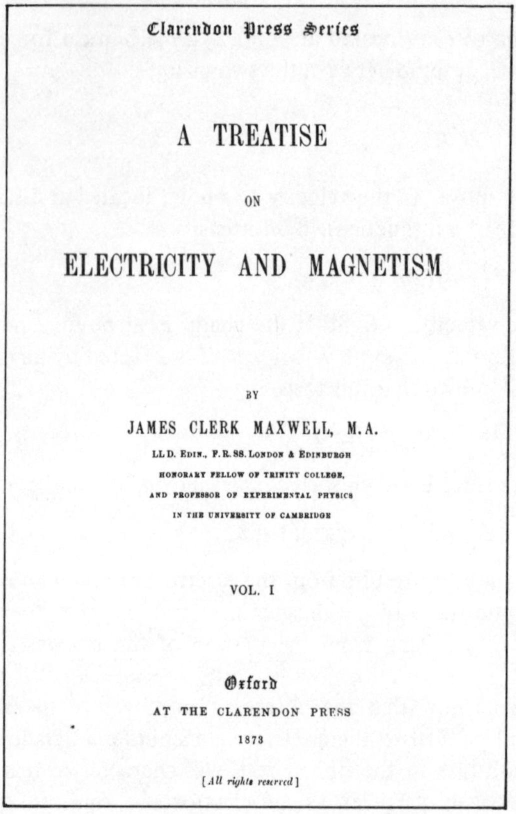

Clarendon Press Series

A TREATISE

ON

ELECTRICITY AND MAGNETISM

BY

JAMES CLERK MAXWELL, M.A.

LL.D. EDIN., F.R.SS. LONDON & EDINBURGH
HONORARY FELLOW OF TRINITY COLLEGE,
AND PROFESSOR OF EXPERIMENTAL PHYSICS
IN THE UNIVERSITY OF CAMBRIDGE

VOL. I

Oxford
AT THE CLARENDON PRESS
1873

[*All rights reserved*]

foundations of mechanics, Maxwell's *Treatise* represents a theoretical generalization of the experimental work of Faraday and can be regarded as the basis of modern electrodynamics.

In connection with this anniversary of the *Treatise* I would like to comment ruefully on the scarcity of its present readers although it is replete with real pearls, not only in the way in which it presents problems that have lost none of their interest in our time but also in the highly original approach to their solution.

The *Treatise* is easily understood today, and it makes aesthetically satisfying reading. In the past, however, it was difficult to assimilate, even 15–20 years after its publication. This is evident from Boltzmann's *Lectures on the Maxwellian Theory,* which were published in 1891 [L. Boltzmann, *Vorlesungen über die Maxwellsche Theorie,* Leipzig, 1891]. Understanding Maxwell's theory of the electromagnetic field was obviously a difficult task even for such a prominent scientist as Boltzmann, who therefore attempted to visualize the dynamics of processes occurring in an electromagnetic field by means of mechanical models. One of these models, used by Boltzmann in his lectures, is shown in Figure 2. Even without a detailed analysis of this model, the very complexity of its design illustrates the difficulties experienced by Boltzmann in understanding Maxwell's theory.

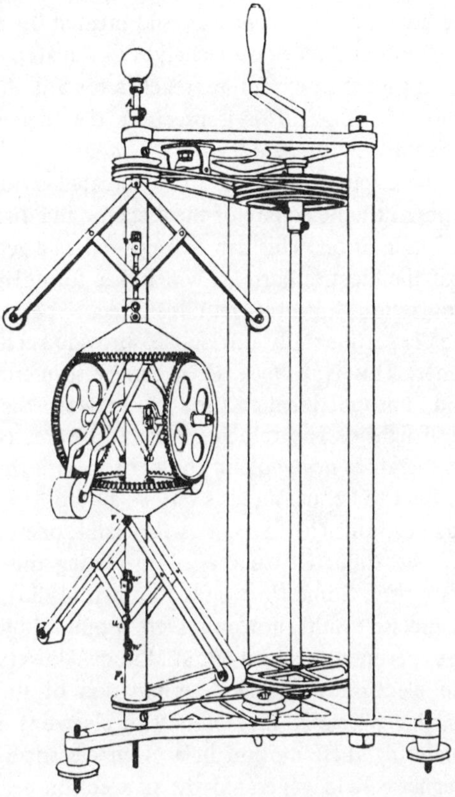

Fig. 2.

Today Maxwell's concept of the electromagnetic field is quite easily understood. From examinations of my students, I have repeatedly seen that they know and understand electromagnetic processes better than, say, the mechanics of gyroscopic phenomena. Perhaps this is due to the fact that from early youth they begin to learn about radio and television receivers. Some teachers today propose an approach opposite to that used by Boltzmann – namely, that mechanics be studied in terms of electrodynamics.

I hope you will agree with me that this anniversary, and in particular the role of the *Treatise*, deserve mention at our conference.

The above example of interaction between charges is also instructive for another reason: it leads directly to the conclusion that the magnetic field H is always created by charges which move with respect to the observer. In other words, the study of the magnetic field created by magnetized bodies can reveal the position and motion of charges in a material medium. This is one of the basic and most powerful approaches toward an understanding of the electric nature of matter. This is precisely the theme of research into magnetic phenomena.

According to the expression (3) the field created by a moving charge is inversely proportional to the square of the distance and directly proportional to the velocity of its motion. This can be regarded as a generalization of the Biot-Savart law of the electric current, which was formulated as long ago as 1820. This generalization, however, requires direct experimental proof. As it turned out, it was experimentally difficult to provide such proof in a direct and precise manner. This is so because, as can be seen from expression (3), the magnetic field is proportional to the ratio between the charge velocity u and the velocity of light c. The velocity u has to be created by a mechanical movement and is therefore negligible in comparison with the velocity of light; consequently the field to be measured is very small too.

The first to achieve this (in 1876) was Rowland, one of the most skilled experimenters of the time. He succeeded in testing this law, though in a rather approximate way. Some time later Röntgen (1885) obtained a better result. The most precise results, however, were produced by A.A. Eichenwald in 1903–1904 in experiments carried out at Moscow University. Following the discovery of the electron and the determination of its mass, it became possible to determine precisely the velocity of electrons moving freely in a vacuum by accelerating their motion in a given electric field. By measurements of the magnetic field generated by an electron beam A.F. Ioffe was able to verify Biot-Savart's law with satisfactory precision in 1911.

The discovery that the carriers of electricity, i.e. the electrons, possess not only a charge but also a definite mass led to the important conclusion that any magnetized body, provided that its magnetic moment results from the motion of electrons, must possess a gyromechanical moment and that the ratio of these moments is $e/2m$. Richardson showed in 1908 that this ratio can be tested experimentally by measuring the gyromoment; the latter magnitude, however, is very small even in a magnetized iron, so that its experimental observation is extremely difficult. Barnett managed to achieve this in 1914. His experiment can be outlined as follows. If an iron cylinder is rotated and it is assumed that the iron atoms represent gyroscopes, the directions of their torques will be deflected in the direction of the rotation axis, as is the case with the ordinary gyroscope. Thus the cylinder must acquire a magnetic moment in the same direction. Barnett managed to obtain this effect. His experiment was very difficult because of the low magnetization, which for the same reason could not be precisely measured.

Another approach – finding the ratio between magnetic moment and gyromoment – was achieved by the Einstein-de Haas experiment. Here the periodic remagnetization of a freely suspended ferromagnetic cylinder causes torsional oscillations from the measurement of which the ratio $e/2m$ can be determined. This experiment likewise requires a great skill; it was first performed in Leiden in 1915, initially with a low degree of precision.

To my knowledge there is no mention in physics books or encyclopedias that the idea of these experiments was expressed by Maxwell in his *Treatise.* Maxwell assumes that the electric current can possess an inert mass, and he presents two experiments performed by him in order to detect it. These experiments are described in § § 574 and 575 of the *Treatise.*

Figure 3 shows the device used for the experiment in 1861. Here the horizontal iron cylinder *A* can be magnetized by sending current through the coil which surrounds it. The cylinder is suspended in the same manner as the gyroscope wheel. The whole system can be rotated around a vertical axis. If the iron cyclinder assumes a magnetic moment upon magnetization, its axis should rotate in the direction of the vertical axis of rotation. This turning of the magnet can be noticed by observing the disk *C,* one half of which is painted red and the other half green. Upon rotation the disk appears to be white owing to the mixing of colors. If the magnetized cylinder begins to rotate, a red or green spot will appear in the disk center, depending on the direction of turning.

While in Cambridge I found this device in the Cavendish Laboratory in a cupboard containing old instruments. Obviously the idea underlying this experiment closely resembles that of Barnett.

The device used by Maxwell for his other experiment is shown in Figure 4. It represents a freely suspended horizontal coil through which current is passed along the suspension. Using the light spot from a mirror attached to the coil, Maxwell tried to detect oscillations of the coil resulting from the passage of current through it. This experiment resembles that of Einstein-de

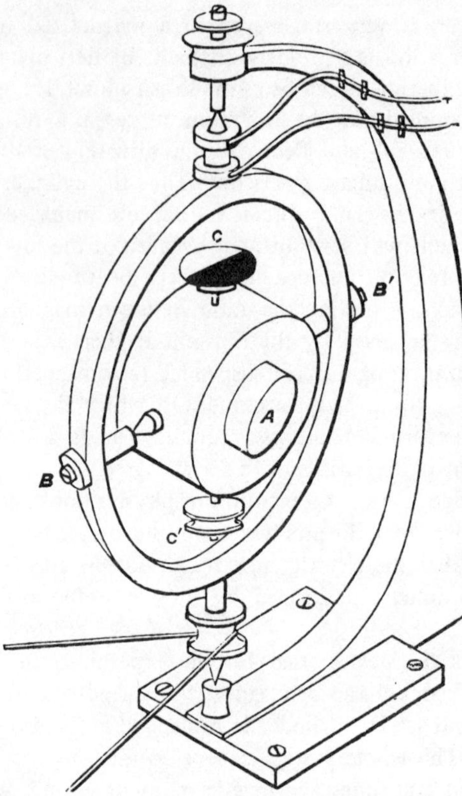

Fig. 3.

Haas in its basic concept. Maxwell writes in his *Treatise* that in both experiments he failed to observe any mechanical moment resulting from the passage of current. Today, knowing the actual magnitude of the effect, we can easily understand that the sensitivity of Maxwell's experiments was too low by several orders of magnitude to reveal any effect.

Because of experimental difficulties neither Barnett nor Einstein and de Haas succeeded at first in determining the ratio between the magnetic and the mechanical moment. Later, when a correct determination became possible, this ratio was found to be twice as large as expected. This surprising result proved to be extremely important in connection with the discovery of the spin nature of the electron itself.

I mentioned these experiments because in my opinion it is only just to

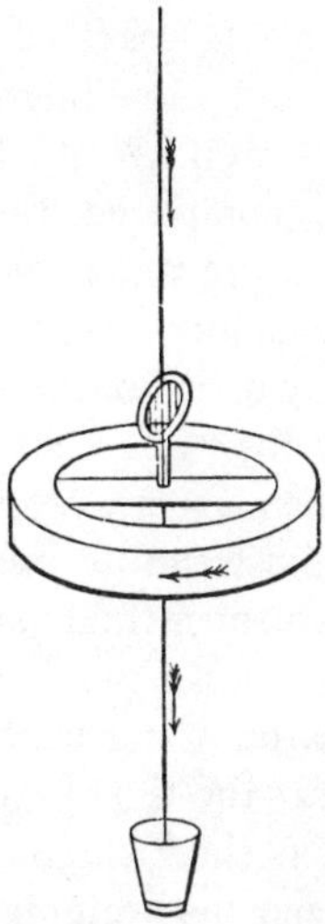

Fig. 4.

draw attention to Maxwell's ideas in these studies at least once in a hundred years.

In 1921 I arrived in England and became a research fellow of the Cavendish Laboratory, which was directed by Rutherford. One of my earliest works arose from the idea of observing the path length of alpha particles in a Wilson chamber in a magnetic field, in order to determine the velocity of each particle from the bending of tracks. Until then the Wilson chamber had never been placed in a magnetic field. For these experiments it was also necessary to find a method of creating a strong magnetic field of about 100 kilogauss in a coil large enough to accomodate a Wilson chamber in its magnetic field. All this seemed technically feasible since only a negligible fraction of a second was necessary for the formation of the tracks, so that the magnetic field would only have to exist for a brief interval of time. This eliminated the main obstacle toward the establishment of a strong magnetic field, namely, overheating of the coil. Calculations easily showed that the winding, in view of its heat capacity, would be heated by not more than 100°C during the brief passage of current through the coil, which lasts only a few hundredths of a second. The main difficulty stemmed from the need for a powerful pulse source of electric energy.

Large-capacitance condensers were not available at the time. We solved the problem by constructing a special storage battery composed of lead plates.

The 1–2 mm gap between the plates was filled with sulfuric acid. This storage battery had a low capacitance and a low internal resistance. It yielded pulses of the necessary output of 200–300 kW. All this could be achieved, and the experiments were successfully completed. Curved tracks of alpha particles were obtained, and their curvature measured. The analysis of the path of charged particles in a Wilson chamber placed inside a magnetic field remains a powerful method for the study of radioactive processes to the present day.

Afterward I began to apply the pulse method for the generation of strong magnetic fields in a study of the magnetic properties of matter. Although this theme did not quite fit the main trend of research at the Cavendish Laboratory, Rutherford took a benevolent attitude and placed everything necessary at my disposal.

When I started my studies on the magnetic properties of matter, the research conducted in France at the time produced a deep impression upon me. Highly interesting work in the field of magnetism was underway in France at the end of the last and the beginning of the present century. Particular contributions were made by a group of scientists composed of Pierre Curie, Langevin and P. Weiss.

It is generally known that the systematic study of the magnetic properties of various elements at different temperatures by Pierre Curie led to the discovery in 1895 of a fundamental law of magnetism which now bears his name.

Langevin's theoretical work on magnetism, performed in 1905 on the basis of Curie's research, provided the foundation for the contemporary theory of paramagnetic phenomena. Langevin's work started from Boltzmann's concepts of statistical mechanics which had led to the well-known distribution law in a force field at a given temperature. Langevin applied this law to the distribution of the directions of the magnetic moments of atoms in a magnetized body. This approach gave not only Curie's law of the effect of temperature on magnetization but also the quantitative relationship between the magnetic moment of the atom and the magnitude of magnetic susceptibility.

Weiss worked at the time in contact with Langevin and Curie. His studies were in the field of ferromagnetism. They can indeed be regarded as classic as they revealed the existence of the molecular field and provided an insight into magnetization processes occurring in ferromagnetic crystals.

Weiss conceived the idea of the existence of the 'magneton', a unit of the magnetic moment of the atom. He attempted to confirm this idea experimentally but did not succeed, since the theoretical foundations were inadequate at the time. Though Weiss was quite correct in this respect, the theoretical proof of his idea had to await the establishment of Bohr's model of the atom on the basis of the quantum theory.

The method used by Weiss for the determination of the magnitude of the magneton proved unsuitable for this purpose. As we know today, Bohr's magneton was observed and determined in the classic Stern-Gerlach experiment which involves the splitting of a molecular beam in a magnetic field.

It was believed at the time that further research in magnetism should proceed in two directions: first, by increasing the intensity of the magnetic field, and second, by lowering the temperature. The basic idea behind these approaches was to arrive at a state in which the energy of the interaction of atoms with the magnetic field exceeds the thermal energy. Obviously the magnetic properties of matter would emerge in a more conspicuous manner under such circumstances. For this reason Weiss incessantly strove to improve his electromagnet: he improved the yoke design, introduced water cooling for the coil, and made the pole terminals of cobalt steel.

The talented physicist Cotton joined this work. In Bellevue he built what was a large electromagnet for the time by making some improvements on Weiss's design. However, he was unable to obtain a field of more than 70 kilogauss even in a very small volume. Afterwards he equipped the coil with an intensive water cooling system, but did not obtain much better results. This method for the creation of a stationary magnetic field is widely used today. It is being applied most successfully at Harvard, in the Bitter Laboratory, where stationary fields of about 200 kilogauss were obtained in this manner. [Actually at Mass. Inst. Tech.]

For the creation of a series of magnetic phenomena, a strong magnetic field lasting for only a fraction of a second was obviously sufficient. Realizing this, I began to develop my pulse method. To increase the field intensity one must first possess a more powerful source of pulse energy. For this purpose we built a special pulse generator where the kinetic energy of the rotor yielded an output of about 200 MW for one hundredth of a second. Rutherford ordered this generator with the Manchester firm Vickers after raising, with great difficulties, the necessary amount of £3,000, an extraordinarily big outlay for research in those days. Having this output at our disposal we managed to obtain fields ranging from 300 to 400 kilogauss within a volume of several cubic centimeters, which was sufficient for our experimental purposes.

The coil strength turned out to be the limiting factor in any further increase of the intensity of the magnetic field. It can easily be calculated that in the presence of the field H, inside the solenoid there appears what could be designated a hydrostatic pressure the magnitude of which is not less than

$$p = (H^2/8\pi) \times 10^{-6} \text{ kg s/cm}^2.$$

It follows from this formula, for example, that the pressure p would reach 10,000 atmospheres in a field of 500 kilogauss. To prevent the coil from bursting with an explosion, which we witnessed more than once, it was necessary to build specially strong coils, replacing the copper with cadmium bronze and reinforcing the winding with steel bands on the outside. In a given volume, the fields obtained by us exceeded those created by electromagnets by approximately one order of magnitude.

Using such strong magnetic fields we magnetized various elements and studied their magnetic properties. We also examined magnetostriction, which we found not only in ferromagnetic substances; moreover, we observed the Paschen-Back effect in the spectra. However, the galvanomagnetic phenomena proved to be of greatest interest.

At the time the electron theory of electric conductivity predicted a quadratic law for the change of resistance in a magnetic field, beyond which the resistance was expected to reach a limit rather abruptly. In fact, however, in all metals used in our experiments, at normal temperature as well as at that of liquid nitrogen and hydrogen, we observed a linear change of resistance without any signs of saturation.

Matters proved different with elements possessing a poor conductivity strongly dependent on temperature and impurities. To this category belong tellurium, germanium and other elements which were later named semi-conductors. Only in these elements does the resistance follow an initial quadratic relationship and a subsequent saturation, as predicted with regard to metals. This phenomenon was first noticed by Gurney, and this particular course of electric conductivity provided an understanding of its nature in semi-conductors.

As for the category of metals, their galvanomagnetic properties remained obscure despite several attempts in this direction over a period of nearly 30 years until I.M. Lifshitz managed to unravel them in his remarkable works. This achievement was made possible by a more thorough understanding of quantum processes in metals, which shed light on the relationship between the galvanomagnetic properties and the structure of the electron energy spectrum, more particularly the form of the Fermi level in metals.

Generally speaking, I did not have much luck in my work on galvanomagnetic phenomena in strong magnetic fields, as it was often interrupted by unforeseen circumstances. In 1934 my work stopped since I had to leave Cambridge. Eventually the Cambridge installation, with the powerful generator for the creation of strong magnetic fields, was shipped to Moscow and reassembled there owing to the courtesy of Rutherford, but soon afterward I had to stop again because of the war and evacuation. When I returned to Moscow after the war I soon had to leave the institute for seven years, until 1954. During all those years my interest shifted to the fields of electronics and plasma. Work on magnetism was continued at our institute, but I no longer took part in it.

The second major trend of research into the magnetic properties of matter was in the low-temperature field. Of course, the most interesting results in this direction came following the liquefaction of helium and the attaining of temperatures close to absolute zero. This, as we know, was achieved in 1908 by Kamerlingh Onnes in Leiden. When I began my work at Cambridge in the 1920's, Leiden was the only place where helium was being liquefied, and the works of McLennan in Toronto and Meissner in Charlottenburg were just beginning. If I remember correctly they used to liquefy not more than 2–3 liters of helium per week at the time in Leiden; yet they used to perform highly interesting experiments with this quantity, including the discovery of superconductivity and a number of other remarkable phenomena. In my opinion, as far as the study of the properties of matter at low temperatures during the last century is concerned, the Leiden laboratory is second to none in the world. Practically all the major discoveries in the field of magnetism at low temperatures were made at Leiden, with the exception of the Meissner effect.

It is remarkable indeed that a single university in a small country should breed such succession of prominent experimental physicists as Van der Waals, Kamerlingh Onnes, de Haas, Keesom, and theoretical physicists of the caliber of Lorentz and Ehrenfest. Leiden retains a leading position in the field of magnetism to the present day owing to the work of Professor Gorter, who is a participant in our conference, and his collaborators.

What are the present avenues of research? I shall list them briefly.

As far as the generation of strong magnetic fields is concerned, the pulse method still yields peak values. However, the use of this method has become much easier owing to the introduction of large-capacitance condensers now in mass production. The problem of coil strength has also been solved; now the coils are made without a special reinforcement, but since the pressure created by the magnetic field has a pulse character it is compensated at present by viscosity forces arising in the coil winding during its deformation. In practical terms this necessitates a drastic reduction of the pulse duration, which is now in the environs of 1/1,000 of a second. Each experiment, however, causes a certain deformation of the coil, which consequently can last for a limited number of experiments only. Even this procedure, however, does not yield more than 500 kilogauss within a volume of a few cubic centimeters.

Much stronger magnetic fields were obtained by an adiabatic compression of a magnetic flux formed in a conducting cylinder. This compression was brought about by causing an explosion outside the cylinder. Fields in excess of one megagauss were obtained in this manner. However, a way must be found to use this field for an experiment taking place, figuratively speaking, inside an exploding bomb. To the best of my knowledge this has not been achieved so far, and one finds it difficult to speculate whether it ever will be.

The generation of strong magnetic fields by means of superconductors of the second kind was found to be highly useful for a number of experiments requiring a stable homogeneous field; here the superconducting solenoid provides great advantages. But despite the existence of alloys where superconductivity persists even in a very strong field, the magnitude of the field is still limited by the strength of the winding of the superconducting coil. It appears that technical factors make the construction of a stronger coil of superconductors rather more difficult in comparison with the conventional copper coil. For this reason one can hardly expect that superconducting solenoids will ever yield a much stronger field than that obtained with a water-cooled coil – that is, about 200 kilogauss.

As for low-temperature physics, the expansion of research caused an increased consumption of liquid helium, which at our institute reaches 350 liters daily. This, naturally, should be attributed to the large-scale application of superconducting solenoids.

Major progress toward the attainment of extremely low temperatures was achieved by the use of He^3 and its mixture with He^4. The method most commonly applied now is that proposed by London and performed by Neganov. Thus the temperature of matter can be brought down to 0.001 K. This value is close to the practical limit imposed by the existence of a temperature jump which almost blocks heat transfer. As I was the first to observe this phenomenon over 30 years ago, I am often blamed for the fact that it largely prevents any further lowering of temperature!

Present research in the field of magnetism devotes an increasing attention to the need for working with highly purified substances and faultless crystals. Progress in this direction owes much to the development of the radioactivation analysis method, which makes it possible to detect impurities at concentrations as low as $10^{-6}\%$ even if they belong to light elements. In the nearest future it will perhaps be necessary to analyze pure isotopes. A more extensive method for the separation of isotopes will have to be worked out for this purpose.

All these trends emerged in the years of my youth, but recently we have been witnessing the appearance of others. The most important of them came in the wake of E.K. Zavoisky's discovery of paramagnetic resonance in 1944. Zavoisky was then a young scientist working quite independently at Kazan University. I distinctly remember the general incredulity when he reported in Moscow on the results of his studies. A few days later, however, he reproduced his experiments in our institute, and there was no longer any doubt that a major discovery had been made. Following Zavoisky's discovery, research into resonance phenomena in high-frequency fields spread to various branches of science.

It was found that in any substance containing elementary magnetic moments, be it atoms, Larmor orbits, electron or nuclear spins, a magnetic field can change the fundamental frequencies of the oscillations of the magnetic moments; and these are studied by resonance. The frequencies involved range usually from radio to microwaves. This wave range is now of great practical importance in radio, television, radar, etc., so that present research in magnetic moments can draw on a considerable technical experience.

An extensive research investigation is now underway not only in paramagnetic but also in nuclear, cyclotron, and other forms of resonance.

Hardly any other approach to the magnetic properties of matter has revealed a wealth of information comparable to that obtained as a result of Zavoisky's discovery of resonance.

Measurements of magnetic fields by nuclear magnetic resonance never fail to fascinate me. This method now allows field measurements with an accuracy of a fraction of one percent, using a very simple device. I still remember the days when Piccard and Cotton spent over a year building a special balance for the measurement of magnetic fields with an accuracy that was lower by one or two orders of magnitude.

Among the more recently discovered phenomena used in the study of the magnetic properties of matter are the Mössbauer effect and the tunnel effect predicted by Josephson. These effects have yielded methods of almost inconceivable sensitivity in the measurement of atomic particles and magnetic fields.

Now a few words about what in my opinion is the conceivable future in the field of electromagnetic phenomena. There is a widespread view that all the basic electromagnetic processes in metals, semiconductors and dielectrics have been essentially understood and quantitatively defined, so that any further research would be necessarily confined to minor details of these phenomena.

I think that this is wrong. Let us consider, for example, polymers, which are composed of atom chains and where many basic processes still remain obscure. The living organism consists of such chains, which make up such formations as muscles, nerve fibres etc. Today we know that all life processes are of electromagnetic nature and involve the passage of electric current along these fibres. But we know next to nothing about the physical nature of these processes, although biologists assert that they understand them. Nevertheless I shall remain a sceptic until the polymer fibres – nerves for the transmission of information, and muscles capable of contraction – are made artificially in the laboratory. Only then will these processes be really understood.

Many new phenomena can be expected in molecular chains. Not only superconductivity but superdielectricity as well. The possible discovery of superdielectrics with a sufficiently high dielectric constant, capable of reflecting a high-frequency field without any losses, would be of great practical value not only for the long-distance transmission of information along wave guides but also in the field of high-power energetics.

Interest in this field of research is just beginning in recent years.

Scientists must remember that the most interesting and important discoveries come unforeseen.

Shakespeare's Hamlet said long ago: "There are more things in heaven and earth, Horatio, than are dreamt of in your philosophy".

Therefore no one should fear that any future conference on magnetism will suffer from a shortage of interesting topics for discussion.

In opening this conference on magnetism I have no doubt that it will prove interesting. I wish you fruitful work!

8. ENERGY AND PHYSICS

It is widely recognized that the main factor determining the development of the material culture of nations is the creation and use of energy sources. The work performed by people now exceeds by many times their muscular work. For example, in the most developed countries the power from different energy sources is up to 10 kilowatt per person and this is at least 100 times greater than the average muscular power of one man.

The role of energy in economics is well illustrated by Fig. 1 (see [1]). Along the abscissa we have the value (per person) of the gross national product of different countries and along the ordinate the energy resources, also per person. Except for the natural fluctuation, it can be seen that there is a simple proportionality between the two. Therefore, if people are deprived of energy resources, their material well-being will undoubtedly fall.

The obtaining, transformation, and conservation of energy are fundamental processes investigated by physics. The main law established by physics is the law of conservation of energy. On the basis of this law a global crisis in the acquisition of energy has been predicted [1]. The main energy resources currently used are peat, coal, petroleum, and natural gas. It has been established that the chemical energy stored in them was accumulated over millennia by biological processes. Statistical data on the use of these resources indicate that in the coming centuries they will be exhausted. Therefore, on the basis of the law of conservation of energy, man, if he does not find other sources of energy, will be faced with the need to limit his requirements, and this will reduce the level of material well-being of mankind.

The inescapability of a global energy crisis is now fully recognized, and therefore, the energy problem for technology and science has become problem number one. In the leading countries, considerable means are now being devoted to the scientific and technical investigations in this region. These searches are generally carried on in a narrow technical approach, without due allowance for the laws established by physics. Life has shown that the effectiveness of investigations is considerably enhanced if they take into

---

Report at the Scientific Session to mark the 250th Anniversary of the U.S.S.R. Academy of Sciences on October 8, 1975.

account more fully the basic laws of physics. In this communication, I wish to mention the laws of physics that should play a leading role in the solution of the energy problem.

The energy used by people is now divided into two parts. The first is the so-called consumer energy. It directly guarantees the cultural way of life. This energy is used for illumination, to operate refrigerators, televisions, electric razors, vacuum cleaners, and a large number of appliances used in everyday life. The power used in day-to-day living is usually measured in kilowatts. The other form of energy is industrial energy, in which high powers are involved.

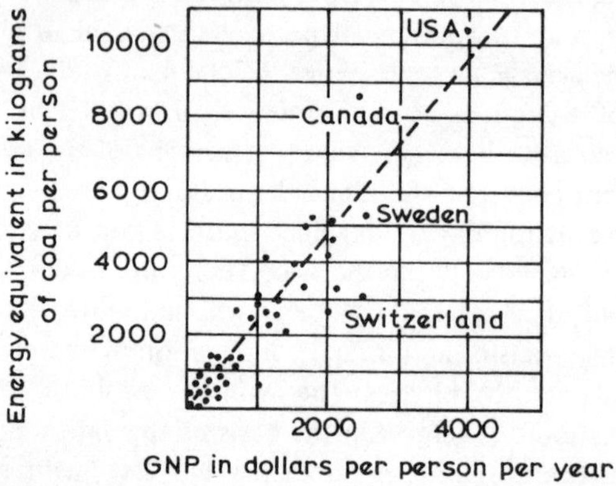

Fig. 1. Energy requirement and gross national product per person. The data refer to 1968 and are based on material of the United Nations Organization and the International Bank for Reconstruction and Development.

It is used in metallurgy, in transport, in mechanical engineering, in mechanization of construction, in agriculture, and in a number of similar fields. This energy is much greater than the consumer energy and is measured in megawatts; its scale and cost determine the level of the gross national product of a country. Of course, the impending crisis will be due to short fall in the energy resources for only the high-power energy supply, and it is the supply of this energy in sufficient quantity that is the main problem now facing science.

I have already said that the predictions of an impending energy crisis are based on the law of conservation of energy. Of course, another law which also plays a large role in restricting the possible use of energy resources is the law which says that entropy must increase in all energy transformation processes. Both these laws impose a 'veto' on the overcoming of the crisis by the creation of perpetual motion machines. The law of conservation of energy 'forbids' a perpetual motion machine of the first kind. Entropy forbids the so-called perpetual motion machine of the second kind. It is interesting to note that this second kind of perpetual motion machine is still being suggested today by inventive engineers and it is frequently difficult to refute them. This question relates to thermodynamics; it has been well studied, and I shall not dwell on it.

I shall restrict my consideration to the laws that govern the development of high-power energy sources and are related to the existence in nature of restrictions on the energy flux density. We shall see that these restrictions are frequently ignored, which leads to losses on projects that are definitely hopeless. This will be the main theme of my report.

All the energy processes of interest to us reduce to the transformation of one form of energy into another, and this takes place subject to the law of conservation of energy. The most widely used forms of energy are electrical, thermal, chemical, mechanical, and the so-called nuclear energy. The transformation of energy can usually be assumed to take place in a certain volume, energy in one form entering the volume through the surface and leaving it in a different form.

The density of the supplied energy is restricted by the physical properties of the medium through which it flows. In a material medium, the power of the energy flux U is restricted by the expression

$$U < vF, \tag{1}$$

where v is the deformation propagation velocity, usually equal to the velocity of sound, F may be any elastic or thermal energy, and U is a vector. In stationary processes, div U determines the amount of energy transformation into a different form. The vector U is very convenient for studying processes of energy transformation. It was first proposed 100 years ago, in 1874, by the Moscow physicist N. A. Umov. Ten years later the same vector for describing electric processes in the electromagnetic field was found by Poynting. In this country we therefore call it the Umov-Poynting vector.

If the expression (1) is used for a gaseous medium, it takes the form

$$U = AT^{1/2} p, \tag{2}$$

where A is a coefficient which depends on the molecular composition of the gas, T is its temperature, and p the pressure.

An expression of this form, for example, determines the limiting power that a hot medium can transmit to unit surface of the piston of a motor or the blade of a turbine. It can be seen that this power decreases with the pressure, and the same expression therefore determines the limiting altitude at which a turbojet aircraft can fly.

Using the Umov-Poynting vector, one can also describe processes in which the energy is transmitted by a belt drive. In this case, the product of the belt velocity and its elastic stress gives the transmission power. In the same way one can determine the limiting power transmitted by a band in a van de Graaff generator.

I have encountered in practice a technical problem when the flux of electric energy restricted implementation of a device. This occurred under the following instructive circumstances.

In the forties, my teacher A. F. Ioffe was concerned with developing an original construction of an electrostatic generator, which fed a small X-ray instrument. This generator was simple in its construction and worked quite well. Then Ioffe had the idea of replacing, on a large scale, electromagnetic generators by electrostatic generators and transferring to them the entire large electrical energy of the country. The main justification was the fact that the electrostatic generators were not only simpler in their construction but could give directly a high voltage for the transmission lines. At that time I had to demonstrate that this project could not be implemented on the basis of an estimate of the flux density of electrical energy in its transformation into mechanical energy.

Let us determine in accordance with the expression for U the flux density of the energy which is transformed from mechanical into electrical energy and vice versa in the gap between the rotor and the stator of the generator. Then v is equal to the rim velocity of the rotor of the generator. For constructional reasons, this velocity is usually taken to be about 100 m/s. The tangential forces of the interaction between the stator and the rotor in an electromagnetic generator are determined by the energy of the magnetic field, and we therefore have

$$F = \alpha H^2/4\pi. \tag{3}$$

The coefficient α is determined by the construction of the generator and it is characterized by the cosine of the angle formed by the force F with the velocity v. Usually, α has a value equal to a few tenths of unity. The magnetic field H is determined by the saturation of iron and does not exceed 2×10^4 Oe. When the flux of the electrical energy is transformed into mechanical energy or vice versa, one obtains about one kilowatt per square centimeter. Thus, for a 100-MW generator the rotor must have a working surface of about 10 square meters.

For an electrostatic generator the Energy F is

$$F = \alpha E^2/4\pi, \tag{4}$$

where the electrostatic field E is restricted by the dielectric strength of air and does not exceed 3×10^4 V/cm or 100 esu. Therefore, to obtain the same power of 100 MW one requires a rotor with a surface $(H/E)^2 \approx 4 \times 10^4$ times greater, i.e., equal to 4×10^5 square meters or half a square kilometer. Thus, a high-power electrostatic generator requires a virtually unrealizable size.

A similar analysis shows that the restriction on the energy flux density means that to obtain high power one must rule out a number of very effective energy transformation processes. For example, in gas elements in which there is a direct transformation of chemical energy of oxidation of hydrogen into electrical energy, the process can already be realized with a high efficiency reaching 70%. However, the possibility of using such elements for high-power generation is restricted by the very low rate of the diffusion processes in electrolytes, and therefore in accordance with the expression (1) its energy density in practice is very small and from a square meter of electrode one can take only 200 watt. For a 100-MW power the working surface of the electrodes reaches a square kilometer and there is no hope that the capital expenditure on the construction of such a power station would be justified by the energy it generates.

Another apparently promising direction in which, however, one cannot repose any hope is the direct transformation of chemical energy into mechanical energy. It is well known that these processes are widely used in nature in the muscles of animals.

To the embarrassment of biophysicists, these processes are still not understood, though it is well known that their efficiency is very high. Even if these processes are in the course of time reproduced outside animate nature, they will nevertheless be of no use for high-power energy sources since here too the energy density is low; it is restricted by the low rate of the diffusion processes that take place through the membrances or the surface of the muscle fibers, and the rate of the diffusion is here no higher than in electrolytes, so that the energy flux density cannot be greater than in gas elements.

The main interest is now devoted to the methods of generating energy that do not depend on the amount of energy stored in the past in fuel of different forms. The principal possibility is here taken to be the direct transformation of solar energy into electrical and mechanical energy – on a large scale of course. Here again a restriction on the practical implementation of this process for high-power energy comes from the restricted energy density. The optimal calculation now shows that the power taken from one square meter of surface illuminated by the Sun will not exceed 100 W on the average. Therefore, in order to generate 100 MW, it is necessary to take electrical energy from an area of one square kilometer. Not one of the hitherto proposed methods of transformation of solar energy can be implemented in such a way that the capital expenditure could be justified by the resulting energy. For the enterprise to be profitable, it is necessary to reduce the expenditure by several orders of magnitude, and as yet one does not even know ways in which this could be done. Therefore, we must assume that the practical direct use of solar energy on a large scale is unrealistic. But, as before, it does remain possible through its transformation into chemical energy, as has been done since time immemorial with the help of the vegetable world. Of course, the possibility cannot be excluded that in time one will find a photochemical process that opens up the possibility of more effective and simpler transformation of solar energy into chemical energy than takes place at present in nature. Such a process of chemical accumulation would also have the great advantage of making it possible to use solar energy irrespective of the change in its intensity during the course of the day or year.

The possibility is also now discussed of using geothermal energy. It is well known that at various places on the Earth's surface where there is volcanic activity this is done successfully, though on a small scale. The advantage of this method for high-power energy supply is undoubtedly very great; the energy reserves are here inexhaustible, and, in contrast to solar energy, which fluctuates not only diurnally but also with the season and the weather, geothermal energy can be generated continuously. As early as the start of this century the brilliant inventor of the modern steam turbine, Parsons, developed a project for its use. Of course, he could not foresee the scale on which energy is now required and his project has only historical interest.

The modern approach to this problem is based on the fact that at any place on the Earth's crust at a depth of 10–15 kilometers one reaches a temperature of several hundred degrees, which is sufficient to obtain vapor and generate energy with a good efficiency. In the implementation of this project in practice we again come up against the restrictions imposed by the energy flux density. It is well known that the thermal conductivity of rocks is very low and to carry heat to the water which must be heated in sufficient amounts with the small temperature gradients that exist within the Earth it is necessary to do this from very large areas, and at a depth of 10–15 kilometers this is very difficult and it is doubtful whether the necessary amount of water can be heated.

In the West a number of interesting propositions are now being advanced. For example, at this depth one could explode atomic bombs and thus produce either a large cavern or a large number of deeply penetrating cracks. The realization of this project would be very expensive, but in view of the importance of the problem and the great advantages of the geothermal method I believe that, despite these expenditures, one should risk the attempt of this project.

Apart from solar and geothermal energy, with inexhaustible reserves, there is also hydroenergy obtained by damming rivers and exploiting tides. The gravitational energy of the water accumulated in this manner can be very effectively transformed into mechanical energy. Currently, the use of hydroenergy accounts for not more than 5% in the energy balance and, unfortunately, a further increase cannot be expected. This is because the damming of rivers is profitable only in mountainous regions in which there is a large potential energy per unit area of the water reservoir. The damming of rivers when the water level is raised to a small height is usually not justified economically, particularly when this entails flooding fruitful earth since the harvest is usually much more valuable than the energy obtained. Once again the energy density is inadequate. The use of wind is also not justified economically, again because of the insufficient energy flux density. Of course, the use of solar energy, small water streams, and windmills can frequently be useful for everyday needs on a small scale.

But it follows from the above analysis that it does not appear possible to find an economic replacement for the depleted reserves of chemical energy in nature for high-power energy purposes. Obviously, one can and must be more careful about the use of energy resources. For example, it is desirable that they should not be wasted on military requirements. But all this only delays the depletion of the fuel reserves; it does not prevent the crisis.

As is already widely known, all hope for solution of the global energy crisis is based on the use of nuclear energy. Physics gives one every reason to believe that this hope is justified.

Nuclear physics suggests two directions for the solution of the energy problem. The first has already been well developed and is based on using a chain reaction in uranium when its nuclei are split with the liberation of neutrons. This is the same process that takes place in the atomic bomb, but slowed down to a stationary state. Calculations have shown that if it is correctly used there are sufficient reserves of uranium for there to be no danger of their depletion in millennia. Uranium power stations already operate and provide electrical energy at an economic rate. But it is also well known that three fundamental difficulties must be overcome if they are to be further developed and supply the entire energy requirements of a country.

(1) Wastes from the decay of uranium are strongly radioactive and their safe storage presents great technical difficulties that have not yet received a generally accepted solution. The best solution would be to dispatch them on rockets into space, but as yet this is thought to be not sufficiently safe.

(2) A large atomic power station supplying millions of kilowatts presents a great danger for surrounding nature and, in particular, man. In the case of an accident or sabotage, the radioactivity dispersed could destroy all living organisms over an area of many square kilometers just as effectively as did the atomic bomb on Hiroshima. The danger is now regarded as so great that in the capitalist world not one insurance company is prepared to take the risk on this scale.

(3) Wide use of atomic electrical energy also results in the wide dispersal of plutonium, which is a necessary participant in the nuclear reaction. Such dispersal of plutonium over all the countries of the Earth makes it more difficult to control the spread of atomic weapons. This means that the atomic bomb could even become a weapon of blackmail for an enterprising band of gangsters.

Under the threat of the energy crisis people will probably find a way of overcoming these difficulties. For example, the last two difficulties could be overcome by siting atomic power stations on small unpopulated islands in the ocean far from densely populated places. These power stations would be under careful control and the consequences of an accident would not present a great danger for people. The liberated energy could be used, for example, to decompose water and the resulting hydrogen could be transported in liquid form and used as fuel, which does not pollute the atmosphere when it is burnt.

It must however be recognized that the best way out of the dilemma would be to obtain energy by the thermonuclear fusion of the nuclei of deuterium and tritium. It is well known that this process takes place in the hydrogen bomb, but for peaceful use it must be slowed down to a stationary process. When this has been done, all the difficulties that result from the use of uranium will disappear, because the thermonuclear process does not produce radioactive waste in significant amounts, does not lead to great danger in the case of an accident, and cannot be used as an explosive material for a bomb. Finally the reserves of deuterium in nature, in the oceans, is even greater than that of uranium.

But the difficulties of realizing controlled thermonuclear fusion have not yet been overcome. I shall speak about them in my talk because, as it now appears, these difficulties are also mainly related to the formation of energy fluxes of sufficient power in a plasma. I shall dwell on this in somewhat more detail.

It is well known that to obtain useful thermonuclear energy the ions in a plasma must have a very high temperature – more than 10^8 degrees. The main difficulty in heating the ions comes about because the plasma is heated by applying to it an electric field, and virtually all the energy is acquired by electrons, which on account of their small mass, do not transmit the energy well to the ions in collisions. With increasing temperature, this transfer becomes even less effective. Calculations of the energy transfer in a plasma from electrons to ions in their Coulomb interaction were described in a satisfactory theoretical manner as early as the thirties. Landau[2] gave an expression for this interaction, which is still regarded as correct.

The power P_a given up by the electrons at temperature T_e to the ions at temperature T_i in a volume V is[3]

$$P_a = Vnk\,(T_e - T_i)/\tau_{eq}, \tag{5}$$

where k is Boltzmann's constant and n is the density of the plasma. The relaxation time τ_{eq} is calculated by Landau on the basis of Coulomb interactions, and it is found from this expression that for the high ion temperatures $T_i = 10^8 - 10^9$ K at which the thermonuclear reaction can give useful power the flux of energy transferred from the electrons to the ions is very small.

The expression (5) shows us that when the ion temperature is $T_i = 0.6T_e$ the transferred power has a maximal value. The maximal power transferred from the electrons to the deuterium ions is [3]

$$P_{max} = 1.57 \times 10^{-34}\ V\,(n^2/\sqrt{T_i})\quad \text{W}. \tag{6}$$

In a plasma at one atmosphere and with electron temperature $T_e = 10^9$ in one cubic meter the energy transmitted by the electrons to the ions is about 400 W. This is a small quantity since one can readily calculate that about 300 s are required in order to heat a cubic meter of plasma to 6×10^8 degrees when energy is supplied in accordance with the expression (6).

The smallness of the amount of energy transmitted to the ions is manifested in particular in the thermonuclear devices that are now most widely developed – the Tokamaks. In them, the ions are kept in a restricted volume by a strong magnetic field and the heating process is performed by electrons, which are first heated by a short current pulse to very high temperatures and then transfer their energy to the ions by Coulomb collisions. Under the conditions assumed in the modern designs of the Tokamak, the time during which the electrons give up their energy to the ions reaches 20–30 s[3]. It is found that during this time the major part of the electron energy is lost through bremsstrahlung. Therefore, one is now looking for ways for the effective transmission of energy from the electrons to the ions over a prolonged period[4]. This may be achieved either by high-frequency heating or injection of fast neutral deuterium atoms, or by the dissipation of magneto-acoustic waves[5]. It goes without saying that all these methods of heating the ions make the construction of Tokamak reactors much more complicated.

It can be seen from the expression for P_a that the efficiency of energy transfer between the electrons and ions increases with the density. We therefore suppose that in the case of heating by a laser pulse of solid condensed tritium or deuterium the initial density will be very high, by several orders of magnitude higher than in the Tokamak, and that it will be possible to heat ions by pulses in a short time interval. But calculations have shown[3] that although the heating time is indeed shortened to 10^{-8} s it is still inadequate since the plasma bunch, which is no way confined, will have expanded to a considerable size during this time.

For a laser-pellet reactor[4] one is now looking for methods of collective interaction of electrons and ions – for example, the creation of shock waves which, by adiabatic compression, can raise the ion temperature faster than the Coulomb interaction.

At the present time, the principal obstacle is that the physical processes in plasmas have not yet been sufficiently well studied. The theory, which is here well developed, applies only to a nonturbulent state of the plasma. Our experiments [6] on a freely evaporating plasma filament obtained in a high-frequency field have begun to show that a hot plasma in which the electrons have a temperature of several million degrees is in a turbulent state in a magnetic field. Even in ordinary hydrodynamics turbulent processes have not yet been given a complete quantitative description, and essentially all calculations are based on similarity theory. In a plasma the hydrodynamic processes are undoubtedly much more complicated, and we are therefore forced to proceed in the same manner.

As yet there are no reasons to assume that the difficulties of heating ions in a plasma will not be overcome, and I believe that the thermonuclear problem of obtaining high power will be solved in the course of time.

The main task facing the physicist is the deeper experimental study of the hydrodynamics of a hot plasma as is needed for the realization of a thermonuclear reaction at high pressures and in strong magnetic fields. This is a large, difficult, and interesting problem of modern physics. Its solution is intimately related to the solution of the energy problem, which has become the central problem of our epoch. Naturally, this is problem number one of physics.

REFERENCES

[1] Meadows, D. H., Meadows, D. L., Rangers, J., and Behrens, W. W., *The Limits to Growth*. New York, University Books, 1972, p. 70.

[2] Landau, L. D., 'Kineticheskoe uravnenie v sluchae kulonovskogo vzaimodeistviya' [Kinetic Equation for the Case of Coulomb Interaction], *ZhETF* 7, 203 (1937).

[3] Kapitza, P. L., 'Poleznoe poluchenie energii ot termoyadernykh reaktorov' [Useful Production of Energy in Thermonuclear Reactors], *ZhETF Letters* **22**(1) 24–25 (1975). [For English translation see *JETP Letters* **22**, 9 (1975).]

[4] Ribe, F. L., 'Fusion Reactor Systems', *Rev. Mod. Phys.* **47**, 7 (1975).

[5] Kapitza, P. L., and Pitaevskii, L. P., 'Nagrev plazmy magnitoakusticheskimi kolebaniyami' [Heating of Plasma by Magnetoacoustic Oscillations], *ZhETF* 67(4), 1411 (1974). [For English translation see *Sov. Phys. JETP* **40**, 701 (1974).]

[6] Kapitza, P. L., 'Svobodnyi plazmennyi shnur v vysokochastotnom pole pri vysokom davlenii' [Free Plasma Pinch in a High Frequency Field at High Pressure], *ZhETF* 57, 1801 (1969). [For English translation see *Sov. Phys. JETP* **30**, 973 (1970).]

9. PLASMA AND THE CONTROLLED THERMONUCLEAR REACTION

The choice of theme for this Nobel lecture presents some difficulty for me. Usually the lecture is connected with work recognized by the prize. In my case, the prize was awarded for work in low-temperature physics, at temperatures of liquid helium, a few degrees above absolute zero. It so happened that I left this field some 30 years ago, although at the institute under my directorship low-temperature research is still being done. Personally I am now studying plasma phenomena at those very high temperatures that are necessary for the thermonuclear reaction to take place. This research has led to interesting results and has opened new possibilities, and I think that as a subject for the lecture this is of more interest than my past low-temperature work. For, it is said, *les extrémes se touchent.*

At present, the controlled thermonuclear reaction is a process for producing energy, which can effectively resolve the approaching global energy crisis which is a consequence of the depletion of the fossil fuels now used as our principal energy source.

Intensive research on fusion is done in many countries and is connected with fundamental studies of high-temperature plasmas. That fusion is possible is well beyond doubt since it actually takes place in the explosion of hydrogen bombs. The detailed theory concerning fusion nuclear reactions is in agreement with experiments. But in spite of the great effort and the large sums of money spent up to now, it is still not possible to conduct or control the process of fusion so as to make it a useful source of energy. This certainly is a cause for some bewilderment.

One could expect that during the decades of experimental and theoretical work on plasma in studying the conditions for fusion we would have reached a sufficient understanding of the various facts that hinder us from setting up a controlled thermonuclear reaction. It could be expected that we should have revealed the main difficulties that bar our progress. In this lecture I hope to clarify what these difficulties are and the chances that they will be resolved.

[This lecture was delivered in Stockholm, Sweden, 8 December 1978, when the author received the Nobel Prize in Physics, a prize shared with Dr. Arno Penzias and Dr. R. W. Wilson. Minor corrections and additions have been made by the author.]

I will also try to explain the divergence of opinions of different scientists on the practical possibilities for obtaining useful thermonuclear energy. Before embarking upon this subject I would like to speak on the practical importance of obtaining energy from nuclear sources.

The reality of the approaching global energy crisis is connected with the unavoidable lack of the raw materials gas, oil, and coal. Also, it is generally considered that the GNP (gross national product), which determines the well-being of people, is proportional to the expenditure of energy. Energy resources depletion will inevitably lead to general impoverishment.

There are two main possible ways of escaping from the approaching energy crisis. The first, maybe the more attractive, is to extensively use the inexhaustible sources of energy, namely, hydroelectric power, the power of wind, solar energy, and geothermal energy. The second way is to use nuclear energy, which was discovered by man less than a hundred years ago. At present, heavy element fission power is already cheaper than energy from some nonexhaustible sources.

The main fuel in these reactors is uranium. As used at present there is enough uranium for only a hundred years. If, in the future, uranium will be more fully used in breeder reactors, it will last about 50 times longer, for a few thousand years. Some consider that uranium dissolved in seawater may also be efficiently used for cheap energy production. Thus it may seem that the processes now used in modern nuclear reactions may resolve the approaching energy crisis. But there are important reasons for not using uranium as a source of energy. These arguments are mainly connected with security.

First, the use of uranium leads to accumulation of long-lived radioactive wastes and to the problem of safely storing a growing amount of these waste materials. This is a problem that at present has not been definitely solved. Second, in a large energy-producing nuclear power plant a vast amount of radioactive material is accumulated, so that in a hypothetical accident, the dispersion of this material might lead to a catastrophe comparable in scale to that of Hiroshima.

I think that eventually modern technology will resolve these two dangers. But there is still a third hazard, even more grave. The danger is that the construction of great numbers of nuclear power stations will inevitably lead to such huge amounts of radioactive material disseminated around the world that an efficient control on its proper use will be practically impossible. In the long run not only a small country, but even a wealthy man or a large industrial organization, will be able to build its own atomic bomb. There is at present no secret of the bomb. The necessary amount of plutonium, especially if breeders are to be widely built, will be readily available. Thus, recently, in India a small bomb was built and exploded. With the present system of international organizations there is no one with sufficient authority that could execute the necessary control of the peaceful use of uranium as a source of energy. Moreover, it is not clear how such an organization could be set up. This is the main reason why it is most important to obtain energy by the third way, through the process of thermonuclear fusion.

It is common knowledge that this process will not lead to generation of large amounts of radioactive wastes and thus to a dangerous accumulation of radioactive material, nor does it open up any chances for a feasible nuclear explosion. This is the main reason why the solution of the scientific and technical problems involved in controlled thermonuclear fusion is considered of prime importance by many physicists.

The conditions for the thermonuclear reaction for energy production have been firmly established. There are two reactions of importance: the D + D and D + T process. The first one is the reaction between two nuclei of deuterium. The second occurs in the interaction of deuterium with tritium. In both cases fast neutrons are emitted, whose energy may be used. Since a small amount of deuterium is present in water and is easy to extract, an abundant source of fuel is available. Free tritium practically does not exist in nature and tritium has to be produced, as it is usually done, through the interaction of neutrons with lithium.

The thermonuclear reaction is to take place in high-temperature plasma. In order to use the energy of neutrons practically the production of energy has to be greater than the power used to sustain the high plasma temperature. Thus the energy obtained from the neutrons has to be much greater than the bremsstrahlung radiation of the electron gas in the plasma. Calculations show that for useful energy production for the D + D reaction the necessary ion temperature is ten times greater than in the case of the D + T reaction. Although the D + T reaction works at a lower temperature, it is hampered by the necessity to burn lithium, whose amount in nature is limited. Moreover, it seems that the use of lithium complicates the design of the reactor. Calculations show that for obtaining useful energy the temperature of ions in a plasma for the D + D reaction should be about 10^9 K and for the D + T reaction about 10^8 K.

From research in plasma and nuclear physics it is thus known that, for practical energy generation purposes, the technical problem of realizing a controlled thermonuclear reaction is reduced to obtaining a plasma ion temperature of at least 10^8 K, with a density of 10^{13} to 10^{14} cm^{-3}. It is obvious that the containment of a plasma in this state by an ordinary vessel is not feasible since there is no material that can withstand the necessary high temperature.

A number of methods for the containment of plasma and its thermal isolation have been suggested. The most original and promising method was the Tokamak which was proposed in the Soviet Union and which has been under development for more than a decade [1, p. 15]. The principle of its operation is illustrated by the design shown in Figure 1. The plasma is confined by a magnetic field, produced in a toroidal solenoid. The plasma has the form of a ring of a radius R and a cross-section a placed in the coil. The plasma is kept at a pressure of a few atmospheres. As it expands in the magnetic field, currents that retard this expansion are excited. The plasma is surrounded by a vacuum insulation, in order to sustain the sufficiently high temperatures at which thermonuclear reactions take place. This method of confinement is limited in time. Calculations show that, because of the low thermal capacity of the plasma, the energy for initial plasma heating, even in cases when the plasma exists for a few seconds, will be small as compared to the generated thermonuclear energy. Thus a reactor of this type may effectively work only in a pulsed mode. The Tokamak is started as a betatron, that is, by discharging condensors through the coils of the transformer yoke. In practice, plasma confinement by this method is not simple. First, there are difficulties in stabilizing the plasma ring in the magnetic field. With the growth of the cross-section radius a and moreover of the torus radius R, the ring loses its proper form and becomes unstable. This difficulty may be circumvented by choosing the appropriate ratio of R to a, and by properly designing the magnetic field, although at present the time for plasma confinement is only a fraction of a second. It is assumed that, by scaling up the Tokamak, this time will be proportional to the square of the size of the machine.

But the main difficulty is attributable to causes not fully appreciated in the beginning. For the thermonuclear reaction, the D and T ions must be heated. The main difficulty in passing heat to them is due to the fact that the plasma is heated by an electric field. In this case all the energy is transferred to the electrons and is only slowly transferred to the ions because of their large mass as compared to the mass of the electrons. At higher temperatures, this heat transfer gets even less efficient. In the Tokamak, the plasma is heated by the betatron current induced through the condensor discharge. Thus, all the energy for heating the plasma is confined to the electrons and is transformed to the ions by collisions. To heat the ions to the desired temperatures the necessary time Δt is much longer than the time during which we may maintain heating of the plasma by an electric current. The usual calculations are complicated, and attempts were made to do them as exactly as possible; even so, they lose in clarity. It is easy to estimate the lower time limit in which the ion heating may be made by the following simple formula [2, p. 24, expression 14].

$$\Delta t > -2.5 \times 10^2 \frac{f}{\Lambda} \frac{T_e^{3/2}}{n} \ln \left(1 - \frac{T_i}{T_e}\right).$$

It is assumed that during heating the plasma density (n),

$$n = \frac{7.3 \times 10^{21} p}{T_e}$$

and that the pressure p (atmospheres) and the electron temperature T_e are constant.

The coefficient f is equal to the ratio of the ion mass to that of the proton; Λ is the well-known logarithmic factor [3], and T_i is the ion temperature. For modern Tokamaks, operating with the D + T reaction and at plasma temperatures $T_i = 5 \times 10^8$ K and $n = 3 \times 10^{13}$ cm^{-3} (with an initial electron temperature $T_e = 10^9$ K), the time necessary to heat the ions to nuclear process temperatures is more than 22 seconds, at least two orders of magnitude more than confinement times in the modern Tokamaks. The plasma confinement time may be made greater only by building a larger machine, since the time Δt is proportional to the square of the size. From this formula it also follows that the time Δt for the D + D reaction is greater by another two orders of magnitude, and then $\Delta t \sim 2 \times 10^3$ seconds. The difficulties with the time for heating the ions is now fully recognized, although one cannot see how to shorten this time and how a Tokamak may work if, before the plasma ions have been heated, all the betatron energy from the condensors will be fully irradiated by the electrons. That is why in the current Tokamak projects extraneous energy sources are envisaged which are greater than the energy of the betatron process, which is used only for initially firing the plasma.

Extra energy must be transferred to the ions by a more efficient way than Coulomb scattering of electrons on ions. There are two possible processes for this. The first [1. p. 20] and already used consists in injecting into the plasma ring atoms of deuterium or tritium, already accelerated to temperatures necessary for the thermonuclear reaction. The second process of heating is through exciting radial Alfvén magnetoacoustic waves in the external magnetic

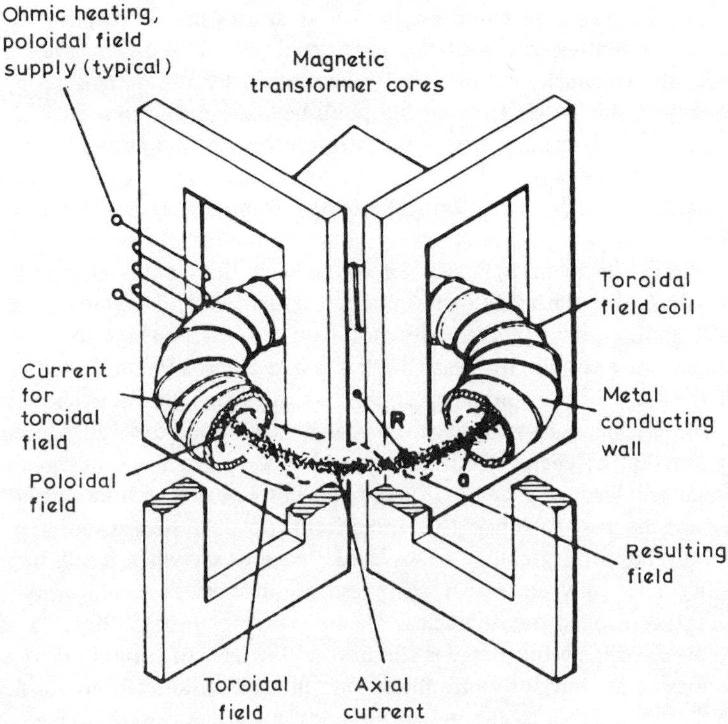

Fig. 1. The main features of a Tokamak.

field by the circulating high-frequency current. It is known [3] that the energy dissipated by magnetoacoustic waves is directly passed onto the ions and the transmitted power may be sufficient to heat the ions and sustain their temperature for a sufficiently long time. Thus the problem of heating the ions may be solved although the mode of operation of Tokamak will be more complicated than at first suggested. The design of the Tokamak becomes more complicated and its efficiency diminishes.

In all nuclear reactors, the power generated is proportional to the volume of the active zone, and the losses are mainly proportional to its surface. Therefore the efficiency of nuclear reactors is greater for larger sizes, and there exists a critical size for a nuclear reactor, after which it may generate useful power. The practically necessary dimension is determined not by scientists but by the engineers who design the machine in general, with proper choice of all the auxiliaries and the technology necessary for energy production. The following development is to a great measure determined by the talent and inventive ability of the design engineers. That is why the critical size of the Tokamak will be mainly determined by the designs proposed. Personally, I think that the existing published design solutions lead us to a critical size for Tokamaks that make them unfeasible. But certainly life does show that the ingenuity of man has no limits and therefore one cannot be sure that a practically useful critical size of Tokamaks may not be reached in the future.

Although the main difficulty for obtaining a thermonuclear reaction in Tokamaks is the heating of deuterium and tritium ions, there are difficulties of still another kind, which does not have a well-defined solution. In a Tokamak, for example, the plasma attracts and absorbs impurities extracted from the walls of the container. These impurities greatly lower the reaction rate. The plasma emits neutral atoms that hit and erode the wall. Moreover, the extraction of energy from neutrons also complicates the design of the Tokamak and leads to a larger critical size. Will we be able to bring the critical dimension of the Tokamak to a practically possible size? Even if it will eventually happen, of course we have no means to say when it will happen. Now we may only state that there are no theoretical reasons why in a Tokamak controlled thermonuclear reactions are not feasible, but the possibility of releasing useful energy is still beyond the scale of current practice.

Among other ways of controlling thermonuclear fusion, serious considerations should be given to the pulsed methods in the absence of magnetic confinement [1, p. 33]. The idea is to heat a D + T pellet about 1 millimeter in diameter in a short time so that it will not have time to fly apart. For this, very high pressures are necessary, to ensure intensive heat transfer between ions and electrons. It is assumed that in this way the thermonuclear reaction in a D + T pellet may fully take place. For this it is necessary to have a very powerful source of focused laser light that should heat the pellet from all sides simultaneously in about a nanosecond. This heating is a complicated process, but with modern computers the necessary conditions may be calculated. If we illuminate a pellet by well-focused laser beams, this may lead to a surplus of thermonuclear energy. But when one considers this process in detail, it is not clear how one can possibly resolve the technical and engineering difficulties. How, for instance, can uniform and simultaneous illumination be assured and can the neutron energy be usefully exploited?

In this case it may also be said that the basic theoretical idea is sound, but the consequent engineering development with current technology is beyond our reach. It is not possible to exclude completely a solution to this problem, although the design for laser implosion seems to me even less probable than the pulsed magnetic methods like the Tokamak.

The third approach to a thermonuclear reactor is based on continuously heating the plasma. Up to now this method has been developed only at our institute. Our work was described 9 years ago [4]. Since then this type of reactor has been studied in detail, and the main difficulties that will be encountered are visible. I will describe here in general terms the problems demanding a scientific solution.

As distinct from Tokamaks and the laser implosion method for producing conditions for the thermonuclear process, our method was not specially invented, but while developing a high-power, continuously operated microwave generator accidentally we discovered a hot plasma phenomenon. We constructed an efficient microwave generator operating at 20-cm wavelength with a power of a few hundred kilowatts. This generator was called the Nigotron, and its principles are described in [5]; full details of its construction and operating characteristics are given in [6]. In the process of its development, beginning in 1950, during tests of our early model, high-power microwave radiation was passed through a quartz sphere filled with helium at a pressure of 10 cm-Hg. A luminescent discharge with well-defined boundaries was observed, but only for a few seconds because the quartz sphere melted through in one place.

These observations led to the suggestion that the ball lightning may be due to high-frequency waves, such as produced by a thunderstorm cloud after the conventional lightning discharge. Thus the necessary energy is produced for

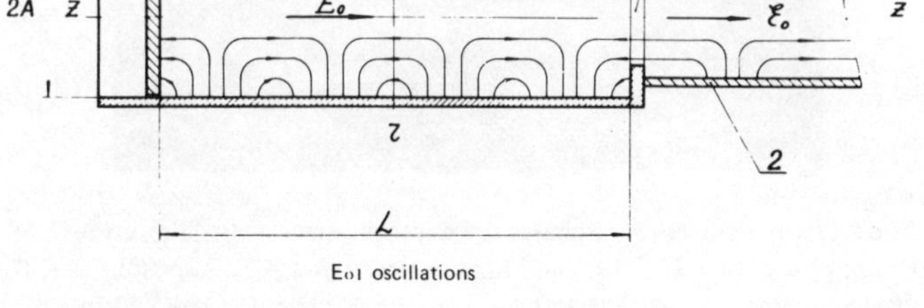

Fig. 2. Structure of the high-frequency field in a resonator for E_{01} oscillations.

sustaining the extensive luminosity observed in ball lightning. This hypothesis was published in 1955 [7]. After some years we were in a position to resume our experiments. In March 1958, we obtained a free gas discharge in a spherical resonator filled with helium at atmospheric pressure under resonance conditions with intense H_{01} oscillations. This discharge was oval-shaped and was formed in the region of the maximum of the electric field and moved slowly, following the circular lines of force.

We started to study this type of discharge where the plasma was not in direct contact with the walls of the resonator. We assume that this plasma may be at a high temperature. For a number of years we studied this interesting phenomenon in various gases and at different pressures, up to some tens of atmospheres at different power levels, reaching tens of kilowatts. We also studied the effect of a magnetic field reaching 2.5 T in our experiments. This work is described in detail in [4]. A sketch of our setup is shown in Figure 2.

The plasma discharge has a cordlike form 10 cm long, equal to half the wavelength. Intense microwave oscillations E_{01} are excited in a cylindrical resonator. The cord of the discharge is situated at the maximum of the electric field, and its stability along the longitudinal axis was due to the high-frequency electric field. In a radial direction the stability was provided by rotating the gas. The discharge in hydrogen or deuterium was of great interest. At low powers the discharge did not have a well-defined boundary and its luminosity was diffuse. At higher power the luminosity was greater and the diameter of the discharge increased. Inside the discharge a well-defined filamentary cordlike nucleus was observed. In our initial experiments the

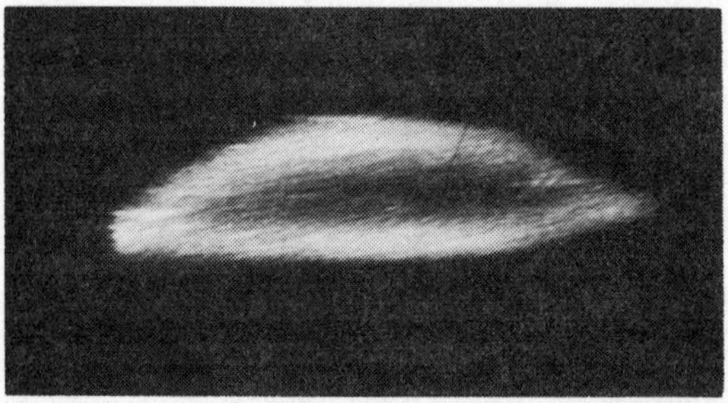

Fig. 3. Photograph of a cord discharge in deuterium with an admixture of 5 percent argon at high power (P = 14.7 kW) and high pressure (p = 3.32 atmospheres). The length of the discharge was 10 cm. The left edge of the discharge is blocked by the window. Oscillations were of E_{01} type.

power dissipated in the discharge was up to 15 kW and the pressure reached 25 atmospheres. The higher the pressure, the more stable was the discharge, with a well-defined shape (Figure 3). By measuring the conductivity of the plasma and by using passive and active spectral diagnostics we could firmly establish that the central part of the discharge had a very high temperature – more than a million kelvins. Thus, at the boundary of the plasma cord in the space of a few millimeters we had a discontinuity of temperature of more than a million kelvins. This meant that at its surface there was a layer of very high heat isolation. At first some doubt was expressed about the existence of such a layer. Various methods of plasma diagnostics were used, but they all, and always, confirmed the extremely high temperature. Later we found out how to explain the physical nature of this temperature jump. It is easy to show that at these high temperatures electrons scattered at the boundary and while freely diffusing into the surrounding gas can carry away a power of hundreds of kilowatts. The lack of such a thermal flux may be explained by assuming the existence of electrons reflected without losses at the boundary of a double layer. The occurrence of a similar phenomenon is well known, as such a layer exists in hot plasmas surrounded by dielectric walls, say, of glass or ceramics.

It is well known that in these conditions, even at high pressures, the electrons may have a temperature of many ten thousands of kelvins and still not markedly heat the walls. This phenomenon may be well explained by the existence of a double layer on the dielectric surface. The mechanism leading to its formation is simple. When the electron hits the surface, owing to its greater mobility it penetrates the dielectric to a greater depth than the ions and leads to the formation of an electric double layer, whose electric field is so directed that it elastically reflects the hot electrons. The low electron heat conductance at the surface of plasmas is widely used in gas discharge lamps, and the method of plasma heat insulation was first suggested by Langmuir. We assume that at a sufficiently high pressure a similar mechanism of heat insulation may take place in our hot plasma. The existence of a double layer in the plasma on the boundary of the cord discharge as a discontinuity in density was experimentally observed by us. This mechanism for a temperature discontinuity may exist only if the ion temperature is much lower than the electron temperature and not much above the temperature at which the plasma is noticeably ionized. But this is only necessary at the boundary of the discharge. In the central part of the discharge, the ion temperature may reach high values. As we will see further, the difference in temperatures inside the core and at the surface is determined by the value of the thermal flux and the heat conductivity of the ion gas. Usually the heat conductivity is high, but in a strong magnetic field the transverse heat conductivity may become very small. Thus, we may expect that in a strong magnetic field the ion temperature in the core will not differ from the electron temperature and may be sufficiently high to obtain in a deuterium or tritium plasma a thermonuclear reaction. This is the basis for designing a thermonuclear reaction to produce useful energy, and this has been worked out [8]. The general outlay and the description of the reactor are shown in Figure 4.

The cord discharge [1] takes place in a confining vessel and resonator [2]. The deuterium pressure is 30 atmospheres, the magnetic field 1 T, produced by an ordinary solenoid. The design shows how the neutron energy is used. The gas heated by the neutrons passes through a gas turbine [5] where it adiabatically expands. Next it passes through a turbo-compressor [6] and is isothermally compressed. The excess power is consumed in the generator [8]. The cord discharge is heated by a high-frequency field as it is done in cylindrical resonators (see Figure 2). The difference is in the coil surrounding the discharge and used to excite magnetoacoustic waves so as to raise the plasma ion temperature [4, p. 1003].

This design and pertinent calculations were published in 1970 [8, p. 200] and demonstrate the expected parameters of our thermonuclear reactor, when working with our plasma cord.

As time passed we were able to increase considerably our understanding of the processes in the plasma. We have mainly improved the microwave diagnostics, and it is now possible to measure with an accuracy of 5 percent the radial density distribution, its dependence on the magnetic field, pressure, and supplied microwave power. The necessary stability conditions have been established. All this has allowed us to raise the microwave power by many times and thus increase the electron temperature up to 50 million kelvins. If we could establish temperature equilibrium between the electrons and ions even without the extra heating of the plasma by magnetoacoustic oscillations, we could have reached the D + T reaction. The design of the reactor is simpler and its size is smaller so that the thermonuclear reactor becomes not only easier to build but also the neutron energy is easier to convert to mechanical power. Thus we escape the main difficulties on the way to building pulsed type of thermonuclear reactors.

But still we have also some unresolved difficulties that merit most serious consideration because they may make the whole problem unsolvable. The main difficulty is the following. Now we can obtain in our installation a high-frequency discharge at a pressure of 25 atmospheres and continuously maintain the electrons at a temperature of 50 million kelvins, and going to a greater size even more. At present the size of our discharge is limited only by the power conveyed to it. Thus we have permanently an electron gas with a record high temperature, even higher than the electron temperature inside the sun. The main problem is to heat the ions to the same temperature for, although the electron gas interacts with the ions in the entire volume of the discharge, it is not easy to raise this temperature in such a way.

The temperature equalization proceeds in two steps. In the first step the energy is passed from the electrons to the ions. This is simply due to the collisions of electrons with ions, and in this case it is obvious that the heat

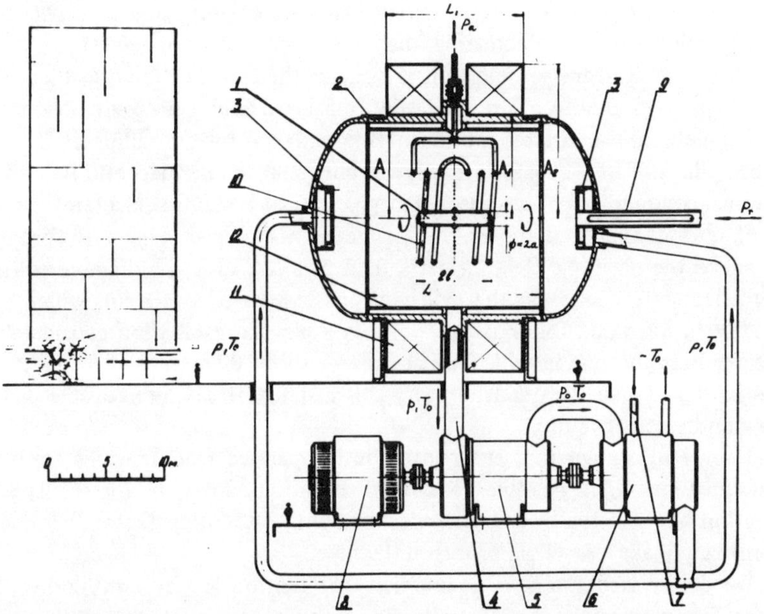

Fig. 4. Drawing of the construction of a thermonuclear reactor operating on a closed cycle. **1**, cord discharge; **2**, cylindrical container of the reactor; **3**, inclined nozzles; **4**, pipe connecting the container of the reactor with the gas turbine; **5**, gas turbine; **6**, isothermal compressor; **7**, cooling water; **8**, generator; **9**, coaxial waveguide; **10**, coil for the alternating magnetic field; **11**, solenoid; **12**, copper wall of the resonator; *L*, length of the resonator; L_1, length of the solenoid; P_a, power of magnetoacoustic oscillations; P_r, high-frequency power; A, radius of the resonator; A_1, internal radius of the winding; A_2, external radius of the winding; $2l$, length of the cord discharge; $2a$, diameter of the cord discharge; h, distance between the wall of the container and the resonator.

transfer will be proportional to the volume. The next stage is the transfer of energy from the ion gas to the surrounding media. This flux will be proportional to the surface of the plasma cord. At a given thermal conductivity of the ion gas the temperature will increase for larger sizes of the plasma cord. Thus at a certain heat conductivity there will be a critical size for the diameter of the plasma cord, when the ion temperature will reach a value close to that of the electrons and the required D + D or D + T reaction can take place. If we know the heat conductivity of the plasma, then it is easy to calculate the critical dimension. If for example, we make this calculation for ordinary ion plasma in the absence of a magnetic field, when the heat conductivity is determined by the mean free path, we will find that the plasma must have an unrealizably large size of many kilometers. One can lower this cross-section only by decreasing the heat conductivity of the ion gas by placing it in a magnetic field as it is done in the reactor shown in Figure 4. The heat conductivity of an ion gas in a magnetic field is markedly decreased and is determined not by the mean free path but by the radius of Larmor orbits, the size of which is inversely proportional to the magnetic field. The thermal conductivity of ion gas in a magnetic field is easy to calculate.

It is thus seen that the critical diameter of the cord is inversely proportional to the magnetic field and at a field of a few tesla the diameter of the cord necessary to obtain thermal neutrons will be 5 to 10 cm, which can readily be provided for. For this we need a plasma installation considerably greater than the one in which we at present study the nature of the electron gas in the plasma. In our laboratory this installation is quite feasible and is now under construction.

It may be shown that the continuously operated reactor which we have described makes it possible to obtain conditions not only for the D + T reaction but also for D + D, if it were not for yet another factor that could eventually make the whole process unfeasible.

We determined the heat conductivity of the ion gas by considering the mean free path of the ion, assuming it to be equal to the Larmor orbit radius, but not having taken into account the effect of convection fluxes of heat in a gas. It is well known that even in ordinary gases the convection heat transfer is much larger than the heat conduction due to molecular collisions. It is also known that unfortunately it is virtually imposible to calculate theoretically the heat transfer by convective currents even for the simple case of random turbulent motion in an ordinary gas. However, usually we can, by dimensional considerations, estimate the thermal conductivity in a similar case and then generalize it for a special case, determining the necessary coefficients empirically. In the case of plasma the process depends on many more parameters and the problem of determining the convectional thermal conductivity is more complicated than in an ordinary gas. But theoretically we may estimate which factors have most influence on the rate of convection. To sustain convection one must supply energy. In a gas this energy is drawn from the kinetic energy of the flow.

In a quiescent plasma there is no such source of energy. But in an ionized plasma there may be another source of energy that will excite convection. This source is connected with temperature gradients and some of the thermal energy flux could produce convection. Quantitatively this process is described by internal stresses and was first studied by Maxwell [9], who showed that internal stresses are proportional to the square of viscosity derivative and of the temperature gradient. In an ordinary gas the stresses are so small that up to now they have not yet been experimentally observed. This is because the viscosity, which is proportional to the mean free path at normal pressures, is equal to about 10^{-5} cm, and hence at low temperature gradients the stresses are small.

In the plasma the mean free path of electrons and ions is of the order of centimeters, and the temperature gradients are high. In this case the internal stresses following Maxwell's formula are ten orders of magnitude greater than in a gas, and we may expect both convection currents and turbulence. The presence of a magnetic field certainly can have an effect on this phenomenon and, with additional effect of an electric field on convection, it makes even a rough theoretical approach to estimating the magnitude of convection very unreliable. Therefore the only alternative is to study these processes experimentally and this is what we are now doing.

However, convectional thermal conductivity will lower the heating of ions and will lead to a greater critical cross section for the thermonuclear plasma cord. Correspondingly, the size of the reactors for useful energy production will be greater.

If this size will be out of our practical reach, then we should consider methods to decrease convectional heat transfer. This may be done by creating on the boundary of the plasma a layer without turbulence as happens in fluids where we have the Prandtl boundary layer. This possibility has been theoretically considered [4, p. 1002].

In conclusion we may say that the pulsed method used in Tokamaks can now be fully worked out theoretically, but the construction of a thermonuclear reactor, based on this method, leads to a large and complicated machine. In contrast, our thermonuclear reactor is simple in construction, but practical means of its realization and size depend on convection heat transfer processes that cannot be treated on a theoretical basis alone.

The main attraction of scientific work is that it leads to problems, the solutions of which cannot be foreseen, and that is why, to a scientist, research on the controlled thermonuclear reactions is so fascinating.

REFERENCES

[1] Ribe, F. L., *Rev. Mod. Phys.* **47**, 7 (1975).
[2] Kapitza, P. L., *JETP Lett.* **22** (No. 1), 9 (1975).
[3] Kapitza, P. L., and Pitaevskii, L. P., *Sov. Phys. JETP* **40** (No. 4), 701 (1975).
[4] Kapitza, P. L., *JETP Lett.* **30** (No. 6), 973 (1970).
[5] Kapitza, P. L., *High-Power Microwave Electronics*, Pergamon, Oxford, 1964.
[6] Kapitza, P. L., Filimonov, S. I., and Kapitza, S. P., 'Elektronika bolshikh Moshchnostai', *Nauka* (No. 6), 7 (1969).
[7] Kapitza, P. L., *Collected Papers*, Pergamon, Oxford, vol. 2, 1965, p. 776.
[8] Kapitza, P. L., *Sov. Phys. JETP* **31** (No. 2), 199 (1970).
[9] Maxwell, J. C., *Phil. Trans. Roy. Soc. London* **170**, 231 (1879).

PART TWO

10. THE CONSTRUCTION AND EARLY WORK OF THE INSTITUTE FOR PHYSICAL PROBLEMS

The Institute was established by a government decree dated December 28, 1934, and was named the Institute for Physical Problems. This somewhat unusual name was intended to indicate that the institute will not deal in any specific realm of knowledge but will carry out research in a variety of fields to be determined by the scientists working there. Thus the institute was to engage in pure science rather than applied science. I have to use the somewhat unpopular expression 'pure science' for lack of a substitute. The term 'theoretical science' was proposed, but in fact every science is theoretical. One must admit that 'pure science' [in English] or *reine Wissenschaft* is a well-entrenched expression. There is only a single difference between applied and pure science: applied science deals with problems arising from daily life, whereas pure science leads to applicable results since no science can remain forever detached from daily life.

Excerpts from a report to a session of the Physics Group at the U.S.S.R. Academy of Sciences, 1937

At the time of its establishment the institute was not intended to engage in diverse physical problems. I was asked to become director of the institute and proceed there with the line of research which I had been pursuing for the preceding 13 years in Britain. My interests there were focused on strong magnetic fields and in connection with the latter, the attainment of low temperatures. These were in effect the main problems studied at the institute at the time.

I shall begin with an outline of the architectural design of the institute with particular reference to the main considerations involved in it.

Selection of a suitable site was the first task. This was an extremely important matter. It is well known that most physical measurements involve highly sensitive instrumentation, so that the institute had to be located away from various sources of disturbance. Cities abound with adverse factors such as shocks from street traffic, induced currents from tram lines and nearby broadcasting stations, etc. Much time was therefore devoted to the selection of a proper site, and I do not know a better location than that occupied by the institute. I would even go as far as saying that the institute has an ideal location for a city: its building lies at a distance of 80 m from the nearest tram line, and street shocks have been reduced to a minimum.

Residential buildings occupy the front line of the institute grounds (in form of a triangle), with the research area located deep inside. To the left of the site is the Neskuchni Park, to the right is also a park, both of which will never be built on; in front of the site is the street, on its other side the Academy of Sciences is to lay out its botanical gardens. The institute will thus be surrounded on three sides by gardens: along the fourth side the Andreyev Canal is to be built so that the isolated position of the institute is likely to be preserved forever [Fig. 1].

The design of the institute was based on the following principles. The working, laboratory part of the building, where the experimental work is done, lies entirely on the ground floor. This is essential for work with liquid helium and hydrogen which have to be transported in the laboratory, as both safety and convenience demand that all these activities be concentrated on a single floor. Moreover, the ground floor is much safer from shocks of any kind [Fig. 2].

The ground floor also contains a hall which accomodates the liquefaction installation. There is no other story above this part of the building, which, moreover, has lightweight roofing in order to minimize the effect of any accidental explosion.

The first floor is intended entirely for administration rooms and various

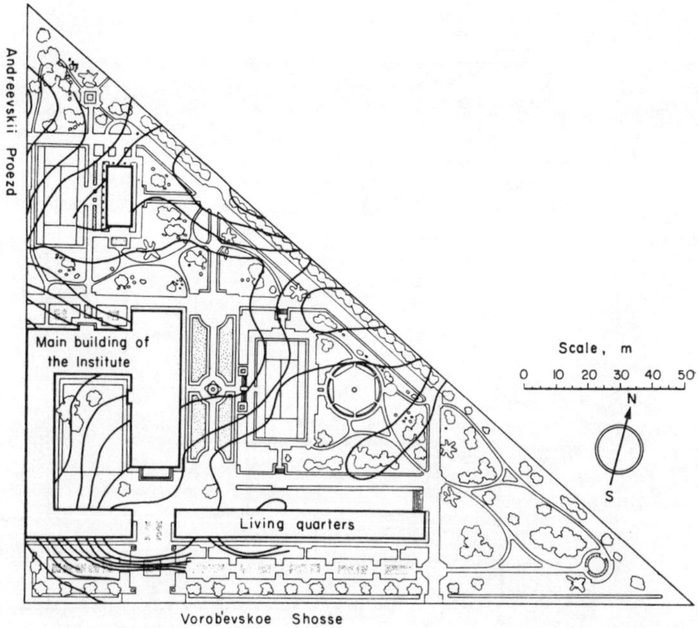

Fig. 1.

services: bookkeeping, recreation and reading room, offices, the 'red corner' auditorium, and the director's room [Fig. 3].

At our institute a pass system does not exist; no one can get straight to the ground floor laboratories as any visitor finds himself immediately on the first floor and is routed to the office room, whence the only way to the ground floor is through the director's study. As for the research workers, they can reach the director's study from their own rooms via an internal staircase. This arrangement provides a direct link between the director and the research staff.

Part of the basement is occupied by the boiler room, and the rest serves for work with highly sensitive devices. A special foundation resting on rubber was prepared there. One barium-plastered room accomodates X-ray apparatus, and another room is used for spectroscopy.

The institute was planned to operate as an independent unit with workshops, libraries, etc., of its own. This concept, however, is clearly incorrect and can only be applied temporarily as our institute must eventually become an integral part of all the central institutes of the Academy of Sciences; once all these are built close to one another, it will be possible to reduce our

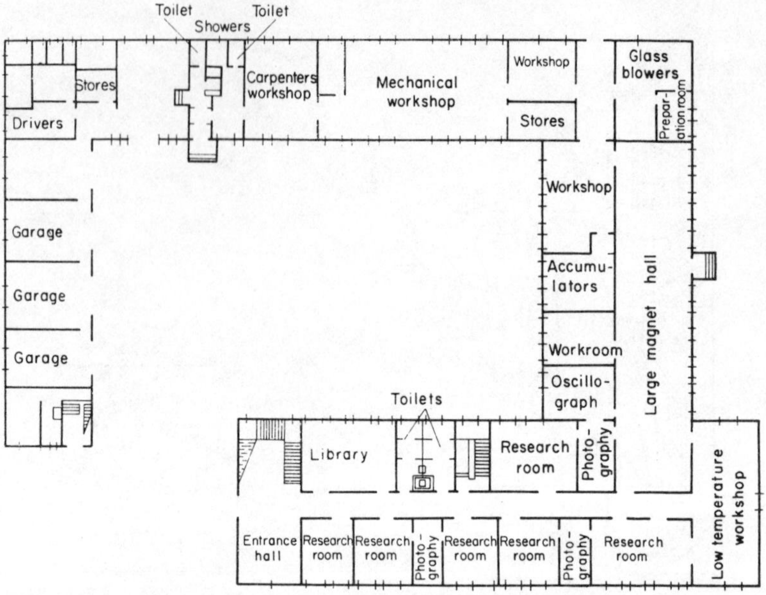

Fig. 2.

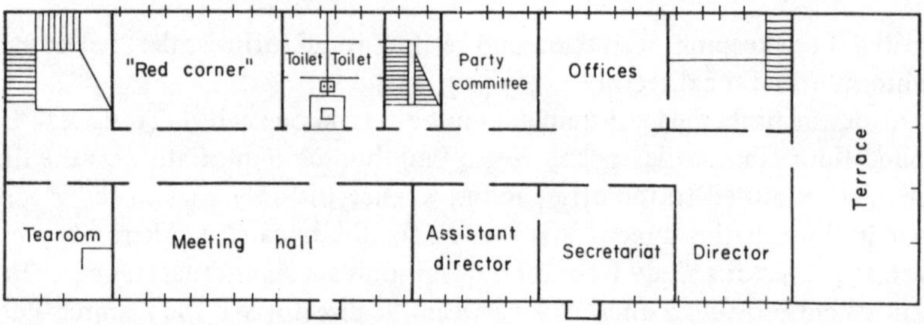

Fig. 3.

administration and services to a considerable extent. Naturally, a complete library of our own will then become superfluous as it will be possible to use that of the Academy of Sciences. The same applies to the workshops; now they may appear excessively large, but meanwhile we must have sufficient facilities of our own.

As for the construction of the institute, it lasted two years though half the time should have been enough. Our builders proved to be unequal to the task.

Despite the smooth, adequate supply of building materials, work on the site was poorly organized and could not be completed in time. As regards the quality of building work, the government committee could only characterize it as 'satisfactory'. This is a pity indeed because the architectural design is in my view highly attractive and should have been implemented with greater dedication . . . .

Now I would like to point out some minor details in the organization of our laboratory facilities.

First and foremost, the entire electric supply system, which comprises storage batteries with a Swiss control panel located in the large hall and providing any desired voltage to any room. The rooms are wired directly to the control panel, so that the simultaneous use of the same batteries by a number of workers causes only a negligible voltage drop which does not interfere with the work.

Next, the generator room. Current from any generator can be supplied to any desired room, and currents as large as several hundred amperes are regulated by a simple sliding rheostat which determines the intensity of current in the exciting winding of the dynamo. For safety reasons the wiring is enclosed in steel and iron pipes, and the whole system is well grounded. A satisfactory grounding system exists all around the institute, so that any room can be separately grounded when necessary.

Every two adjacent laboratories have a common darkroom. A possibility of blackout is provided everywhere. Electric lamps can be connected not only in parallel but also in series to provide a subdued lighting in the rooms. Each room is fitted with a chromium-plated faucet with a finely controlled water output. Moreover, compressed air and gas facilities are present in each room.

There is at the institute a well-equipped workshop capable of building almost any complex physical device. We can proudly say that most of the precision lathes at the workshop are Soviet made.

A separate spacious and well-ventilated room accomodates the glassblowing workshop.

Finally, there is a carpentry workshop equipped with planing machines.

Now to the crux of the matter, namely, the special equipment present in our institute and obtained from Britain. The question arose whether we should make this special equipment here in the Soviet Union or have it shipped from Britain. To manufacture this equipment in the Soviet Union was of course quite possible because our Soviet industry can produce anything made abroad. Yet manufacturing this equipment in the Soviet Union proved extremely difficult. The trouble is that our industry is not suited to the service of research; everything is done on a large scale, and a non-mass-production machine production is almost nonexistent. In such cases orders must be placed with the tool or instrument workshops of the respective factories, but then the prices of custom-made equipment becomes prohibitive and production times impossibly long. We scientists would like to see a more forthcoming industry, a greater interest and assistance from that direction.

So the special equipment came from Britain. For this I am deeply obliged to my friend and teacher Rutherford, who persuaded Cambridge University to sell the equipment of the Mond Laboratory to the Soviet Union. The equipment arrived and is already installed. It includes, most importantly, a machine for the generation of strong magnetic fields. This machine was built some 12–14 years ago. It consists of a single-phase generator which provides a great power output within a brief time. I shall not dwell on the details of this machine. At any rate, it operates on a rather simple principle. A limiting factor in the production of strong magnetic fields in an electromagnet is the phenomenon of magnetic saturation of iron. Once the iron has been saturated, any further progress can only take place at a very slow rate. The magnetic field strength is proportional to the logarithm of the linear size of the magnet. By building a magnet weighing more than 100 tons, as Cotton did in Paris with an investment of several million francs, we obtain a magnetic field gain of only 25% in comparison with a conventional-type laboratory magnet. Clearly we had to seek an alternative course of action.

Our approach can be outlined as follows. We had to discard permanent fields in favor of brief transient fields obtained by the passage of very strong currents through a coil, so as to create a strong magnetic field within the coil, but for an interval brief enough to preclude any overheating of the coil. The drawing in Figure 4 shows a coil which can be traversed by tens of thousands of kilowatts within one hundredth of a second. During this short interval the coil heats up to 100°C. Any attempt to continue the experiment for a whole second would result in a coil temperature of 10,000°C, which is high above the melting point of the coil. Yet within one hundredth of a second it is indeed possible to observe all phenomena which take place in a static magnetic field. This is so because the brevity of a strong magnetic field does not prevent the phenomena from reaching a magnitude large enough to be observed even during such short intervals. Naturally, special instrumentation had to be designed for this purpose.

In Britain I had already studied many phenomena in such strong fields, and found a number of new physical effects.

The appearance of this machine is shown on p. 5 of this volume. The

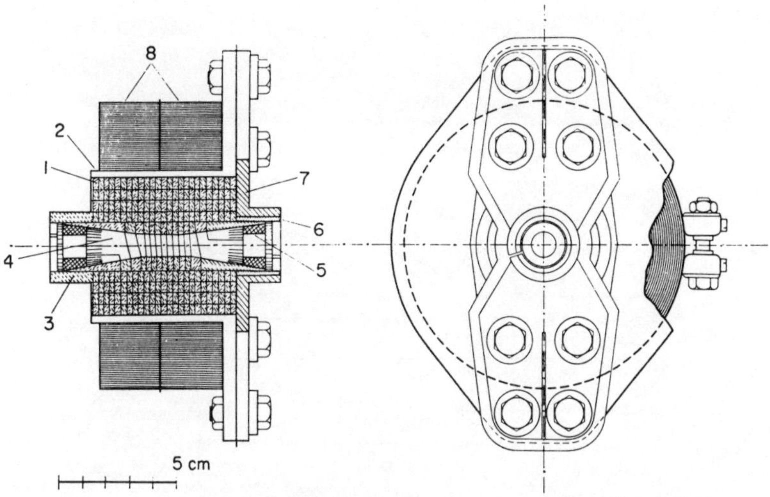

Fig. 4.

rotor weighs 2.5 tons and rotates at a velocity of 1,500 rpm. The machine was constructed by the Metropolitan-Vickers company with the participation of the Soviet engineer M. P. Kostenko and the British engineer M. Walker. To build such a machine is not an easy job because of the colossal forces which are generated inside during the brief closing and which can literally break it apart. Therefore the machine had to be extremely sturdy in construction; its terminals are specially fixed, the entire stator is made of steel, to say nothing of the heavy-duty bearings and many other details which distinguish the machine.

By a brief closing of the circuit the machine yielded an output of 220,000 kW within 0.01 s, which corresponds to 72,000 A at 3,000 V. This is a vast power indeed, over a quarter of the output of the Dnieper Hydroelectric Power Station. Though tested at this output, the machine was actually used at one quarter of it, for safety reasons.

The second important part is the switch, which makes it possible to cut out a single wave, a half-period, from the current generated by the machine. Figure 5 shows an oscillogram of the coil current (9,400 A current, 277,000 gauss); the break in the line 1 indicates the moment of spark formation in the magnetic field for observing the Zeeman effect.

One of the main obstacles was the construction of the coil for the passage of current. Though the coil does not manage to heat up during the passage of

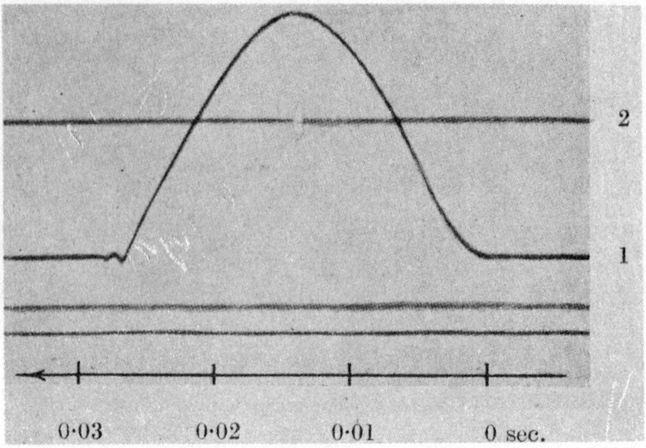

Fig. 5.

current, it does generate colossal forces which tend to break it apart. These forces reach as high as 150 tons in the cross-section of the coil, whereas the pressure on the copper reaches 5,000–6,000 atm.

The first coil we tested broke apart. Its diameter, originally 1 cm, became 2 cm. The whole circuit disintegrated with a big noise and various components of the coil flew off in all directions. Therefore we began to reinforce all coils by enclosing them in steel bands. This, however, proved insufficient as the copper continued to fly off. The whole secret of coil design lies in selecting the coil shape and winding in such a way that the stress on the copper becomes equivalent to the hydrostatic compression. Under such circumstances the copper is effectively subjected to pressure on all sides, which cancels the disintegrating stresses. Cockcroft was able to achieve this by special calculations, and now the coil operates uneventfully.

It is noteworthy that the coil and machine occupy opposite ends of the hall. The reason behind this arrangement is as follows. During the brief closing, the machine generates a strong couple which brakes the rotor and creates a reaction in the foundation. Since the current lasts only one hundredth of a second, something like a minor earthquake takes place at the time of the experiment. If photographs are taken during this interval, in studies of the Zeeman effect for example, the spectrograph will vibrate and the tracing will be blurred. How to avoid this jolt? It is obviously possible to absorb it, but placing the machine and coil at opposite ends of the hall would be a much simpler solution. Since the experiment lasts only 0.01 s the shock wave will reach the coil when the work is already over, and no disturbance will result. Hence the considerable length of the hall.

Brief experiments are very tempting. The short duration of the experiment may seem a limiting factor; if correctly used, however, it yields great advantages indeed. As noted previously, the great force of the magnetic field brings all phenomena to such dimensions that they can usually be observed even within such a brief interval as one hundredth of a second, provided the necessary instrumentation is available. On the other hand, all time-dependent disturbing factors, such as thermal disturbances, often remain unnoticed. One example is the study of magnetostriction in bismuth. During the 0.01 s of our observation the changes of length caused by random thermal phenomena are too small to distort the lengthening produced by the magnetic field. In stationary magnetic fields, thermal disturbances represent the main factor which limits the accuracy of observation. Such cases are numerous indeed.

In the course of extensive experimentation with magnetic fields we found that many phenomena, especially galvanomagnetic ones, become especially interesting at low temperatures. Consequently we began work on the production of liquid hydrogen and liquid helium, which included the construction of suitable equipment. This was rather difficult at the time (8 years ago) as the only laboratory with the necessary experience was that of Kamerlingh Onnes in Leiden. Success in this field depends largely on the ability to organize the work, select the suitable manpower, train people to observe a certain degree of cleanliness, pick out the right materials for work, etc. All these seemingly minor details are crucial as any oversight may doom the whole experiment.

I started with the construction of a hydrogen liquefier, which resembled in design the one now in operation here. The only difference between this liquefier and the conventional one lies in its double cycle, one cycle for pure hydrogen and another for impure (industrial) hydrogen. Cooling takes place with pure hydrogen whereas the impure, technical-grade hydrogen undergoes liquefaction. Thus we managed to overcome the greatest obstacle in the production of liquid hydrogen, having overcome the need for using pure hydrogen as the starting material for liquefaction and making do with the industrial product. This liquefier had an output of 7 liters/hour and a start-up time of 20 minutes, so short because of the correct calculation of all details of the heat exchanger and the reduction of excess metal to a minimum in the liquefier. When we were building the hydrogen liquefier, our intention was to use hydrogen for the production of liquid helium. Once we started with the preparation of liquid helium, however, we found a new procedure for the production of the latter without the use of hydrogen. Work with hydrogen is hazardous and should be avoided, especially on a large scale. The new helium liquefier developed by us allows for the production of liquid helium without a preliminary cooling of the helium with liquid hydrogen, in contrast to the installation used by Kamerlingh Onnes.

The device works on a rather simple principle. Liquid helium was usually obtained as follows. Helium was cooled in hydrogen boiling under a reduced pressure in order to obtain a temperature below the inversion point of helium; then the Joule–Thomson effect caused a cooling, and part of the helium was liquefied. Such liquefiers have a thermal efficiency of only 0.5–1%. In Leiden, the procedure for the production of liquid helium is as follows: liquid air is obtained on the first day, about 20–30 liters of liquid hydrogen on the second day, and a small quantity of liquid helium on the third day.

The difficulties involved in this complex manipulation for the production of liquid helium can be circumvented only by using an adiabatic process for cooling the material, i.e. by making it expand adiabatically at a low temperature. The problem was to design a machine capable of removing heat from the expanding helium. To construct such a machine is very difficult, mainly because of the expander (cylinder with piston), which has to work at a low temperature. The piston needs lubrication, but at such low temperatures all substances become quite solid. Initially the idea of using a turbine crossed our minds, since turbines can operate without lubrication. Here, however, an amusing difficulty emerged. At such low temperatures helium has an extremely small specific volume, whereas turbines are only efficient when they work large masses; engineers know all too well that only a large-sized steam turbine can work efficiently.

A turbine of feasible dimensions would have an output of not less than several thousand liters of liquid helium per hour. For our laboratory requirements of 1–2 liters of helium per hour a properly operating turbine would have to be of an impossibly small size (1–2 cm in diameter). The turbine option should be kept in mind whenever large quantities of liquid helium are required. As far as we were concerned, however, the idea of using a turbine had to be discarded in favor of a piston-type machine. Here we had various possibilities at our disposal including the use of a vibrating diaphragm, but the simplest solution we could think of was as follows.

Our piston cannot be tight-fitting. Obviously it must be surrounded by a clearance of a few hundredths of a millimeter in order to move freely. Under such circumstances when the cylinder is filled with helium at high pressure, most of the material evidently will escape through the clearance because of its low viscosity. If the piston can perform a quick expansion, however, a condition can be reached at which the helium losses become negligible. The parameters necessary for this purpose are easy to calculate. We were able to build such a machine while still in Cambridge, and it works quite satisfactorily to the present day. Its thermal efficiency is about 60%.

The second difficulty in the construction of the machine lies in the selection of material. All materials become extremely brittle at the temperature of liquid helium, which makes them clearly unsuitable for this purpose. After a long search we found the necessary properties in austenite steel, which does not lose its plasticity even at the lowest temperatures.

Our helium liquefier is shown in the photograph on p. 10 of this volume. We apply the following working cycle: at first the helium is cooled down with liquid nitrogen boiling at a reduced pressure, until a temperature of 65 K is reached; then the expander cools it further to 10 K, and the final stage is one of liquefaction using the Joule–Thomson effect. This is the first working machine; it yielded 1.7 liters of liquid helium per hour, consuming 1.5 liters of liquid nitrogen per liter of helium. As I have already pointed out, this machine can be improved to a considerable extent. Now we are in the process of constructing a two-cycle machine which will yield 6–8 liters according to our calculations.

In establishing the institute our objective was to make it a center of advanced research. I believe this aim has been achieved and today our institute is indeed one of the best, not only in the U.S.S.R. but in Europe as well.

The question often arises, in the U.S.S.R. and in the West alike, as to whether the idea of establishing such modern, fully equipped research institutes is basically correct. We know that nearly all major scientific discoveries were made with the simplest and most elementary of means. Why should we build such institutes if all outstanding discoveries were made with extremely simple equipment? This topic has been extensively discussed in our country and abroad. I have been reading *Recollections and Reflections*, the recently published autobiography of J. J. Thomson. Here Thomson also dwells on this question. I would like to cite an excerpt of particular interest because of all physicists of the late 19th and early 20th centuries, Thomson made the most fundamental discoveries. Indeed, he discovered the electron, and the existence of isotopes, and all this by using only very simple means. What is Thomson's opinion on this topic? He writes:

It is not, in general, the first discovery of some quite new physical phenomena that is costly. For example, the discovery of X-rays by Röntgen, of radium by the Curies, the long-continued experiments of C. T. R. Wilson on the formation of drops on particles charged with electricity, cost quite insignificant sums. Discoveries such as these were due to what cannot be bought: to keen powers of observation, to physical insight, to an enthusiasm which does not falter until the difficulties and discrepancies which attend on pioneering work are overcome.

When first the discovery is made, the effects observed are generally very small, and require a succession of lengthy experiments to obtain trustworthy results. It is the attempt to get larger effects that is so expensive. It may mean spending many thousand pounds in making powerful magnets or in producing electromotive forces of hundreds of thousand volts, or in obtaining large supplies of radium. Such money is, however, well spent, for it enables us to obtain new knowledge much more quickly and with greater certainty.[1]

Thomson's words are quite correct. When Columbus sailed off on his expedition which eventually culminated in the discovery of America, all he had at his disposal was a small frigate, a mere rowboat in modern terms. But for the actual 'acquisition' of America, its 'rediscovery' and the realization of its potentialities it was necessary to build bigger ships such as the *Lusitania* and the *Titanic*, and the investment paid off handsomely. I am totally in favor of building modern research institutes with the latest in equipment.

---

[1] Thomson, J. J., *Recollections and Reflections,* Bell, London, 1936, pp. 126–127.

11. THE ORGANIZATION OF RESEARCH AT THE INSTITUTE FOR PHYSICAL PROBLEMS

When I came back to work in the U.S.S.R. in 1934, I was deeply interested in the organization of science generally and of research at my institute in particular. I was well acquainted with the organization of science and research abroad. For many years I was director of an institute in Cambridge, the focus of scientific research in Britain. On the basis of this experience I felt that Western organizational patterns of research were not fully applicable here. I was convinced that we had to work out our own organizational patterns of scientific research at the institute and even more so for our science generally.

This is so mainly because science occupies a particular place in our socialist country. Naturally, science abroad is likewise recognized as a major factor in the development of culture and technology. In our country, however, science is a foundation of culture, a guiding force in the progress of technology and the development of our national economy. Therefore the organization of science in the U.S.S.R. must be more purposeful than in capitalist countries, where it is largely accidental and spontaneous in character. We strive toward a close link between science and life. The organization of science is a matter of particular importance to us, the workers of the U.S.S.R. Academy of Sciences.

In my present report on the organization of research at our institute I shall outline first the principles which guided us at the beginning, and then the results obtained by us.

I believe a few additional explanations are necessary. Here I shall speak mainly on the organization of our institute in peacetime. At the end of my report a few words will be devoted to the impact of the war and the manner in which we adapted our institute to military requirements. All this, however, was but a brief episode in the life of our institute. Today we are mainly concerned with the setup of the institute in peacetime. The sounder the peacetime structure, the easier the adaptation to war needs, whenever necessary.

Let me point out that our institute is very young, having been founded only 7–8 years ago. Though I arrived here as a relatively mature scientist,

---

Report at the enlarged session of the Presidium of the U.S.S.R. Academy of Sciences, 1943.

I soon found out how difficult it was to establish a new institute in the absence of a school and a team of collaborators. Consequently the institute grew much more slowly than if it were branched off from an existing research body. Further obstacles in the selection of personnel arose from the specific nature of our work on strong magnetic fields and low temperatures, fields poorly developed in the Soviet Union at the time. The early years were spent in the training of basic research personnel and auxiliary staff at the institute. Only after the formation of a working core of manpower were we able to grow and expand normally. In other words, the future development of our institute will be much more rapid.

After these preliminary explanations I can begin my report proper.

One of the first questions I asked myself concerned the tasks of an institute belonging to the Academy of Sciences. Naturally, in asking this question I had in mind an institute of physics or, generally speaking, natural sciences; the organization and objectives of an institute engaged in any other realm of knowledge would obviously be different, so that the views expressed here are not universally applicable.

Let me also stress that we are dealing with the organization of an institute belonging precisely to the Academy of Sciences. What is the Academy of Sciences? It can be characterized as the headquarters and the major manpower of Soviet science. In my view, the Academy of Sciences must provide the guiding ideas for all of Soviet science from the very top to the bottom and steer it on the right course. This is what it must do, or at least strive for. For its part, each institute of the Academy of Sciences must follow the same line and assume a decisive role in science within its specific field, which it has to bring to the front line of research.

Therefore the main task of an institute belonging to the Academy of Sciences is to deal with pure science, that is, a science engaged in the study of fundamental phenomena necessary for a more profound understanding of nature. Pure science must provide the knowledge necessary for transforming nature into a tool for the progress of human culture. Hence the importance of the proper selection of basic themes of research at the institute. In effect, the scientic work at the institute must correspond to the principal and most promising trends of research of the respective field at the time, taking into account the existing capacity for fruitful work at a satisfactory pace.

I think there are three such basic trends in physics, namely, low temperatures, the atomic nucleus, and the solid state. I cannot enter now into a detailed justification of these choices, and many colleagues will perhaps disagree with me. Our institute is engaged in the study of phenomena which take place at low temperatures, near absolute zero. This is one of the most rapidly developing fields of physics in recent years, and many fundamental discoveries still lie ahead.

Research at our institute is carried out by a small team of leading scientific workers. This makes our work more streamlined and focused on a limited number of major topics. Nothing is as dangerous for a research institute as getting bogged down in minor themes which only lead away from the principal tasks and trends. The main program of the institute is worked out by a small group, only three to four scientists, which makes the endeavor all the more purposeful.

Once the general trend of research has been set, the next step is the selection of the scientific staff. Only a person of great creative talent and a creative approach to work can aspire to success in pure science. There is only a limited circle of such persons in science, just as there can only be a few major writers, composers and artists in the country. Especially for this reason we must give these people the necessary conditions for fruitful work in pure science. Consequently, the nucleus of the institute must consist of only a limited number of very carefully selected scientists. This nucleus must devote all its energies to research.

Organization and working conditions at the institute must be such that the research scientist must spend at least 80% of the time in scientific work in the laboratory, leaving not more than 20% for social and other activities. Only under such conditions will the researchers be able to stay in the laboratory and work by themselves.

The only way of obtaining actual results in science is to perform the experiment by yourself, with your own hands, even including the routine parts. Good work is never done with someone else's hands. A person who spends about half an hour in instructing others to do research cannot possibly be a major scientist. For my part I have never seen or heard of a great scientist working in this manner, and in my opinion this is clearly impossible. I am convinced that even the greatest scientist stops his own growth, and even ceases being one, the moment he abandons his own work in the laboratory. These principles are of crucial importance, but only in a time of peace; war dictates a different approach.

These principles must be made quite clear to young scientists. For this purpose it is my policy to organize their work in a somewhat rigid framework. For example, a scientist should not deal with several matters at once, especially at the beginning of his career. Later, at a more mature age and with greater scientific caliber he might be able to cope simultaneously with two or three themes as a rare exception; as a beginning, however, he must be satisfied with one.

Another rule of successful work states that the researcher must spend only a limited number of hours at the laboratory. Working by fits is harmful: it exhausts the person and drains his creative power. In our institute, for example, all work in the laboratory must end by 6:00 p.m. The researcher must go home to reflect about his work, read, study and rest. In exceptional cases, with the permission of the Deputy Director, work may be continued until 8:00 p.m. Night work is subject to the approval of the Director himself and must be justified in terms of technical requirements dictated by the specific condition of the experiment. Such are the working rules for our research workers.

An institute, even if endowed with the scientific resources that make it a center of pure science, runs the risk of sinking to the level of an isolated unit incapable of achieving the original objectives of becoming a factor of major influence on the science and culture of the country.

How can the institute exert its influence on the progress of science in the country and how can it communicate with other research centers in the country? This can be achieved in a number of ways, of which we shall name only a few.

First of all, the institute must make use of its advantages as an institute belonging to the Academy of Sciences. These advantages include rich and modern equipment and the availability of suitable manpower, circumstances which allow research in fields unthinkable elsewhere. In our institute, for example, the availability of a special installation for the large-scale production of liquid helium opens wide horizons for experimentation in the field of low temperatures, which in other institutes simply cannot be done. Therefore we offer researchers of other institutes the opportunity of working with us as guest scientists in the field of low temperatures, as ours is the only place where such work can be done. These works, however, are not of a leading nature and lie outside the main trend of research at our institute.

Visits by colleagues from other institutes are usually organized as follows. The colleague who wants to work with us is invited to one of our scientific sessions or seminars for a report on the experiment he intends to make. A discussion follows, and if the idea proves to be of scientific interest and its author sufficiently qualified, he is allowed to conduct his work. In order to prevent any interference with the mainstream of research at the institute the number of such works is kept low, rarely more than two or three at a time.

Applications for working in our institute have so far exceeded the available facilities, a clear indication of the reputation of our institute as a leading research body.

The constant flow of workers from other research organizations to our institute has enabled us to establish a lively active link with the external scientific world. Apart from the experience acquired in the course of their work with us, the returning guest scientists acquaint their colleagues with various other aspects of our work, which thus penetrates more and more deeply into various research institutes all over the country. This active connection enables us to learn about research being carried out there. Indeed, the active link is the most efficient one and a powerful tool for the progress of science in the country.

A similar active living link should also be established with foreign scientists. Research workers from abroad did visit our institute in the early years of its existence. Lately, however, the political atmosphere has become so complicated that foreign links have been severed, though not for lack of potential visitors. Under these circumstances any connection with foreign scientists can only be a matter of the future. Yet it must be realized that such connection is a normal and healthy requirement for the work of any research institute, since science in the world consists of an integrated whole.

The leading position of a research institute is measured by the desire of scientists from various countries to visit it and work there. This represents a real proof that our institute is indeed a center of advanced research.

Another area in which advanced academic institutes contribute to the culture and science of the country is the training of scientific cadres.

Only the institute is in a position to train its own future scientific staff, which must be carefully cultivated from among the young generation. Our postgraduate program should therefore be welcomed and encouraged in every possible way. In this area, however, there are certain difficulties on which I would like to comment.

The main difficulty is the selection of postgraduate students. Regrettably enough, the links between our universities and research institutes are rather unsatisfactory. In my opinion the trouble lies in our organization. Many of the major scientists have gone to work in research institutes, leaving the universities largely in the hands of teachers for whom research work represents a minor occupation. Their way of educating the students, and what they demand of them is not designed to pick out the most creative youths. Hence the poor condition of creativity in our universities.

While participating in postgraduate examinations I have seen that in the majority of cases the university professors give top marks to students who *know* most, rather than to those who *understand* best. Yet science needs people who *understand* better. It is therefore very difficult to select university students for postgraduate courses on the basis of examination data along. For a correct selection of promising postgraduate students one must observe for some period of time while they are engaged in a work which provides an opportunity for independent, creative thought.

In my opinion, this gap between universities and research institutes is responsible for the present inadequacy in the selection of young scientists cadres in comparison with my time, when research took place mainly in universities. I remember the time when Academician A. F. Ioffe was head of the physics department of the St. Petersburg Polytechnic Institute. It is not by chance that a considerable number of young scientists of his research team at the time eventually became very prominent (including four Academicians).

Today our universities are undoubtedly rich in promising, talented youth, but the sieve we use to pick them out for research work is riddled with such holes that most of the talent and promise are lost as far as the research institutes are concerned. If we truly want to pick out the most talented scientists we must find a way to link our research institutes with the universities, so as to be able to detect and cultivate the creative capacity of youth.

With this objective in mind we must seek new ways for the selection of postgraduate students among the university youth. Since we possess liquid helium for low-temperature research in greater quantities than all other cryogenic laboratories in the world combined, we were able to set up at the institute a laboratory program for all students of the physics faculty of Moscow University. Initially this course could only cover the best students, but in the last two years all students of the faculty have passed it, each student performing 2–3 laboratory works with liquid helium. From the viewpoint of cryogenics institutes this is a veritable luxury; in Leiden and other laboratories, for example, work with liquid helium is still almost inaccessible even for research workers, while here every Moscow University student can study superconductivity, magnetic phenomena at temperatures close to absolute zero, etc. Not surprisingly, the university welcomed this opportunity and readily sent students to us.

This work at our institute was organized as follows. The best students, those with a superior performance during their course, were allowed to do more than the three mandatory works. Institute researchers serving as instructors used to talk with these selected students and the best of them were given an opportunity to talk with me. Thus we were able to pick out the most promising youths, become acquainted with them from their 3rd or 4th year onward, and keep an eye on them. Then we invited the best of them to work in the institute as probationers. As such they were allowed to participate in our research work as technical assistants to our staff, making records, setting up some elementary works, etc.

Postgraduate students were selected among these assistants not only on the basis of examination grades but also according to actual performance at the institute. Such selection of young scientists obviously increases the circle of candidates and eliminates the element of chance.

This policy has been interrupted by the war. Were we to continue it, the outcome would be as follows. Upon completion of the postgraduate program, the young scientists are awarded the degree of Candidate and join the staff of various other research institutes where they apply the knowledge acquired with us. Moreover, one out of ten or fifteen such postgraduates would be found talented enough to join the permanent staff of our own institute. Thus the institute would grow.

In my view, this thorough and continuous follow-up of youths from the university bench onward is the only correct way for the selection of fresh academic cadres. No effort should be spared in this direction, as the young scientists represent our future. Even more so, they also constitute our present. As we grow older, only our students can save us from a precocious stagnation of the brain. A student working in a given field must naturally know more than his teacher does in the same field. And who will teach the teacher, if not his own student?

By virtue of his own experience the teacher sets the course of work, but in the final analysis it is the students who educate their teacher: they deepen his knowledge and expand his horizon. In the absence of students the teacher ceases his own progress and his creative talent quickly withers away. I shall never forget the words of my great teacher Rutherford: "Kapitza" – he used to say – "you know that only my students make me feel young". And as I approach old age it is becoming increasingly clear to me that communication with the young must be a *modus vivendi* which keeps senility away, preserves freshness of mind and interest in everything new and progressive in science. Scientific conservatism is a lethal danger to the scientist and an impediment to the development of science.

Now I shall consider another important link between the institute and the outer world, one unjustly ignored not only by research institutes but also by the Academy of Sciences as a whole. I am speaking of propaganda of science.

The popularization of science, which means making it intelligible to the masses, is widely spoken of in the Soviet Union. However, little attention is given to the propaganda of science. Every major scientific achievement can not only be popularized – and this, of course, is not necessarily the task of scientists; what we must do is propagandize it – that is, explain its significance to our fellow scientists, point out its contribution to science, its impact on the development of scientific thought, on our philosophic concepts, our technology, etc.

The propaganda of science does not mean the retelling of scientific thoughts in a simpler language. It constitutes a creative process, since there are real difficulties in perceiving the impact of a given scientific advance on the development of science, technology and culture as a whole, let alone explaining it to others. We tend to devote little attention to the propaganda of science as defined here, and such matters are largely ignored by many of our scientists. We in our institute share the blame. Our activities in the field of science propaganda have been confined to isolated lectures in research institutes, invitations to colleagues from other institutes to attend our scientific sessions in fields of joint interest, etc.

Such ties between science and life are rather accidental and unorganized in our country. This state of affairs impedes the influence of one field of science on another and the implementation of scientific achievements in daily life. We must do some thinking about the training of science propagandists and the organization of their work. I have always striven to encourage a broad discussion of scientific works and have never tried to prevent scientific arguments at research sessions; to the contrary, in my opinion people should even be goaded a little into a healthy intense discussion. A broad discussion of a scientific work should only be welcomed, leading to contradictions and arguments which provide a powerful stimulus for the healthy development of scientific thought. It seems to me that our institute has given eloquent proof of this policy at sessions of the Physical-Mathematical Division of the Academy of Science.

Now I would like to consider one of the crucial expressions of scientific work on culture – namely, the influence of science on the progress of technology and industry.

What should be the form of organization of this influence of science on the technology and economy of our socialist country?

This topic, often debated here, has lost none of its urgency. Let me say quite frankly that I do not share the prevailing attitude to this problem as expressed by persons holding responsible positions with whom I have had the opportunity to talk. It seems to me that the relationship between science and technology is interpreted in an oversimplified and vulgar way in our country. Many people are convinced that any scientific work must yield an immediate application in technology. Such people judge the performance of a research institute merely in terms of the concrete advantages to be derived from it by a given branch of industry. This, of course, is wrong. Such a naive, simplistic approach can only bring harm. Even a superficial review of the history of science and culture shows quite clearly that any pure science inevitably exerts its influence not only on technology but also on our way of life as a whole.

Only the fundamental work and discoveries of Faraday have made possible the creation of such totally new tools of human culture as the dynamo, the telephone, etc. Yet it would be absurd to expect of Faraday and the likes of him to make the telephone and the dynamo all by themselves. Faraday did not possess any engineering background, and the industry of his time was not yet prepared to put his ideas into practice in daily life. Such first-rate engineers as Bell, Siemens, Edison and others were able to achieve this a few decades later. Many similar examples can be cited. The fact that Faraday did not translate his ideas into technological language does not in any way detract from the monumental importance of his discoveries of the laws and properties of electric current. In our country, we tend to judge the achievements of science strictly in terms of practical results; the man who picks the apple is given all the credit, whereas those who have planted the tree and cultivated it are largely ignored.

Such an attitude, which I reject, minimizes the significance of pure science; more particularly, it greatly underestimates the bulk of the work done at the Academy of Sciences.

The ties between science and technology are complex indeed. When an ordinary engineer calculates the brake action of a truck or the strength of a construction he uses Newton's laws of mechanics. When a patent expert rejects yet another 'highly promising' proposal for a *perpetuum mobile,* he does so on the basis of Mayer's law of the conservation of energy, etc. When an engineer comes to a scientist for advice or for an explanation of a poorly understood phenomenon in the production process, or for guidance on the calculation of a mechanism, all these cases embody important links between science and technology. Such instances are an everyday occurrence under various circumstances in hundreds of places. We have grown so accustomed to them that we hardly mention them, and there is a prevailing tendency to ignore them. Yet such ties constitute a major channel by which science exerts its influence on technology and industry. Such influence, however, can only take place in the presence of pure science and of scientists engaged in it.

Our military technology, for example, equals that of our enemies and in many respects is even better. Who is to be given credit for this? First and foremost, our pure science and our researchers, who exert their fruitful influence on technology in a variety of invisible ways.

Further, how has our metallurgy attained its enviable present level? No doubt largely because of the works of Chernov and all his co-workers with the entire tradition of a scientific approach to metallurgy which they have established over many years. Engineers undoubtedly have great merit: they have managed to assimilate and implement everything that is pertinent out of the pure science established by the fathers of our scientific metallurgy. But without such people as Chernov, Kurnakov and their followers our metallurgy would not have been able to produce such good steel for the weapons of our army, or the excellent armor we have. In the absence of such steel no designer could ever construct any first-rate tanks.

Our aviation provides yet another example. What is the main factor in its progress? Obviously no development would be possible without the works of Zhukovsky, Chaplygin and their school. But Chaplygin could not even draw a profile, let alone design an airplane. He was a great mathematician like his brilliant teacher Zhukovsky, the founder of flight aerodynamics. Zhukovsky became world famous for the discovery of the fundamental theorem of the calculation of airplane wing profiles, which shed light on the mechanism of wing lift. Should we expect of Zhukovsky that he should have engaged in airplane design on top of that? His theorem is the wonderful apple tree which will yield abundant fruit to all airplane designers for centuries to come.

Naturally, the influence of pure science on technology should be better organized than it is now in our country, and science communication should be used for this purpose. It is also necessary to improve the consulting services rendered by scientists to industry. Our scientists must show more interest in those fields of technology where their knowledge can prove most fruitful. If any planning of science is possible at all, it should consist in a stimulation of the development of those fields of knowledge which can exert the greatest influence on the progress of technology. However, one should never demand of a major scientist any direct, obligatory involvement in technology by personally deriving a practical result result from his ideas.

After these general considerations let me dwell briefly on the ties of our institute with technology. What I intend to say may seem at first glance to contradict the ideas developed previously. This discrepancy, however, reflects the accidental fact that apart from my research work I am also engaged in engineering problems. My exceptional case, however, should not be regarded as the rule. Perhaps the best approach would be to recount how our institute applied our oxygen research in industry.

Our technological literature of the early 1930s contained extensive discussions of the wide application of oxygen in industry and its possible influence on modern technology. Many interesting articles and calculations by prominent Soviet engineers emphasized the great potential role of cheap oxygen in industry. Especially promising was the intensification of the iron and steel industry: blast furnace smelting, oxygen blast in steel production. Other possible applications included underground gasification, the speeding-up of various chemical processes, etc. All these interesting and tempting prospects rested on the possibility of a cheap production of oxygen on a large scale. Accordingly, various methods for the large-scale production of oxygen were proposed and discussed.

These works attracted my interest, and I drew attention to some papers which contained obvious errors. A discussion began on the cheapest method for the production of oxygen. It was shown on the basis of modern physical concepts that the cheapest source of oxygen is air, which contains this element in a free state. I was also able to demonstrate, as I reported before the Academy of Sciences, that the cheapest way for the production of oxygen with present technological means was by liquefaction and subsequent fractional distillation of air.

Liquid air can be fractionated into oxygen and nitrogen in much the same way as the fractional distillation of alcohol from its liquid mixture with water. Next, likewise on the basis of general scientific considerations it can be shown that present installations for the production of liquid air have an efficiency of not more than 10–15% and that the existing cycles of liquefaction and distillation are rather complex. Further, it is possible to show how to design a cycle that is closer to the ideal one. It can be proven that the best way for the simplification and cost reduction of these processes for the large-scale production of oxygen is by discarding the piston refrigerators in favor of the rotary type, namely, turbines. It is noteworthy that the idea of building a refrigeration turbine, though expressed as long ago as the 1890s by Rayleigh, has not yet been successfully achieved despite numerous attempts. We were able to show theoretically what appears to be the error of these attempts and how to overcome it. All this theoretical work had to be done, and naturally, this was the task of a scientist.

Having obtained these results, I explained them to engineers and specialists, indicating what in my opinion should be done for developing a new technology for the production of cheap oxygen. They told me frankly that the professor was indulging in fantasies, that all these matters were unreal and detached from their present concepts. In other words, our technical thought was not mature enough to assimilate these new ideas.

As a scientist, I could have stopped there, having published my results, in expectation of the day when the prevailing technological concepts would be mature enough to absorb them and convert them into reality. Today I know that this creative research provided the foundation for my own work as an engineer during the last four years, a work I initially considered the task of our industry. I was fully entitled to stop upon ending the theoretical part; but I am also an engineer, and was burning with engineering fervor, I must admit. They tell me my ideas as a scientist are unreal. I decided to take yet another step forward.

Within a year-and-a-half or two, I built, at the institute, a machine for the production of liquid air on the basis of these new principles. The above-mentioned general theoretical considerations proved to be true. The machine was checked by a government commission. The State Economic Council [Ekonomsovet] instructed a factory to adopt our expert scientific and technical knowledge to develop the project. I thought I could rest there. The factory began the design and development of new intallations in the indicated direction. At first I hoped that the industry would manage to work out a new technology for the production of cheap oxygen on the basis of our laboratory model, which provided all the necessary parameters and confirmed all our basic theoretical considerations. However, the actual course of events was different. Although the government gave the factory rather strict instructions, they were not carried out.

In considering what happened at the factory, one finds it easier to understand the causes for the obstacles to be overcome in the development and implementation of a novel installation. At the factory there were young, gifted engineers and designers who showed great interest in the job. The overall attitude of the factory team to the new task was not hostile. Everyone recognized the value of the innovation, but they simply could find no time for it, being preoccupied in everyday work and most importantly, in fulfilling the production plan of the factory. Our project was clearly in the way, tapping their strength and interfering with the production plan; moreover, its negligible magnitude as a small-scale production deprived it of any significance in the annual production figures.

I think the attitude of the factory to any new, creative initiative can be best characterized by paraphrasing a few lines from *Faust*. Perhaps you remember them; in our case they would sound as follows:

Upon the highest concepts of the mind
There grows an alien and more alien mould;
When we but fill our annual plan completely,
That which is better is labelled a fraud, a blind.*

Our factories have an honest, positive attitude to scientific innovations. The trouble is that their objective conditions make the annual plan the highest goal. After a year of futile attempts it became clear that the factory would never manage independently to solve the problem of cheap oxygen production under such circumstances.

Then we changed our policy. The task was entrusted to another factory, where a special department and design office were established for handling our installations. By order of the Economic Council, our institute was charged with the selection of manpower and the technical management of the project.

To save time, the institute did what we thought was a job of industry. From the installation for the production of liquid air we went on to the development of new cycles, the construction of a liquid oxygen plant. We proceeded with verification of our theoretical considerations and used turbines for the production of liquid oxygen. We also tried to find out how long our plant can work without interruption, and to determine the conditions of its operation in the factory. Although the oxygen installation of the institute worked satisfactorily it was obvious that our project had not yet reached the industrial stage.

Matters went much better at the second factory. Yet the advance was sluggish, and our position as managers of the project was sometimes interpreted as interference by strangers. Within a year or so a few installations were built and delivered to industry. Then the war broke out, and this new tie with industry was severed.

We have learned a great deal from our joint work with industry. In the factories we met creative engineers and saw their striving for new technology. From our earliest steps in the production of oxygen we received much help, support and interest on the part of the government in all our initiatives. We were readily given all the assistance we needed. Otherwise our work would have failed. Then what was the cause of our troubles? Organization, no doubt. In effect, our industrial organization is inadequate for a quick, smooth implementation of new ideas of technology. Yet I have no doubt that our

* [Paraphrased in the *Collected Papers* of P. L. Kapitza, Vol. 3, p. 115.]

economic system can find and create the organizational forms necessary for a rapid and smooth introduction of advanced ideas in technology, for making the industry more receptive to the influence of science. These forms, however, are yet to be found.

The national demand for oxygen increases in wartime. Every effort must be made for the construction of industrial-type machines and solving the problem of durability and service life. This we did in Kazan where the institute was relocated after the wartime evacuation. At the same time urgent construction of large-scale industrial plants began, and these are being introduced to production work, everything on the basis of the Kazan experience, using our drawings and with the cooperation and guidance of our institute.

The war and the postwar economic problems make the oxygen problem especially urgent. We must act quickly to utilize the full industrial potential of our method for the production of oxygen. I cannot enter into any details on the steps taken; suffice it to say that a separate Oxygen Board has been set up for the development and application of plants of our type. This board has a factory of its own. I was placed in charge of this organization.

The scope of my present report leaves no room for a detailed discussion of the tasks of the Oxygen Board and the simultaneously established Technical Council for Oxygen Applications. This organization can be regarded as an attempt to create a tie between pure science and industry, to use oxygen for the intensification of our metallurgy, chemical industry, power supply, etc.

Perhaps a contradiction with my basic assertions can be detected here. However, the discrepancy can be easily solved by assuming the existence of 'two Kapitzas', one of them a scientist, the other an engineer. As an engineer I made every effort to establish an industrial body capable of adopting and implementing new scientific ideas. The outcome of this attempt remains unknown; at any rate, war conditions make it imperative to spare no effort in achieving our goal.

This, of course, does not contradict what I said earlier. It is only a coincidence that I can work both as a scientist and an engineer. Other persons have been known to possess two professions. Borodin, for example, was a chemist and a composer. This, however, should not be made a rule or a model to be imitated. You do not expect of a singer to accompany himself. Neither should the scientist be compelled to take an active part in the introduction of his work into industry.

Some scientists have the necessary engineering inclinations. Such fortunate cases are a pure blessing for all. But forcing a person to act beyond his capabilities can only bring harm. The case of Academician N. N. Semenov provides a good example.

The works of Academician Semenov on chain reactions and combustion represent some of the most brilliant and leading scientific achievements in the Soviet Union. The theories of combustion, explosion and detonation developed by Semenov and his school are generally recognized as a major factor in the design of modern internal combustion engines, explosives, and in various other fields of technology. Semenov enjoys an enviable reputation in the field of combustion processes both in the U.S.S.R. and abroad. But were Semenov to attempt to build an internal combustion engine by himself, or event to manage such a project, the result would only be a waste of precious time and energy best spent in pure science where he is nothing less than a virtuoso.

We value Semenov as a prominent Russian scientist, as the pride of our theoretical thought; his works on theoretical chemistry will be appreciated by future generations. But as an engineer Semenov is just mediocre. And if the singer cannot possibly accompany his own songs, why force him do it? Is it not better to train accompanists for this purpose? Admittedly we devote insufficient attention to the training of suitable cadres for the application of new scientific processes in our industry. We must overcome this major shortcoming of ours. But it is just as harmful to place this burden on the shoulders of our major scientists.

This burning question of the links between science and industry must be widely discussed in order to determine the best course of action for ensuring a rapid cultural growth. Any vulgarization in formulating the problem should be avoided, – notably the requirement that all scientists become actively involved in the application of their work, as is frequently stated, even by the Presidium of the Academy of Sciences. I shall always protest against such policy. Pure science has always been and will always be the moving force of technological development. We in the Soviet Union have all the necessary tools for promoting this influence. But these matters should not be simplified to such a primitive level.

Allow me to consider now the last subject – namely, the organizational setup of the institute. The tasks of the institute, which I outlined here, are naturally reflected in its organization. We have at the institute a small staff of permanent research workers and a number of other scientists and postgraduate students employed on a temporary basis. Only one third of the personnel have permanent positions, which means that about two thirds constitute a fluid work force. This state of affairs has left its imprint on the whole structure of the institute. Since the temporary personnel are not paid by the institute, our auxiliary services are clearly out of proportion to the number of permanent employees. From a bookkeeping viewpoint this means excessive expenditure for services at the institute, but there is no discrepancy if one takes into account the ratio of auxiliary staff to the total number of research workers.

Moreover, the presence of scientists on temporary work at the institute necessitates the availability of a highly qualified auxiliary personnel. Post-graduate students, if left unsupervised, often break various instruments and devices until they learn how to work. Arriving workers may also break various items of equipment if left unassisted by experienced laboratory technicians. Such technicians can also help speed up the work by assisting in the construction of special equipment for work at low temperature, assembling the rather complex thermometry for this purpose, demonstrating how liquid helium should be handled, etc.

Apart from the staff of trusted, experienced technicians, our institute employs highly skilled craftsmen for the preparation of special equipment. Nothing can be as frustrating to research work as the sluggish preparation of equipment for experimentation. Hence the value of a good workshop at the institute.

The financial structure of the institute is an important factor in its organization. The financial system adopted for U.S.S.R. research institutes is essentially identical to that of any other economic enterprise. Obviously this is wrong. The tasks and organizational patterns of economic bodies differ radically from those of research institutes. Yet for some reason even the Academy of Sciences has made no serious attempt to change this state of affairs and adapt it to its own requirements, rather than resort to endless gimmicks and tricks in trying to adapt itself to the system.

Our institute was allowed much freedom to alter and simplify its financial procedures. Recognition was given to the fact that a research institute requires a considerable measure of administrative flexibility. In the course of creative work, it is nearly impossible to foresee the status of research next year and even next month, so that little can be said in advance about the expenditure and organizational patterns required for a successful outcome. Under our flexible organization we are, for example, exempt from registering our staff lists in advance. Staffs are determined by the director of the institute as needs dictate. Without entering into further details regarding our financial setup I shall merely note that we are given more freedom to dispose of our annual budget than ordinary state institutions.

Some perseverance was required to convince the People's Commissariat of Finance of the necessity for such an arrangement. At that time the Finance people wanted to introduce so-called thematic accounting whereby expenditure was reported in detail for each line of research. In my correspondence with the Treasury I asked them: "When looking at a Rembrandt picture are you really interested in the great artist's expenditure for brushes and canvas? Then why are you so interested in the cost of equipment and materials when you consider research?" A fruitful research is worth incomparably more than the expenditure involved in it. The cost of research in money terms is incommensurable with its cultural value. I asked: "How much funds, in the Commissariat's opinion, should Isaac Newton have been given for his work which culminated in the discovery of universal gravitation?"

Yet the Commissariat remained adamant. The arguments lasted over six months, and I think they would have gained the upper hand were it not for the intervention of the U.S.S.R. Council of People's Commissars. Finally our institute was given a simplified financial system which saved the director from everyday troubles and endless 'combinations' of work. Under the new system, for example, the institute employs a single bookkeeper, and when we are shorthanded he finds time to help us in testing equipment, taking records and making measurements. All this helps a great deal toward an efficient and healthy operation.

Now I arrive at the last topic – wartime changes in the work of our institute.

In effect, we did not have to introduce any major alterations. The oxygen problem turned out to be no less urgent in wartime. True, the war forced us to apply all our experience and expert knowledge in this field as quickly as possible. We did our best to provide all these to industry as part of the war effort. Moreover, some other fields of research at the institute turned out to be of military value. For a number of reasons I cannot give any details on these matters. Having focused all our energies on these objectives, we had to make deep cuts in the fields which I discussed at the beginning of my report. A great deal was sacrificed for oxygen, our main objective, in order to obtain rapid and definite results. We were of the opinion that research not completed prior to the war had to be discontinued so as not to consume resources better spent for more urgent matters.

In concluding my report I would like to point out that I have tried to cover only the most general and basic aspects of the organization of scientific research work. Some of the problems involved are indeed far from being solved. Regrettably enough, the question of the organization of science is still largely being ignored in our country. Therefore I hope that many of the solutions found so far will be improved considerably in future. But there is no doubt in my mind that many possibilities still remain to be explored in the field of research organization in our country. Even at the inadequate present status of research organization, our pure science is exerting a major influence on technology and daily life as a whole, in fact more so than we tend to realize. This influence is brought about by the growing traditions created by pure science, by way of a multitude of invisible links with our industry and everyday life. Without the great scientific traditions established by our scientists since the days of Lomonosov, we would be left without good rifles, strong armor and fast airplanes, even if none of our academicians can personally design an airplane or fire a rifle.

We have not yet realized the vast possibilities at the disposal of our country, the great power which arises from the close ties between our science and daily life, the inexhaustible resources which the Soviet government places at the disposal of scientific research. We have not yet managed to make proper use of the enormous freedom for scientific thought in our country. Many mistakes have yet to be corrected, there is sometimes little understanding of the powerful role which the Academy of Sciences is called upon to perform, and the Academy itself tends to minimize this role. Nevertheless our achievements are impressive indeed.

My scientific colleagues! We are called upon to make a major contribution to a great endeavor in a great country. It is our obligation to honor and respect our own task, and we must spare no effort to fulfil it.

PART THREE

12. THE UNITY OF SCIENCE AND TECHNOLOGY

The life of the socialist state is being built and developed on a scientific basis. For this reason the Soviet government has been devoting so much attention and care to science from the earliest days of its existence. As a result we now have a science capable of solving the most advanced and intricate problems. We also have an industry which makes the most complex and delicate mechanisms. Yet the link between science and technology in our country is still weak, which prevents us from utilizing all our vast resources of creative power.

Only a live, healthy unity of science and technology can bring about a fruitful cooperation between them: technology readily absorbs the new prospects which science reveals, while the new technical possibilities acquired by the growth of technology provide science with new tools and expand its fields of research.

Nowhere in the world is the ground as suitable for the extensive application of science as in the U.S.S.R. Therefore the unsatisfactory union between science and technology can only be attributed to the lack of efficient organizational patterns for achieving it. At the present level of our science and technology, one could expect much greater creative technological advances even with the existing shortcomings. I do not think that this lag is due solely to such obvious and gradually vanishing obstacles as the stagnation which prevails among some of our creative cadres, the red tape, the irregular supply etc. It seems to me that we are also afflicted with more severe organizational defects to which we have devoted too little attention.

Success in the struggle for the conquest of nature requires a strategy and tactic of its own. Here as in real war the most important thing is a correct deployment of forces on the front and a clear understanding of the task on the part of the soldiers. In reality, however, not only is the headquarters of our scientific army below the required level but also our soldiers often have to fight without any commanding officers. On many fronts of our science and technology an organized army simply does not exist, and only isolated 'guerrilla' units do the fighting. We also lack concentrations of elite forces in specific sectors of decisive importance. Moreover, there is a pronounced

---

From an article in the newspaper *Pravda*, June 17, 1941.

hesitation to deal a concentrated blow at definite points, and a tendency to postpone the handling of major problems until the last moment. Thus our creative forces remain scattered over a long front line, and even the greatest general cannot achieve victory under such circumstances.

Our scientific army must be inspired with the challenge of the tasks ahead and made to feel the significance of victory on the frontiers of research. Fighting units must be suitably formed and trained by responsible commanding officers. Life teaches us that progress in science and technology can only thrive with the support of a healthy public opinion. Such a creative atmosphere can be established in scientific meetings and by the press. These major organizational prospects remain largely neglected.

The general sessions of the Academy of Sciences provide a fitting example. Generally well attended, they include some interesting reports, but in most cases the agenda consists of professorial lectures which enrich the education but leave no room for scientific discussions. Debates are scarce, and no attempt is made to seek a public appraisal of problems and works in specific fields of science and technology. Debates are not encouraged. Yet they do not arise spontaneously but have to be cultivated. For this purpose it is necessary to select the most crucial and acute problems in advance, and opponents have to be chosen and prepared. The chairman must steer the discussion, formulate the contradictory views, and prevent unnecessary digressions from the main topics. If a scientific report does not by its very nature give rise to a discussions, as may be the case with a report on a scientific discovery, the meeting must hear the opinion of leading specialists in order to allow for a correct public evaluation. As a matter of fact, to read reports is pointless in the absence of a discussion, as it would be much cheaper and simpler just to print them.

It seems to me that we in scientific circles are often afraid of discussions and appraisals. This may be due to our false pride, based on the wrong idea that a good scientist cannot err and that any 'mistake' is bound to discredit him. We forget that only those who do nothing never err. The scientific truth of today may well be modified or supplemented tomorrow as we advance toward an understanding of the true nature of things. Only by revealing the contradictions involved and overcoming one mistake after another can we solve problems in science. Errors do not constitute a pseudoscience. To the contrary, pseudoscience is the non-recognition of errors and therefore a major impediment to the healthy growth of science.

Our press often avoids a discussion of controversial problems of science and technology.

The lack of a solid public opinion on a number of topics is responsible for the development of some pointless and even harmful research in our country; moreover, it prevents the concentration of our best forces on problems of crucial importance.

This lack of discussion is totally alien to the general setup of our country. With us any scientific or technological achievement becomes the property of the whole country, and people show a marked desire to understand and appraise it. A just public evaluation can only inspire the worker and fire his enthusiasm which is so essential for creative work. Any major scientific or technological problem is a matter of public impotance here, which means that it must be discussed and judged.

In my opinion, the army of creative industrial workers is even less organized than that of research scientists. Our scientists are more or less united in associations of one form or another, but the creative workers of technology are so thinly spread all over our vast industry that any attempt to gather them into an effective strike force becomes extremely difficult.

I do not have in mind those skirmishes which lead to various technical improvements such as the selection of more suitable materials, the rationalization of processes, the reduction of labor expenditure, the improvement of product quality, etc. Such forms of technological progress, though often achieved by 'guerrilla' tactics in the hands of single enthusiasts, are of considerable importance and thrive fairly well in our country. The remarkable thing is that here as nowhere else, such endeavors involve wide circles of technological workers, from ordinary workers to engineers. I am speaking about major battles which are carried out by large units and effectively widen the horizons of our country. Achievements of this category may culminate in something comparable to the invention of the turbine or the Diesel engine.

In such major battles, victory over nature cannot be guaranteed in advance even by committing a vast amount of money, labor and the best creative forces of enthusiastic, hard-working scientists and engineers. But victory on the battlefront brings us to a new stage in the development of industry. Achievements in this field will help us overtake all. The socialist regime, no doubt, must be the best organizer of such major victories in technology. Under no circumstances can capitalism mobilize such vast human and material resources for such battles.

To make use of this extraordinary possibility, we must rapidly make an organized army of our best creative workers from all leading sectors of our technology and coordinate their work with our scientists.

In my experience the most severe organizational shortcoming is the fact that we do not attach the necessary importance to the introduction of technological advances into daily life, as we usually do not consider such matters to be creative. Conditions for creative work do exist in the various institutes for applied research and in factory laboratories. Despite this, the final stage – the actual introduction of new ideas – is usually done by the rank and file together with their current production tasks. But just as a good citizen is not born ready-made but has to be educated over a long period, a new idea about a machine or a process can only be developed side by side with their growth in everyday life. The successful 'rearing' of new ideas should not be entrusted to rank-and-file engineers, craftsmen or workers; such matters are the task of a carefully selected, highly qualified, creative team.

That our industry is suffering from a shortage of organized cadres of such creative technical workers I realized only too acutely in my efforts to start the manufacture of our installations for the production of liquid air and oxygen. These installations, based on new liquefaction and separation principles, were adopted by laboratories only. The industry faced the task of converting them into a reliable, workable machine.

We found our greatest difficulties in trying to organize a base for such an introduction into the production stage and in the selection of suitable cadres of engineers, technicians, designers, craftsmen and other workers capable of a creative adoption of such machines. Of course, such people are numerous in the industry. But they are anonymous, scattered, unrecognized, being engaged in a large variety of tasks. One has to grope along to find them out, and getting them is a tough job indeed. Our economic managers, still largely unaware of the importance of opening new horizons for technology, often object to any concentration of first-rate people and to the establishment of the necessary conditions for the solution of advanced technological problems.

The importance of having a sound base for the introduction of new technological achievements is evident from the following example of a foreign firm which by a random course of events set up an experimental factory for the development of new types of machines. This is the Brown–Boweri factory in Switzerland. The factory was built in the town of Baden, located near the University of Zürich. Using the advice of Stodola and other well-known professors, the factory manufactured very good electric machines and turbines. But when the neighboring countries erected customs barriers all around, the firm had to build independent factories in such countries as Canada, Germany, France, etc. However, work on the design of new machines was concentrated in Baden, where the firm had its best manpower. Thus there arose in Baden what is perhaps one of the most advanced experimental factories.

The causes of the establishment of this factory are, of course, of little interest to us. What matters from our viewpoint is the remarkable success of a factory engaged mainly in the design and introduction of new types of machines. How advanced was the factory is evident from the fact that it managed to build the first successfully working gas turbine.

This example reveals quite clearly the great power of a concentrated team of creative scientific-technological people having a base of their own.

I therefore suggest that carefully selected teams comprising creative workers of all categories be set up in each sector of our industry. Depending on the scale of production, such organizations could range in size from separate production departments to entire factories with a specially selected personnel. Thus we shall create a much better atmosphere for the introduction of new technological trends into our industry. Given a caring, flexible management, such teams will undoubtedly open new technological horizons for our whole science, introducing the most advanced and daring ideas. Under such circumstances it would be much easier to establish a connection between scientific and technological teams of creative workers. And this would indeed be a major step forward toward bridging the present gap between theory and life. This gap owes its existence mainly to temporary organizational shortcomings. In reality, however, as we realize more and more clearly, it is becoming increasingly difficult to tell where science ends and where everyday life begins.

13. PLANNING IN SCIENCE

It is evident from the discussion which took place that we still lack a clear and generally acceptable idea of planning and reporting at the Academy of Sciences. This state of affairs can be attributed to the divergence of opinion about the objectives of planning and reporting. I believe that four possible objectives can be clearly formulated.

To begin with, planning and reporting are tools of financial control. Funds are allocated according to the plan, and the report shows how these funds have been used.

Secondly, follow-up of the work performed. In other words, the performance reported is checked against the plan.

Thirdly, a largely neglected matter in our country, the plan and the report are the basic means for coordinating the research activities of institutes and individual scientists all over the country.

Finally, the report and plan are major instruments for coordinating the research work of institutes and scientists with the needs of the country, its economy and culture.

Contrary to the established attitude I attach particular significance to the last two objectives of planning, which are of a more general nature.

The advocates of planning as a tool for exercising a control over scientific work would rather have it on a strictly thematic basis whereby the scientist must declare in advance that he will be working within the framework of a given plan and state his needs in terms of money and materials, so that by the end of the year he will be able to justify the expenditure by producing specific results. By matching the report against the plan it would thus be possible to assess the work of an individual scientist or research institute as unsatisfactory, satisfactory or good, as is commonly done with regard to factories or industrial enterprises generally.

Such requirements concerning plans and report come usually from officials of the Commissariat of Finance and the administrative staff of the managing organization. Among scientists they can only arouse irritation and discontent.

Address to the Conference of the Directors of the Moscow Institutes of the U.S.S.R. Academy of Sciences, 1943.

Many major scientists oppose planning of any kind. They regard research as a creative activity, which by its very nature is incompatible with planning and requires complete freedom for the scientist to create in accordance with his own inclinations at the time. Many convincing examples are cited in support of this view. Newton, for instance, could hardly have discovered the law of gravitation according to a given plan; the idea came to him quite spontaneously as he was looking at the famous falling apple. It is obviously impossible to plan the moment when the scientist will see a falling apple, to say nothing of the possible effect of this, even on him. Scientific breakthroughs are impossible to plan as they constitute the culmination of a creative process, the succcess of which depends on the talent of the scientist.

A compromise was devised in order to bridge these contradictions. At our conference this view was represented by Academician O. Yu. Shmidt. In his opinion scientific discoveries are indisputable facts and no one can ever force a scientist to plan his discoveries in advance, so that scientists need freedom in this respect. But there is a lot of work other than the creative kind, and this work has to be planned. I do not share this view. For example, a Rubens picture, a masterpiece of inestimable value, can hardly be assessed in terms of the expenditure for such items as frame, paints, etc. Such an approach is totally meaningless and can only satisfy a bureaucratic administration. In my opinion Shmidt's and all other proposals leave the problem unsolved and we should seek a broader approach based on creativity as the most valuable element of science. Planning and reporting must accordingly stimulate the freedom of creation rather than restrict it.

However, I disagree with those scientists who regard research as totally spontaneous and therefore unamenable to planning of any kind. Even a glance at the work of such a major scientist as Newton shows quite clearly that his achievements in the field of celestial mechanics were not unrelated to the objectives of world science and the British interests of the time. His research in celestial mechanics, for example, was related to the needs of the expanding colonial policy of Britain with its intensive shipping, which required a reliable navigation based at the time on celestial mechanics. Faraday's works were closely connected not only with the developing science of electromagnetism but also with the rapid development of industry in Britain. From Faraday's biography we know how serious was his attitude toward scientific problems related to the industrial needs of the country and how he strove to solve them.

We know from the history of science that even a major creative work always has a certain orientation which reflects not only the problems of world science but also the pressing needs of the country.

Under capitalism the scientist does not have to determine these trends in advance. But in our socialist country, where science must be organized on a national scale, we must know in advance the direction of research in various fields of science. The work of our scientists must be organized on the basis of comprehensive, thoroughly worked-out plans by which every effort is made to coordinate the creative activity of research scientists with the objectives of world science and the interests of our country. Planning and creativity are not necessarily contradictory; to the contrary, the plan must contribute toward a free and successful development of pure science in our country. Once this has been recognized as the main task of planning, we must cleanse it of all petty elements of bureaucratic control. Only the broad organizational function of the plan should be retained. In my opinion, planning in the Academy of Sciences must be built on these principles.

They will tell us such a plan cannot possibly be prepared and at any rate no one has succeeded in doing so. True, we are going to be the first because science has never been organized on a national scale as it should be in a socialist country. We shall be indeed the first to have achieved this; therefore we must seek out the way for ourselves.

In summing up along these lines, I believe that the plans and reports must achieve two major objectives. Firstly, a coordination of research all over the country. Secondly, a development of pure science in accordance with the requirements of world science and the cultural needs of the country.

For this purpose we do not have to work out in detail any standard pattern for planning or reporting; general guidelines will do. Instead of tethering the scientist in a standard form, one must allow him complete freedom to describe the main trends of his work and justify his choice of subjects in a concise manner. Plans and reports based on these principles can be both interesting and useful. Success in this direction requires also that the plan and report be united and submitted simultaneously, instead of the present procedure of handling them separately. The reader can only understand the significance of the research performed when he considers it in connection with the tasks formulated by the scientist himself. Moreover, the plan must possess an element of orientation – it should be concerned with the main trend only, leaving all secondary themes aside. Such plans and reports will show how our research at the Academy of Sciences meets the requirements of world science. This is of particular importance because we are a part of world science.

An objection might be anticipated here: what about the financial aspect of the plan and how are the allocated funds to be determined?

As a matter of fact the planning of funds in research can be reduced to a rather simple pattern which is actually being applied in our country, though in a form concealed by programmatic or thematic planning. Let me describe this pattern briefly.

First, the premises where the research is being carried out within the institute. It is well known that a given area can accomodate a definite number of research workers. Consequently, only a definite number of researchers can work in a building having a given area. Another established fact is that a given number of researchers in a specified field of science require the assistance of a definite number of technical personnel, servicemen, administration clerks, etc. All of this scale of operations has already been established to an accuracy of ±5–10%, and it can be used for planning the magnitude of work at any research institute.

Now with regard to the planning of funds. We know that every researcher spends roughly the same amount of money for his work. Physicists spend a given amount, chemists a somewhat lower amount, etc. These figures, with variations of only a few percentage points, will reflect the financial parameters of institute funds spent for a given research.

Such parameters can characterize research expenditure much more accurately and correctly than do all the detailed sham estimates which are so widespread here and bear so little resemblance to actual work.

If a given research work requires special expensive equipment such as a spectrograph or a big electromagnet, the expenditure involved can be entered in an additional, separately explained estimate.

The organization of research at an institute can be judged not in terms of absolute figures but from such relative parameters as the percentages formed by salaries, administrative expenses, equipment purchases etc., out of the total expenditure. These figures reflect the course of research at the institute. I introduced them long ago for supervising the expenses of our institute.

The financial-accounting establishment of the Academy of Sciences must likewise learn to exercise control in terms of relative rather than absolute parameters. A detailed allocation of funds for creative research under a thematic plan is both absurd and harmful to the progress of science. Far from helping steer research toward the goals of world science and the cultural growth of the country, such an approach can only result in a waste of money for idle bureaucratic scribbling.

14. ON LEADERSHIP IN SCIENCE

I shall not dare speak of my own field but of the organization of science at the Academy as a whole.

Several colleagues have just made a critical review of the obstacles on the way of development in their specific fields. However, we must also analyze our entire work as a large team, and point out the shortcomings of our work as a whole.

The fact that each of us criticizes the state of affairs in his own field reflects the healthy growth of our science. Naturally, we shall always do so, 10 years and 100 years from now, never failing to point out the shortcomings that must be eliminated. Such is the natural dialectics of our growth.

What is the present status of our science?

We can say quite objectively that today our researchers can work fruitfully in any field of world science. Perhaps our pace would be slower or the work more expensive, but we are now capable of conducting research in any field, and this is indeed a major achievement of our science as a whole. This statement is no mere bragging; it can be confirmed by vast evidence.

On the other hand, there are only very few fields of research where we occupy a leading position.

Leadership in science has its own specific traits. Let me give the following example. In a convoy of ships there is one leading vessel, with the next one trailing just a little behind. But if leadership in science is to be compared to a

---

Address at the session of the active members of the U.S.S.R. Academy of Sciences, 1956.

This report was read 20 years ago during a discussion of the reorganization of the U.S.S.R. Academy of Sciences. It may prove of interest to the modern reader because it shows the changes that have taken place since in the scientific and organizational activities of our Academy. By the way, those years were marked by such achievements of U.S.S.R. science as the launching of the first artificial Earth Satellite and Gagarin's flight in space.

The report deals with the role of the Academy of Sciences as the headquarters of our science, a widely discussed topic in our time. The XXV Party Congress has decreed: "The U.S.S.R. Academy of Sciences must be elevated to the status of a center of theoretical research and a coordinator of research throughout the country".

[This note was added by the author for the 1977 reprint of this paper for this volume.]

convoy of ships in the open sea, then the waters are frozen and the leader must break ice to make way for the convoy. The leading vessel must be the strongest of all and have the best navigator. And although the leader is sailing only a short distance before the next vessel, its role is totally different.

A review of various branches of science shows that the number of fields where we do not have a leading position is actually increasing. In other words, we have lost our lead in many fields. We often conceive new ideas in science but fail to give them birth. This means that the organization of science in our country is afflicted with essential shortcomings. Such defects can be overcome if they are indeed of an organizational nature. But even the best organization would be helpless in front of a scarcity of ideas.

Let us consider the development of science in the U.S.S.R. during the last 30–40 years.

During this period applied science at institutes belonging to the various ministries and administration has moved far ahead of the pure research conducted at the Academy. With regard to such major fields of science as aviation, aerodynamics, nuclear physics, radio technology, optics, etc., it can be easily shown that the ministries have progressed further than the Academy. Research in these fields under the respective ministries is being conducted on a scale beyond the resources of the Academy of Sciences.

Can we assume the responsibility of leadership in these fields of science? Of course not.

But the matter can also be viewed from a different angle. Perhaps some day the Academy will have to surrender many branches of science and entrust them to our growing industry. In such circumstances the Academy would dwindle to nothing within 20–30 years or so. And the same science in the hands of industry would grow, being closely linked to everyday life. And then the Academy will no longer be criticized for its poor performance in applied sciences.

This, of course, is a fictitious projection. As we all know, the Academy of Sciences will exist forever. It is essential under any regime, capitalist or socialist, and especially under communism. The problem is to determine more clearly the main tasks of the Academy – that is, what the Academy research teams must do, and what cannot be achieved by research institutes belonging to the industrial sector. These tasks must be clearly formulated and worked out. And the Academy must not be distracted from these objectives, if it is to operate efficiently at all. It is essential that the Academy give up everything that can be placed in the hands of industry. Let us consider these major tasks in greater detail.

Science, as we all know, comprises a number of fields. These are comparable to the division of the armed forces into infantry, aviation, artillery, etc., each of them performing its part of the overall strategic task. Science has a single frontline and but one strategic task – the conquest of nature, the understanding and taming of nature in service of human culture. Animals adapt themselves to nature; man, on the other hand, conquers it and fits it to his own needs. This is the main distinction.

Though somewhat artificial, the division of science into separate fields allows a more efficient organization of research.

Pursuing our analogy with the military, we can say that the conquest of nature is impossible without a headquarters engaged in general command and strategic planning. And in science there can be no headquarters other than the Academy. Here are gathered the best specialists in all branches of science, people who can organize and guide the whole campaign for the conquest of nature.

If the main job of the Academy is to serve as such headquarters, we must examine critically its performance and qualifications for this task.

In my view, the lack of Soviet leadership in science, of which I spoke earlier, results from the fact that we do not cope well enough with the strategy of science or more specifically, with the ideological guidance of research as a whole.

Let us consider where and how a new concept in science appears. It is well known that breakthroughs in science usually take place as a result of contact between two distinct fields of research, or figuratively speaking, by a process of mutual fertilization.

We just heard Academician V. V. Vinogradov. Though a philologist, he spoke of computer translation. This new branch of electronics has attracted the attention of philologists who have started to work in this field together with physicists and mathematicians.

Many examples of the application of physical methods in biology can be cited. New trends of research, notably in genetics, emerge in this manner.

Where else can scientists from different fields get together for joint research in such fields, if not at the Academy? One of the main objectives of the Academy of Sciences must be the creation of a suitable atmosphere for a mutual fertilization between different fields. Thus, chemists must be allowed freedom of action in biology, physicists in chemistry, mathematicians in economics and technology, etc. New branches of science and trends of research will thus emerge at the various institutes of the Academy.

Life shows that breakthroughs in science are often the work of scientifically minded youths or young research workers. These must be given all possible support for their fruitful work. Such people are often hampered by their own timidity, which prevents them from overcoming the scepticism of a conservative environment. This is why moral support is so important to the young scientist, even more so than financial assistance. And only the leading scientists of the Academy, by virtue of their own authority, can provide an effective moral support.

It is even more important for the Academy to deal with research in new directions involving a risk of failure and an element of uncertainty as to the outcome of the work. Such endeavors usually require a hothouse atmosphere. In such matters the Academy must assume the risk because of the great advantage in the event of success.

But is the Academy really doing all this? Is it functioning as a creative public organization rather than a bureaucratic establishment?

Authority is essential for any organization engaged in the ideological guidance and management of science. The authority of a scientist is totally unrelated to his administrative-bureaucratic position. A tradition of two centuries of the annals of the Academy does not by itself impart any authority to our Academicians. Authority comes only by way of public recognition. It is therefore quite right that Academicians are elected by secret ballot, because elected bodies wield the greatest ideological authority. But for the elections to be effective at all it is necessary to elect the truly best scientists, and this can only be done if the eligible candidates are known as scientists. And if we are to know the scientists standing for election, they must be given the opportunity to appear before the Academy, show their knowledge and gain the respect of the scientific community. In this respect our performance is poor indeed.

A few words on the duties of the Academy members. The title of Academician is not granted merely for satisfying the personal ambition of people; it places specific public burdens on the shoulders of scientists, notably the obligation to contribute to the progress of science. No Academician can be satisfied with work in his laboratory or at his desk. He is morally bound to take part in the public life of the Academy and contribute to the healthy progress of science. Otherwise we would not be able to create an advanced science in our country. This is the only way by which the Academy can assume the ideological guidance of research on a national scale and become a real headquarters in the struggle for the conquest of nature.

Obviously the Academy is still far from being able to cope with the ideological and creative guidance of science, as it is still working after the pattern of a bureaucratic ministerial body.

Let us consider, for example, the bureaucratic contents of the papers dispatched from the Presidium to all of us at the research institutes.

Thus, on February 20, 1956, the institute directors were ordered to submit their own timetables for approval. But I, for example, can never tell my timetable in advance, as only the course of research can dictate the number of working hours spent at the laboratory and in my study. In the same order, the Secretary-Academician is instructed to supervise the matter and check whether the scientists indeed follow the predetermined timetable.

What should be the principles of reporting on the course of research at the Academy of Sciences?

The pattern to be adopted in a leading research organization should be as follows: first, a review of the achievements of world science in recent years; second, the contribution of our science to this progress and the measures taken by the Presidium to prevent us from lagging behind. But in reality there is a purely bureaucratic evaluation of performance versus plan in terms of mere percentage points without any consideration of specific achievements and failures; or that a given institute has failed to meet the norm by so many printer's sheets. Such reporting is totally unfit for creative work in science; it can only satisfy a ministerial establishment.

Article 31 of the Academy statutes stipulates that the report is subject to the discussion and approval of the General Meeting of the Academy. But we never bother to analyze the contents of the report; instead, we approve everything in a purely formal manner, thus converting the Academy into a bureaucratic body incapable of assuming a guiding role in the ideological organization of our science or of ensuring its leading position.

The unhealthy character of our scientific communications is evident from the disregard shown by U.S.S.R. Academicians toward our own scientific sessions. Thus, A.N. Nesmeyanov went to Britain to present a report to the Royal Society (the British equivalent of our Academy of Sciences).

I have received the Royal Society publication. It contains the report read there by A. N. Nesmeyanov. Aleksandr Nikolaevich never told us, the Fellows of the U.S.S.R. Academy, of his interesting work, but he did report it before the British audience.

I. V. Kurchatov did the same in Britain with the leading research of his institute in the field of thermonuclear phenomena, which he did not share with us. Many other examples can be given, but these two are sufficient.

Now the President of the Royal Society is coming to us on a return visit.

Can we manage to organize an interesting discussion in the field of biology for him?

Lord Adrian is a major scientist; we ought to organize a discussion, if we can manage it, in order to criticize his idealistic concepts, as we are so fond of doing.

My criticism is of course intended to eliminate our shortcomings. Therefore I shall propose the concrete steps to be taken in order to improve the work of the Academy and strengthen its influence on the progress of our science.

If the Chairman permits, I would like to tell briefly about the coordination between the ideological guidance and administrative management of science at the British and American academies. Obviously their patterns cannot be copied here, but it would be worthwhile for us to become acquainted with the ideological guidance of pure science in these countries.

As I noted, the British Academy of Sciences is the Royal Society. I am a Fellow of this society, which sends me annual reports of its activity. Our own Academy, on the other hand, never sends any reports of its work to me or to any other of its members. Moreover, the Royal Society regularly sends me timetables of its scientific sessions. They have some 20 sessions each year on a large variety of themes. These themes are selected in advance. Here are some of the subjects discussed in 1955: 'Viability of Hybrids', 'Magnetic Hydrodynamics', 'Radiation Balance in the Atmosphere', etc. Such are the subjects of the general sessions of the Royal Society. The reports there cover current problems of potential interest to the scientific community. Moreover, the Royal Society has the tradition of inviting at least two other leading scientists to serve as opponents to the lecturer; this arrangement ensures a lively discussion.

It seems to me that our Academy of Sciences must adopt this tradition and raise the level of scientific discussions not only in our general meetings but also in the sessions of the Presidium of the Academy.

The British Academy of Sciences has no administrative organs of its own, no research institutes, no garages, no building activities – just a number of scientific journals which are sent free of charge to its fellows. Thus its main activity is the ideological guidance of science. The organization and development of research in Britain is in the hands of the Department of Scientific and Industrial Research. This department is organized like a ministry; it finances and controls a considerable portion of the research in the country. It controls only a few research institutes; mostly it organizes and supports research on specific subjects at universities and by individual major scientists.

We never do such things, being concerned little or not at all with research at the universities, which thus remain largely detached from the work of Academy institutes. University research, being essentially of the same nature as the Academy work, must be united with the latter and placed under a single ideological guidance.

Now a brief mention about the U.S. National Academy of Sciences. Its activity closely resembles that of the Royal Society. I am likewise a Fellow of this Academy, which also sends me its reports and journals. The structure there, however, is somewhat different. The U.S. Academy of Sciences similarly exerts an ideological influence on science, but administrative matters are entrusted to a different body, the so-called National Research Council. There, as in Britain, administrative affairs do not interfere with the ideological guidance of research.

Let me point out that administration in capitalist countries is much simpler than in the U.S.S.R. There it is unnecessary to have building services, garages, etc., – just money, and all these things are done by private firms under contracts.

I think we should adopt this division between the ideological steering of research and administrative matters. Scientists cannot possibly form an Academy Presidium capable of coping just as successfully with construction, housekeeping, financial matters and the ideological guidance of the entire Soviet science. These tasks should be separated.

Therefore I propose the establishment of a Presidium charged exclusively with the ideological guidance of science and if possible, such tasks as the selection of cadres and scientific information. Nothing else should be added to the burdens of this Presidium. A Presidium of this kind must consist of scientists capable of exercising ideological guidance of science and the scientific life of the Academy in a most authoritative manner by way of discussions in general meetings or sessions of the Presidium. I do not think it necessary to split up the Academy and set up a separate Ministry of Science in our country. Such a step would be precocious and perhaps even totally unnecessary. But at the Academy there must be an administrative-economic council responsible to the Presidium but operating as an independent organ for the handling of financial, administrative, construction, housing, and other matters – briefly, the whole scientific and domestic economy of the Academy of Sciences. Such an organ does not have to be composed entirely of Academy members.

In specific terms I propose that the Presidium be converted into a true authoritative research organization exercising an ideological control over science in our country and providing advice to our government on all matters requiring scientific evaluation; in addition, that a separate administrative organ be set up. But to prevent the bureaucrats from breathing down the necks of our scientists, this organ must be under the responsibility of the Presidium, which will also form it.

15. COMPLEX SCIENTIFIC PROBLEMS

In the working-out of new organizational forms of research in science and technology, and in the implementation in everyday life of the results achieved by science, it is necessary to learn from the experience gained in the solution of various major technological problems arising in connection with the rapid growth of science and technology during recent years. This experience teaches us, I believe, that new, unorthodox approaches are necessary for the scientific solution of major technological problems in a most efficient manner. These new organizational patterns are still little known or spoken of; therefore I shall consider this matter in somewhat greater detail.

To begin with, a brief chronological outline of the principal stages of research on a national scale.

The importance of science for technology and the national economy was realized long ago. Research was initially confined to laboratories belonging to universities or other institutions of higher education. By the end of the last century, however, research expanded to such an extent that independent research laboratories and institutes had to be established. Scientific work there was divided into specific fields of science, such as physics, chemistry, biology, astronomy, etc. With time it became necessary to set up research institutes of a more specialized nature; physical institutes, for example, gave rise to institutes of acoustics, crystallography, optics, etc., and in chemistry, institutes of physical chemistry, organic chemistry, and so on.

The need for establishing a closer link between science and practice emerged in the early years of the 20th century and especially after the First World War. At that time the major metallurgical, electric engineering and other enterprises began to set up factory laboratories of their own for the solution of current problems concerning the industrial implementation of scientific and technological achievements. The necessity for a continued growth of these factory laboratories eventually converted them into large separate research institutes engaged in specific fields of technology. Abroad, these institutes are usually administered by major manufacturing associations, and in the U.S.S.R., by the respective ministries. We call them 'branch institutes.'

From an article in the newspaper *Pravda,* 1957.

The scientific institutions and branch institutes of any developed country employ a work force of thousands of selected, talented and capable people. Vast amounts of money are spent for the needs of these institutions, whose numbers and size are constantly growing. An essential feature of these bodies is that research there is organized according to the respective fields of science. Until recently, this arrangement proved more or less satisfactory.

But major interdisciplinary problems began to arise at an ever increasing rate during the last decade. One example is the design of an airplane powered by atomic energy. This task requires the combined efforts of atomic physicists and specialists in heat engineering and aerodynamics, to say nothing of designers, metallurgists, etc.

The name 'complex problems' is applied to such tasks in our country. But if we consider the present problems of science and practice it will become clear that nearly all of them can be placed in this category. Obviously the solution of an interdisciplinary problem is beyond the capability of a single specialized institute. In the U.S.S.R. such problems are usually tackled in the following manner. There is invariably a most interested party which assumes the initiative for the management and organization of the whole project. The work is split up into separate tasks which are entrusted to various institutes, designers, etc.

Given adequate support by the central administrative organs, such an organization proved workable in our country. But it is cumbersome and afflicted with grave shortcomings. One of its major faults is that the participants in the project are scattered, and there is no steady contact between them.

Such working conditions stifle the enthusiasm and cloud the final objective, which are so essential for an intensive creative endeavor.

Life itself teaches us how to find a way out of this difficulty. Many examples show that interdisciplinary scientific or scientific-technological problems are best solved by setting up a single, independent organization composed of scientists and engineers who differ in background but share the common task of solving the problem and introducing the obtained result into daily life. One case in point is the organization established in the U.S.S.R. to find ways for the application of oxygen for the intensification of metallurgical and other processes. True, in this case life itself dictated that a project of such magnitude be entrusted to a separate, specially established organization. But in principle nothing prevents us from applying this approach to problems in the fields of semiconductors, heat-resisting steels, polymers, etc.

In addition to the existing specialized research institutes and design offices, it is therefore necessary to set up scientific-technological organizations totally engaged in specific whole projects. Each organization of this kind will be entrusted with a single specific project. If necessary, it will comprise research institutes, laboratories, design offices, pilot plants, etc.

To be efficient, such an organization should be regarded not as a permanent establishment but a temporary body existing for the duration of the project, a few months or years as the case may be. Because of their temporary nature such organizations can easily be set up in areas where the specific problem is most urgent and related to practical needs.

Such an approach to scientific-technological problems will not require a greater expenditure. But the main difficulty lies in the need for re-educating some of our scientific-technological workers who prefer the stability of their old jobs and recoil from any transfer. We shall also have to cultivate a feeling of mobility among our scientists and top engineers. Such an organization for the execution of scientific-technological projects can be visualized as a mobile army group composed of different fighting units and given a specific task; once the objective has been achieved, the army group is split up and reorganized in accordance with the requirements of the next task.

Naturally, the formation of such interdisciplinary scientific organizations will not in any way be at the expense of existing specialized institutes of the Academy of Sciences or the branch institutes at major industrial and economic enterprises. As a matter of fact the interdisciplinary scientific-technological organizations must ease the burden on our specialized research institutes by relieving them not only of subjects that have nothing to do with them but also of superfluous personnel. This will enable the research institutes to concentrate their efforts on matters of fundamental scientific importance.

In my opinion, this principle of work by projects will very likely be applied in the future not only to problems of applied science and technology but also in pure science, thus inevitably exerting its effect on the organization of research at the Academy of Sciences. The promise which the interdisciplinary approach holds for pure science is evident from the fact that today the most interesting and advanced problems of science emerge at the boundary between different branches of knowledge: biology, chemistry and physics; radiophysics and astronomy, etc. The need for a comprehensive treatment of major scientific problems is evident from the recent integration of research and scientists into interdisciplinary subjects at the U.S.S.R. Academy of Sciences. This integration proved not only feasible but smooth and it has been favorably accepted by our scientific community.

16. EXPERIMENT, THEORY, PRACTICE

I would like to point out the discrepancy between the development of theoretical and experimental research and the lack of the necessary link between theory and practice.

Naturally, I shall dwell on physical-mathematical sciences in particular. In the U.S.S.R., experimental physics obviously lags behind its theoretical counterpart. Consider, for example, the works submitted in 1962 for the Lenin Prize Competition: a multitude of papers on mathematics and theoretical physics, but not a single one on experimental physics. As editor of the *Journal of Experimental and Theoretical Physics* I know all too well that most of the papers received deal with theoretical physics. The ratio is about 1:4 or 1:3.

Comparison between Soviet and foreign physics shows how far behind we are in experimental matters.

Let us take nuclear physics as an example. Although we are well equipped technically, our experimental performance in this field has been less than satisfactory. Of the whole multitude of new particles discovered, the lion's share belongs to foreign physicists.

This is opposite to what takes place in the U.S.A. where it is the experimental sciences that are developed at the expense of the theoretical sciences; this was already evident enough twenty years ago with the development of nuclear physics, when they started developing that field in the U.S., and it became necessary to create nuclear weapons. To fulfill this task, Americans had to 'import' many theoretical physicists from Europe.

It is well known that when we were confronted with the same tasks in nuclear physics, our Soviet theorists proved to be well-prepared and sufficiently qualified to solve those tasks quickly and independently. Recent years saw a strengthening of theoretical physics in the U.S.A. and a reduction in their lag.

It is significant that young people on leaving college tend to go for theoretical work, and not into experimental research.

Experimental physics must not be allowed to lag behind, for this would

---

From an address at the General Meeting of the U.S.S.R. Academy of Sciences, 1962.

greatly hold back the normal growth of our physics and would obstruct its gateway to advanced positions in world science over the entire front of most important research. The gap between theory and experiment, between theory and practice, is the symptom of a grave disorder in the normal development of science. That there should be a gap between theoretical physics and practice on the one hand, whilst the quality of experimental research is insufficiently high on the other, upsets the harmonious development of our science and, it seems to me, this applies not only to physics, but also to many other fields of natural science.

We have to implement the necessary measures to raise our experimental physics to a new high stage. Of course, for this it is first necessary to discover what is holding back experimental research and upsetting the normal relationship between theoretical science and practical life. It takes a correct diagnosis of a disease to make ways of curing evident.

From the history of physics it is well known that the division of physicists into theorists and experimenters occurred comparatively recently. Previously, not only Newton and Huygens, but also such theorists as Maxwell, usually verified their theoretical deductions and constructions experimentally themselves. Now, however, only in exceptional cases do theorists attempt to check their theories themselves.

This is happening for the very simple reason that experimental techniques are becoming much more complicated. It requires great effort to carry out tests, which is not usually within the capacity of one person, and it is done by a collective of scientific workers. In actual fact such equipment as accelerators, liquefiers, elaborate electronic circuits, reactors etc., require a large staff of scientific workers to conduct the experiments. Therefore a theoretical physicist cannot possibly check his own theoretical conclusions in practice and he is obliged to rely on the favours of experimenters and to wait until they decide to test his conclusions and proposals in practice.

This is how the disproportion arises between the amount of theoretical work and the possibility of subjecting it to experimental investigation: a theoretician may well produce several papers per year, say four, but to do the experimental checking, a year or eighteen months is usually required and a group of, shall we say, five persons has to work on each. Clearly the theorist will have to have between 20 and 30 experimenters. This, of course, is a simplified roster, but in general it shows the correlation between theoreticians and experimenters necessary for scientific development.

At present the number of theorists and experimenters is approximately equal, and the result is that the majority of theoretical conclusions are not being tested in practice. Theorists lose the habit of thinking that none of their work acquires a valuation until after experimental verification. The theory begins to work of its own accord, and in the best case the value placed on it depends upon methodological and aesthetic considerations.

For harmonious development of science, of course, it is necessary for theory not to lose touch with experimentation, and this can only be the case when theory rests on a sufficiently large experimental basis.

Why do so few of our people go into experimental work and why is it with us so inadequately organized? The answer to this question is simple: in these times the work of the experimenter is rather more arduous and less 'profitable', and not just because the experimenter in the event of failure loses, not two to three months like the theorist, but a year or eighteen months, i.e. the time which at present is usually necessary to complete an experiment. The work of the experimenter requires rather greater effort; not only does he need to understand theory, but also to have a number of practical skills in handling instruments, and he needs to create a well-harmonized team. Often the experiment requires continuous work day and night. All this leads to the fact that recognition of an experimenter as a scientist who has achieved scientific status comes much later than for the theoretical physicist.

To submit and defend a thesis from work done collectively, he has to separate the part which is supposedly his independent contribution, and this must be confirmed by his director. It is easy to see that this requirement is radically at variance with that healthy spirit of collective work when persons are continually exchanging experience and ideas, helping each other and taking over for each other. The separation of 'personal ownership' for defence of a thesis is an unnatural factor holding back the development of collective creativeness. This repels many people from experimental work.

The director of an experimental group is also placed in a difficult position. He bears the responsibility for the work, but since often he himself takes no part in many of the divisions of work, it is usually considered that his name should not appear in the list of authors. Young people underestimate the Director's role, though he selects the team, divides the duties between its members, and separates the good ideas from the bad. Naturally, the Director's function is extremely important. In modern conditions the director of scientific work is like theater or film producer – he creates the show, but he himself does not appear on stage.

Modern theater and film recognize the decisive role of a producer in staging productions. However, the fact that in modern scientific work, the role of the organizer is usually decisive has not yet been fully realized, and therefore the conditions necessary for the productive work of an organizer have not been created; his work has not been taken care of properly.

At present it is very difficult to attract capable scientific workers into supervisory posts – heads of laboratories, directors of institutes. These positions are often filled by personnel with administrative skills, without the creative scientific qualification. And this lowers the quality of the scientific experiments about which we have been speaking.

In such large fields of experimental physics as space research, plasma-physics, nuclear-physics, design of accelerators, scientific teams are very big, and the role of the director is decisive. Only the right appointment can ensure success. When a theorist does his own work, his only apparatus is pencil and paper, and even this is unnecessary to some. Thus, when Euler was blind, he did his fundamentally mathematical work in his head.

But the experimenter needs a good material base, room for every possible kind of equipment, a large selection of instruments, special materials, workshops, a well-trained staff of laboratory assistants and so on. Progress and success in the work is conditional upon the perfecting of this material base. Ten years ago our material base was very poor, since then it has considerably improved, but all the same its condition is still holding back experimental scientific work, and making it less attractive for the scientist.

From all this it becomes understandable why our young people are tending towards theoretical scientific work and why such a disproportion has grown up between theory and experiment. And this leads to theory being divorced from practice.

It is necessary to give the experimenter and director of experimental research working conditions which are at least as attractive as for theorists. It is also necessary to go on improving our organizational system of scientific work and, more particularly, to encourage its collective forms. Such incentives may be based on material and moral factors. For instance, one could arrange for thematic prizes for experimental work, easier methods of obtaining degrees on the basis of one and the same collective dissertation.

What has been said of physics, may also apply to other fields of natural science. The separation of theory from experimental work, from experience and from practice, above all harms theory itself.

I would like to say that separation of theory from experimental work and from reality has also been characteristic of those philosophers who were occupied with philosophical problems of natural science. A striking example of such a separation of some philosophers from experimental practice and reality was their attitude toward cybernetics in the late 40's and 50's when cybernetics was labelled 'reactionary pseudo-science' in our philosophical dictionaries.

If our scientists then had listened to the philosophers and accepted this definition as a guide for the development of their science, one may say that the conquest of space, of which we are all justly proud and for which all the world respects us, could not have been carried out since it is impossible to guide space vehicles without cybernetic machines.

Here is another example which shows the consequences of an insufficient understanding and knowledge of physical experimentation. It is still fresh in the minds of many how some of our philosophers, dogmatically applying the dialectical method, proved the untenability of the theory of relativity. Most criticism was centred on the conclusion that energy is equivalent to mass multiplied by the speed of light squared ($E = mc^2$). Physicists had already subjected this law of Einstein to tests with elementary particles. To understand these tests required a profound knowledge of modern physics, which some philosophers did not have. And by now the physicists have produced nuclear reactions and verified Einstein's law, not on individual atoms, but on the scale of the atom bomb. A fine thing it would be if physicists were to have followed the conclusions of some of the philosophers and stopped working on the problem of applying the theory of relativity in nuclear physics! In what a position physicists would place the country if they were not prepared for the practical use of the achievements of nuclear physics?

These examples most vividly characterize the chasm between some philosophers and practice, though there are still other instances, for example, their incorrect evaluation of the principle of uncertainty in quantum theory, and their incorrect evaluation of the theory of resonance in the study of chemical bonds. It seems to me that such incorrect generalizations have been made by our philosophers not only in the field of physics, but also in biology.

This shows that the use of dialectics in natural science requires an exceptionally deep knowledge of experimental facts and their theoretical generalization. Without this, dialectics is unable to give the solution to problems. It is, as it were, a Stradivarius violin, the most perfect of violins, but to play on it, one needs to be a musician and know music. Without this, it will play out of tune the same as an ordinary violin.

The work of scientific naturalists who are making a great contribution to modern natural science, and whom Lenin called 'spontaneous materialists', has unfailingly progressed in the close tie of theory and experience. Therefore for the developing of natural sciences on a sound materialistic basis, any theoretical generalizations must unfailingly be verified by experiment. Harmonious development of theory and practice is absolutely necessary in all fields of natural science.

Concerning the 'mechanism' of linking theory and practice, I would like to recall an illustration given long ago by Kelvin. He compared theory, primarily, of course, mathematical theory, with millstones, and the test data with the grain which is poured into these mills. The product – flour – is a benefit to people. Quite clearly, some mills, however much they grind, cannot produce anything beneficial – this is theory operating of its own accord. The quality of the flour is determined by the quality of the grain, and rotten grain cannot give edible flour. Therefore good quality data are required in an experiment, both for constructing an advanced theory, and also for obtaining practical results.

High quality of experiment is a necessary condition for a healthy development of science.

To conclude our discussion, I would ask you to remember the meaning and role of a good experiment in the aphoristic phrasing of the novel *Gentlemen Prefer Blondes,* that classic American work: "Love is a good thing but a golden bracelet is forever".*

I think that we scientists may say: "Theory is a good thing but a good experiment is forever".

---

*[Adapted from the lines in Anita Loos, *Gentlemen Prefer Blondes* (New York, Boni and Liveright, 1925), p. 100: " . . . kissing your hand may make you feel very, very good but a diamond bracelet lasts forever."]

17. EFFECTIVENESS OF SCIENTIFIC WORK

The question which I wish to pose is: how efficiently are our scientists and institutes working? Should we not direct our main efforts at improving the working conditions of the existing institutes and of their research workers so as to improve their productivity, rather than towards further growth of the Academy? Perhaps with the same expenditure in physical terms we may be able to gain more in this way. One may enquire by what factor the productivity of our scientists can be increased. It is of course impossible to give a precise answer, but I think there is still very great scope.

There are three main ways of influencing the work of scientists in order to increase their effectiveness. For the sake of brevity I shall call them morale, finance, and establishment. Undoubtedly the first is the most important. The attitude of scientists is largely determined by the choice of a right direction for his scientific work and its proper implementation. Unless a scientist has the right to disagree with one or more of his own colleagues as to the line which his own work should follow, he is no longer able to disregard the opinion of the scientific community as a whole. We must pay more attention to cultivating and maintaining social concourse amongst scientists in the Academy. This is one of the most effective ways not only of raising the level of scientific work, but it is also the main way of concentrating our forces on the more promising and chief lines of development. It is essentially this which is the real way to plan scientific work.

One means of social influence is to hold discussions on 'problematics' and particular themes at scientific meetings. It needs to be acknowledged straightaway that we are still not working hard in this direction. Meetings on narrower themes and conferences on special topics go well, but broader discussions do not usually proceed so well; the pulse of social life of the scientists attending them is very weak. The art of discussion and of scientific arguments has faded away in our country.

I have said many times that we need a club for scientists where we could gather and talk over urgent matters at ease. When I was in England the most

---

From an address at the session of the Presidium of the Academy of Sciences of the U.S.S.R., 4 December, 1964.

interesting conversations on the burning topics in science at the time took place at dinner in the colleges. We discussed questions which embraced many fields of science, and this was the best way of widening one's horizon and so understanding the significance of some particular line of scientific enquiry.

It is necessary to develop social life in the Academy. In the epoch of the cult of personality this was difficult. But you see today there is again an opportunity for this! There is a question, then, of why 'fraternity' is therefore still lagging, because after all this lag is by no means inherent in our nature.

An important factor in morale is to take part in scientific life abroad – at conferences and similar gatherings. As yet we still take little part in them. Our delegations are still only one quarter or one fifth as many as those of the U.S.A. and other countries, and often the delegates are selected by bureaucratic methods, and without strict regard to the criterion of scientific qualifiation and interest. This matter needs to be improved and without begrudging the means.

Another factor of morale which we do not take into account sufficiently, concerns our journals. A journal should be selective and provide constructive criticism. When an article is received, it is necessary to subject it to competent appraisal and notify the writer about the critique; in other words the Editorial Board of a scientific journal should carry on its work along with the author. We are engaged on this in the *Journal of Experimental and Theoretical Physics,* and although this is troublesome and sometimes difficult, yet I consider that this work is of great benefit for the direction of scientific research.

Let us now consider how the development of science is affected by the 'establishment'. When we in the Academy arrive at the conclusion that some field of science is lagging in our country, at once the question is raised about material support for some laboratory or even about the construction of institutes and so on. But it should be understood that it is impossible for us to maintain all fields on the same high level, so it is rather more correct to concentrate our efforts wherever we are powerful and where there are already good scientific traditions. Science needs to be developed in those directions where we are lucky to have a great, bold and talented scientist. It is well known that no matter how much you support an ungifted person, all the same he will do nothing great and purposeful in science. In the development of any particular field our first duty is therefore to proceed from a consideration of the creative forces of the person who is working in this field. You see, our science is a creative vocation, like art, music, and so on. It cannot be thought that by setting up a department for writing hymns and cantatas we shall get them: unless there is in this department of the conservatory a great composer equal in power, for instance to Handel, nothing will be produced. The lame cannot be taught to run, no matter how much money you spend on this. It is the same in science as well. The governing body of the Academy should seek out, attract and support the most talented people, and it should be engaged on this even more than on thematics.

In an institute the leading role in the selection of cadres should be played by the Director, who himself should be a great scientist. Yet this implies that he should enjoy great rights in this matter. But in general today the Director is largely fettered hand and foot both in financial and establishment questions. As I have pointed out, finance is the third way of influencing the development of science. Here at home there is money for research, the State is not stingy, and it is easier for us to obtain it than, for instance, for American scientists. But perhaps you remember the children's game about the Lady of the Manor. "She sent a hundred rubles, buy what you want", but then afterwards the rider follows – "buy nothing black or white, and do not say anything either", and so forth. Hence it appears impossible to spend the hundred rubles without tricks. A Director finds himself in this same position too: they give him money, but greatly restrict him in deciding how it is to be spent.

It is also generally known that a Director is greatly restricted in the choice of cadres, he cannot himself dismiss anyone, nor can he provide any incentive. The fettering of the Director, his restricted role in this important matter, is one of the reasons for the low efficiency of our institutes.

Unless the work of the Director of an Institute is made easier and simpler, for he is a scientist directing scientific research, as previously it will be hard to attract scientists to this work. Today many outstanding scientific collaborators of existing institutes (for instance, of our Institute for Physical Problems) would be quite able to head independent scientific collectives, but they do not wish to do so. If the work of the Director were made easier and simpler, this would of course be instrumental in a more rapid promotion of capable young scientists into directorships. It should be added further that the level of our scientific youth is higher than, shall we say, in America, where many creative and talented persons do not go into scientific institutions, but are attracted to commercial work which has much higher material rewards to offer there. Here at home this does not take place, but we are insufficiently using this advantage.

And the last thing which I should like to mention concerns the building of new institutes outside the limits of Moscow. From my own experience it is more productive and quieter to work 50 kilometres from a large town. But to put this into effect, it is necessary to provide really good conditions, better than today in Moscow, both for scientific creativity and, of course, amenities. We have still not managed to achieve the latter. Scientific workers are therefore very reluctant to leave Moscow, and in this respect success cannot be achieved by compulsion.

In setting up institutes outside large towns one should never begrudge expenditure on good living conditions, even if this is more than the outlay on the institute itself.

18. APPLYING THE ACHIEVEMENTS OF SCIENCE AND ENGINEERING

It is well known that the main index of economic progress is productivity, and increases in productivity are achieved mainly by successfully applying the new achievements of science and engineering.

When the growth of productivity slackens, the reasons need to be sought in the failure of industry to adopt scientific and technical achievements.

From official statistics it will be seen that the rise of productivity in the U.S.S.R. has in the past reached 13 per cent per annum, but the pace has since slackened and in recent years it has dropped to 4–5 per cent.

This shows that the process of applying new techniques in the U.S.S.R. is not now satisfying the demands of U.S.S.R. industry.

I want to dwell on the reasons why the application of achievements of science and engineering is lagging behind. It is generally known that in the U.S.S.R. the application of new techniques by industry always proceeds slowly and with difficulty. This will be seen from the word *vnedreniye* which we usually use in this connection, meaning 'introduction'. We speak of the 'introduction of new methods' and 'introduction of the achievements of science'. The word for 'introduction' in the Russian language signifies that movement forward meets with resistance from the environment. We are thus accustomed to new techniques meeting resistance in their adoption and for a long time we have been using the word 'introduction' without noticing that we thereby characterize abnormal conditions in the adoption of new techniques. When we begin to use the word *dsudeniye* (adoption) one may suppose that normal conditions will have been achieved for applying them.

My experience over many years in this field shows that six conditions have to be fulfilled for successful application of achievements of science and new technique by industry.

I shall enumerate them and then look at what is needed for them to be implemented in our country.

Adopting and mastering new techniques means that industry must learn to do what it has not been doing before. So adopting new techniques should be

---

Address to the general assembly of the Academy of Sciences of the U.S.S.R., 13 December 1965.

regarded as a process of study and therefore it needs to be carried out by the same pedagogic methods which are usually used whenever we teach somebody something new.

When we are teaching students or schoolchildren, the main essential precondition is the person's wish to be taught. If there is no such strong desire for knowledge, it cannot be driven in by the cane. Does our industry always want to be taught what is new? Are the conditions here at home always created such that industry actually feels it is worthwhile to learn new techniques?

Obviously, for this wish to be present one needs to create favourable morale and material conditions. It is, of course, necessary to create those conditions under which industry and also the factories would have an interest in being taught what is new. They must feel that this is worthwhile to them, useful and honourable. It is this which is condition No. 1.

Condition No. 2 is that when we teach a person something new, this person must always have appropriate training. It is impossible to teach advanced mathematics if the pupil does not know algebra and trigonometry. Therefore when teaching something new the person being taught must have been trained sufficiently. In our country this is often not taken into account. I know of a number of cases where management has authorized the production of new apparatus at a factory which has not been prepared at all, and so despite every effort the factory executive has not been able to complete the order. For instance, one good factory was authorized to make special metallic vacuum Dewar flasks even though the works had no experience with vacuum engineering. With difficulty the factory met this order, but it produced many poor quality Dewars and the process of mastery took several years. At a factory prepared for this order, the work could have been done more quickly and more easily.

Hence the second condition is adequate training of the pupil.

The third condition, also well known from the teaching profession, is that the pupil must never be overloaded with studies. Each factory, each industry, can only assimilate and master annually a definite quantity of what is new, even if there is adequate training and the desire to learn. But in our country it often happens that as soon as a works shows interest and starts to adopt something, they start to overload it out of proportion. One needs to remember that there are also limits to how much new knowledge can be assimilated by industry.

Fourth condition: when you teach somebody something, sufficiently favourable material conditions must be created. It is irrational to teach either a person or a factory entirely at their own expense in terms of internal resources. For all training and education it is always necessary to provide a good material basis appropriate to the task posed. Simply speaking one needs to allocate sufficient means to whoever is being taught something new.

The necessity of these four conditions is easily understood, and it is comparatively simple to realise them in practice.

The fifth condition, less obvious and rather more difficult to implement, is the following. From practical teaching experience it is well known that if you are educating someone, it is always necessary to produce a clear program as to the course of instruction. Similarly, if something new is being adopted in industry, for speedy success it is necessary to have a well-developed program indicating the best way of proceeding. But, as a rule, little attention is paid to this here at home; we often leave adoption to *laissez-faire* and even do not consider any kind of program whatever to be necessary.

In forming such programs it is necessary to fulfil two conditions. First, the program must take into account the production possibilities of the works and, second, it must have regard to the specificity of the new process which is adopted. At the works there is usually no single person able to see both these sides of the picture. So when a scientist or an inventor, or even a scientific institute, is entrusted with drawing up the program, the specificity of production is not taken into account in this program. When the factory itself draws up the program, the special requirements of the new technology are not taken into account. In either case the program is deficient.

How do we escape from this situation? In practice one can find versatile engineers who are able to grasp both sides of the program. At present there are few such engineers, and so one needs to value them highly. These highly-skilled engineers are just as necessary to us as design engineers, so one needs to encourage them and also give to them the opportunity of playing an active part in production in the capacity of adjusters of new techniques. At Ministries and Central Boards, bureaus must be organized with such specialists. It should also to be the task of these bureaus to draw up the required programs and organise their implementation. It needs to be laid down as a general rule that there should always be a well-developed program when starting up new production. But here at home little attention is being paid to this important condition.

Finally, the sixth condition concerns the teacher. If he is a scientist, then he should be a teacher. It is generally known that success in teaching requires good friendly relations between the teacher and the pupil. Furthermore, any creator of new technique who is passing on his work to industry, whoever it may be – scientist, inventor or a collective of a scientific institute – should be as interested in its successful adoption as the works executive. Are our scientists and our inventors and engineers interested in their achievements being mastered by industry? How are they linked with industry?

I shall dwell only on the case where the scientist is a teacher.

As is generally known, a scientist who is engaged in 'introduction' of technology in industry, receives no material recompense for this. In our country it is accepted that a scientist should work with industry as a social obligation. This is completely different from the position in capitalist countries. When I lived in England and received the academic rank of Doctor, as a scientist I joined the professional association of scientific workers and was obliged to sign an undertaking according to which I had no right to serve as a consultant to industry without payment, and what is more there was a definite scale of fees corresponding to my academic rank below which it was forbidden to charge. In capitalist countries this is done so that members of the associations cannot undercut each other. Naturally, our conditions at home are quite different and so measures of this kind can have no place here.

I do not want to insist that material recompense in our country is of decisive importance, but undoubtedly conditions must always be favourable for the morale of scientists working with industry. It is necessary for it to be to the interest of the scientist to do this work, and an atmosphere needs to be created in which his work would have widespread social recognition and in which collaboration with industry would be considered a useful activity for the state. Unfortunately, today, whenever one has to turn to a Ministry, rarely are you met with a pleasant attitude. All this, of course, is not instrumental in developing good relations between teacher and pupils.

Often in our country they assume it is enough to get an order prescribing the adoption of something, to assume that the process of 'introduction' is thereby provided for. Yet from the foregoing analysis it follows that the process of adopting new techniques can never be regarded simply as an administrative act; rather this process needs to be approached as one would approach a 'pedagogical poem'. Hence in organizing the adoption of new techniques by industry, one needs the individual approach, no rigid pattern, and one needs to allow for the nature of people and the conditions prevailing in each particular case. It is by government order, of course, that staffing and financial matters mainly are determined on the spot, but success in adopting new techniques is founded on good relations between the pupil and the teacher and on their common personal interest in success in carrying out their program, but this in our country, unfortunately, is not always the case.

Now I want to touch upon another topic which is no less important: is our science delivering sufficient goods of new technological output for the economy to adopt, is the productivity of the scientists high enough? To these questions we need to pay serious attention. To look into them, I think the best thing is to compare some data of our scientific work with American data.

It is of interest that today in the U.S.A. they are seriously concerned with problems of the development of science in their country and with its relation to industry, and in this connection they are publishing much statistical material.*

I quote some figures which are interesting for us. This year the United States of America are spending on all scientific work 21 billion dollars, of which two thirds comes from the federal budget, and one third from patrons and industry. Of this total academic science receives 11 per cent, or 2500 million dollars, hence the bulk of the expenditure is on science which directly serves for adoption by industry, or, as we say, goes on scientific research on applied themes.

The Americans further state that with their high cultural level and available capital the existing stock of science is not enough to meet the industrial demand for new techniques. Industry in the U.S.A. mainly needs developments in new directions, such as television and cinema in the past, and now, for instance, cybernetic machines and synthetic fibres. New applications of engineering are not only the most profitable investment, but they also have a social significance since they absorb unemployed workers and they raise the standard of living.

The Americans consider that the inadequate scales of development of theirs are the consequence, mainly, of a shortage of highly-qualified scientists and engineers. They consider that scientific work could produce still more money, but at present they lack highly-gifted people able to direct their own scientific work and so largely determine the development of scientific work in the direction which they need. Therefore in recent years they have started bringing in large numbers of scientists from Great Britain and West Germany. In the last ten years 53,000 scientists have emigrated to America, mainly young scientists, of whom 30,000 had engineering training, 14,000 were physicists, and 9000 scientists in other specialization. This means more than 5000 persons every year.

Assuming that one university produces on an average 500 persons each year, this means that in Europe in the last ten years at least 10 universities have gratuitously trained cadres for America. Insofar as the Americans took

---

* [See, for example, *Scientific American,* July, 1965, p. 19.]

the best people, this means that they took the cream of approximately 50 universities. At the Royal Society a special commission was set up to discover the reason for such a large drain of young scientists to America and to decide what measures needed to be taken to stop this bloodletting of British science.

Roughly the following was discovered: although the English and Germans pay their leading scientists quite well, the Americans pay twice as much. But it appears that this is not the decisive reason for the emigration of scientists. The commission discovered that young scientists were attracted not only by high salaries. The conditions under which scientific work is set up in America attract them.

The Americans allocate their funds to scientific work slightly differently from other countries. The main funds for science are not allocated to scientific institutions as such. The Americans prefer to allocate their funds for a definite topical theme, or to support the work of an outstanding scientist, who is usually given the opportunity of a free hand as to the choice of theme. Only a small percentage of budget funds is allocated to scientific institutions as such (apparently no more than 2–3 per cent). It is of course very attractive to a scientist to be an independent boss of his own material base. They feel that their work can be completely assured, providing great independence and freedom of action is retained.

Let us now contrast American statistical data with ours. Not only is this difficult, it is almost impossible since there is a real difference between the organization of science and its finance in a capitalist and socialist country. The difficulty is still further aggravated by the fact that, unfortunately, we have not as yet properly undertaken the collection of statistics in connection with the organization of science.

We compare, first of all, the number of scientific workers. The Americans consider that in their country today, starting from the lower posts, from engineers and technicians, there are 800,000 persons occupied in scientific work. In the U.S.S.R. from official statistical material, counting all scientific workers, starting with the juniors there are about 700,000. Hence it will be seen that within statistical reliability limits we differ little from each other as regards the number of scientific workers. This conclusion is corroborated also by data taken from a French source. They consider that for every 10,000 persons in the population in America there are 23 scientific workers with high qualifications, in the Soviet Union the figure is 18.5 and in England it is only 9.7, whilst it is much less in all other countries. Since our population is greater than in America, the total number of highly qualified scientific workers in the U.S.S.R. is therefore the same as in the U.S.A. – about 400,000.

To determine the productivity of scientists, one needs to evaluate scientific production. It is, of course, hard to do this precisely. The Americans try to do it in the following way. They have counted the number of scientific papers in the leading branches of natural science and of technical sciences published by scientists in all languages in different countries in the main scientific journals. Judging from this American data we find that today they are producing one third of all world science. We produce one sixth of all world science, i.e. only half as much as the Americans. Every other country is doing less than we are. Thus as regards scientific production we are the second country in the world. But if we take these figures, we find that with approximately the same number of scientific workers we are only doing half as much scientific work. Therefore, no matter how sad it may be, it has to be recognised that the productivity of our scientists is approximately half that of scientists in the U.S.A. Moreover, as will be seen from further data, recently the rates of growth of science in the U.S.S.R. have started to drop slightly. It is therefore timely to pose the question: how, under the conditions which have been created, are we to develop our science so as to increase the productivity of our scientists?

Labor productivity in science is determined, first, by the material potential at the disposal of the scientist and, second, by the quality of education and the selection of cadres in scientific institutions.

Let us consider the first factor. American budget expenditure on science is rapidly growing. Over the last 20 years it has been increasing on an average by 14 per cent annually, and in the present year the growth has reached its greatest yet: 20 per cent. Expenditure for science from the federal budget alone amounts this year to a sum of 14 billion dollars.

In the newspaper *Pravda* in the leading editorial of 17 December 1965, our budget expenditure on science for the coming year was given as 6.5 billion rubles, an increase of 9.9 per cent over the previous year.

Thus with the same number of scientific workers, our material base is much weaker and this, of course, greatly influences productivity. Insofar as today there is no possibility of strengthening the material base, we have no alternative but to reduce the number of new cadres considerably, roughly by half, in order to improve their quality, of course, i.e. it is necessary to screen out the people who cannot justify by their work the advantages which come to them due to their positions as scientists in the country.

Unfortunately, one can see no other possibilities, except, in the near future, to transfer a whole row of workers who are not working sufficiently effectively in science, into industry where they can be of real benefit to the country. Of course, such a serious step cannot be taken at once. But such should be the trend in the development of our scientific institutions. One could, for instance, transfer 15–20 per cent of the cadres each year from our scientific institutions into industry and take in 7.5–10 per cent well-selected and trained young people, thus raising the quality of the cadres without closing the doors to the flow of fresh forces into science. But here it should be pointed out that even if it were decided to follow such a path, we shall be unable to put such a measure into effect so long as all directors of institutes in the Academy of Sciences continue to have the same powers and rights as are prescribed for them.

One needs to be unafraid of saying that during the last few years the gap in science between our country and America has not only stopped becoming narrower, it has increased, and we must quickly look for a way of making up for the lost ground. Unless in the near future we increase the productivity of our scientists, unless we improve the conditions for the adoption of the achievements of science and engineering in industry, the problem of catching up with the Americans cannot, of course, be solved. If we decisively and boldly use the advantages which socialist construction gives us in the organization of our science and industry, then this lag in growth will only be a temporary hitch. I sincerely believe that if we are not afraid of speaking the truth about our shortcomings and errors and if we amicably look for ways of eliminating them, we shall soon again be leading in the growth of our scientific work.

It is necessary gradually and continuously to improve the organizational forms for the development of our science: improve the material base, raise the quality of cadres and increase the productivity of scientists. We, as members of the Academy of Sciences of the U.S.S.R., need to apply ourselves with particular seriousness to these problems. We are the leading scientific institution in the country and therefore we more than anybody else are responsible for the development of science and the implementation of its achievements.

19. THE CENTENARY OF THE *JOURNAL OF EXPERIMENTAL AND THEORETICAL PHYSICS* AND THE ROLE OF JOURNALS IN THE DEVELOPMENT OF SCIENCE

It has already been noted by the historians of science that science began to develop on an international scale only after the appearance and spread of the printing press in the 15th and 16th centuries. This is easily explained by the fact that the fundamental factor determining the collective work of scientists is the organization of information exchange. The more effectively this is carried out, the greater its scale and the more intensively science develops. The most effective method of scientific information up to date appears to be its dissemination through periodicals, since one can most widely and quickly communicate the scientific achievements in this way to a large number of interested scientists.

The first scientific journal in the world appeared in 1665, but their number began to grow uninterruptedly only after 1750 when a regular postal

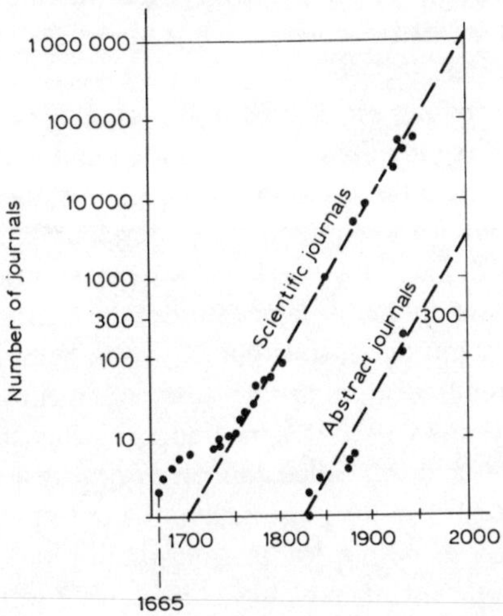

Fig. 1. Growth of number of scientific journals since 1665.

Report delivered at the session of the Presidium of the U.S.S.R. Academy of Sciences, March 15, 1973.

network was established in Europe. Figure 1, composed by the historian of science D. J. de Solla Price[1] shows a plot of the growth of the total number of scientific journals published in all countries. The number of scientific journals is given in a logarithmic scale; it is seen that during the last three hundred years their growth has been steadily exponential.

Every 10–15 years, the number of journals doubles and it has now reached the imposing number of 200 000. The number of abstract journals now published is also shown in Figure 1. These journals were aimed at facilitating familiarity with a great quantity of scientific information. They appeared back in 1830 and, as follows from Figure 1, their number grows in parallel to the main curve and therefore follows the same exponential law as the growth in the number of the main journals. There are now about three hundred abstract journals.

Statistical data, which are at present widely available, show that the number of scientific workers and the funds allotted to scientific work also grow according to the same exponential law, doubling every 10–15 years. Thus it is obvious that a close connection exists between the growth of the number of scientific journals and the scale of development of scientific research work. This points to the possibility that useful data can be had about the organization of scientific work from a study of the number and character of the scientific journals.

For example, it follows from Figure 1 that science has been developing uniformly over all three hundred years, and the jump which one could expect in accordance with recent and current scientific-technical revolution cannot be observed. The intensive contemporary development of science can be explained only by the well-known fact that every process that follows an exponential law always sooner or later assumes the character of an explosion.

The first physics journal in our country was published by the Russian Physics Society a hundred years ago; its title page is shown in Figure 2. After the October revolution, in 1930 it came under the cognizance of the U.S.S.R. Academy of Sciences and was called *Journal of Experimental and Theoretical Physics,* as can be seen from the photograph of a recent cover page (Figure 3). It is now considered to be our leading journal in physics. The fundamental stages of the development of our physics have been reflected in the journal for a hundred years. Therefore if one studies the matérial contained in the journal one can not only construct a picture of the development of the physical sciences, but one can also visualize their present status.

A special article was printed in the jubilee edition of the journal by Yu. M. M. Tsipenyuk[2], in which a picture of the development of physics over the

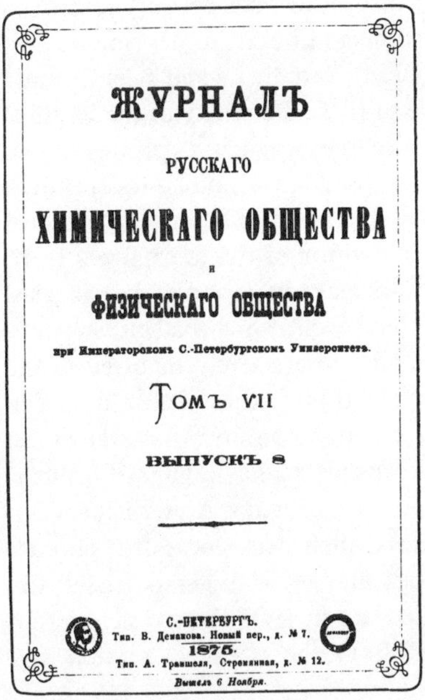

Fig. 2. Title page of the 'Journal of the Russian Chemical Society and Physical Society' for 1875.

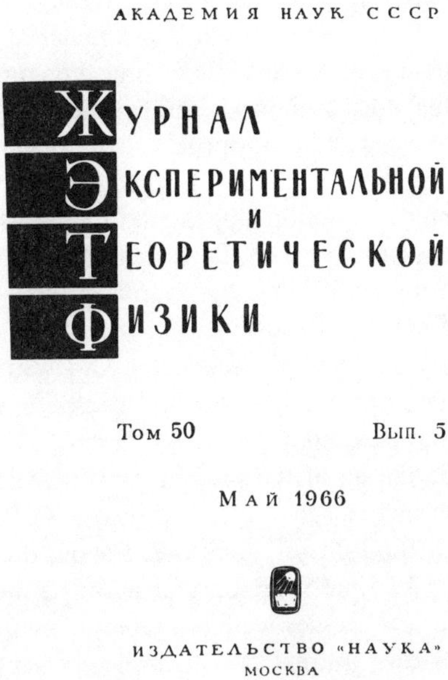

Fig. 3. Front cover of Zhurnal Eksperimental'noĭ i Teoreticheskoĭ Fiziki, 1973.

last hundred years is drawn, compiled from the material which had been printed in the journal. Here we can note only a few stages of the development of physics in our country over that period of time, focusing our main attention on how the journal reflects the contemporary status of the physical sciences in our country.

It is customary to think that the natural sciences began to develop in Russia at the beginning of the 18th century when Peter I founded the Academy of Sciences, to which great scientists like Euler and Bernoulli were attracted from the beginning. It was there that Lomonosov, the first Russian scientist of genius, was educated. After Lomonosov, however, and in spite of this brilliant beginning of physics and chemistry, toward the end of the 18th and the beginning of the 19th century our physico-mathematical sciences developed much more slowly than in Western Europe. Effective progress in these sciences called for material and technical bases which did not exist in our still largely agricultural country. These sciences started to develop properly in the second half of the 19th century, after the peasant reform in 1861. The country became richer at that time and began to develop its own industry. A scientific-technical community came into being and scientific and technical societies began to be created. One of these societies was the Physical Society and our journal was founded a hundred years ago as its printed organ. The journal of the Chemical Society was published first, and an independent physics section appeared in it only in 1873. The founder of the Physical Society was S. F. Petrushevskiĭ, and the first editor of the journal was D. K. Bobylev. At first the journal printed both original papers and review articles. Since the basic scientific activity in those days went on abroad, it was only natural that our scientists who wanted to be more actively involved in the development of the world's science, preferred to print their works in foreign journals, either German or French. It is interesting to observe that even at the beginning of the present century, when we already had a number of brilliant scientists, such as N. A. Umov, A. G. Stoletov, P. N. Lebedev, A. A. Eichenwald, B. B. Golitsyn, and A. S. Popov, they printed their best works first abroad.

As time went on and with the increasing number of original articles, the review articles were relegated to an appendix, and in 1907 the physics section of the journal was completely separated from the chemical section and became an independent publication.

It is enlightening to see what happened to the journal after the October revolution. Despite the civil war and the harsh ordeals, scientific work in the country resumed immediately in all fields of learning, especially in the natural sciences. Our journal received steady material support from the state so that publication never stopped, and in the worst years of the Twenties, when it was impossible to print it in Petrograd because of the devastation, it was printed in Germany.

At that time only original scientific papers were printed in the journal. A journal *Uspekhi Fizicheskikh Nauk* for review articles was founded in 1918 and organized by E. V. Shpolskii who is still today its editor.* This journal steadily publishes highly qualified information on the most important achievements in physics in all countries.

The Russian Physicochemical Society was dissolved in 1930, and its journal came under the cognizance of the Academy of Sciences and its name changed to *Journal of Experimental and Theoretical Physics* (*JETP*). Its editor and editorial staff was nominated by the Academy of Sciences on the basis of elections. After its transfer to the Academy, A. F. Ioffe and L. I.

* E. V. Shpolskii died on August 21, 1975 while still Editor-in-Chief of *Uspekhi*.

Mandelshtam remained the chief editors; S. I. Vavilov was editor from 1939 to 1952; until 1956 it was N. N. Andreev; and since that year the editing of the journal has been entrusted to me.

After the establishment of the socialist system in our country, an intensive growth of the natural sciences took place and the previously lost time was made up. This is well illustrated by the materials introduced by D. de Solla Price in his cited paper, where data are given (p. 40) on the number of journal publications in chemistry in a number of countries between 1910 and 1960 (unfortunately, no such data are given for other sciences). Before the October Revolution the greatest number of papers in chemistry were published in Germany (34%), followed by the USA (20%); Russia's contribution was only 2%. In 1960 the picture changed sharply: German journals – 6%, American – 28%, Soviet – 18%.

In these years physics developed no less intensively, as reflected correspondingly in our journal. Although *JETP* has increased in volume, it could no longer cover all the scientific papers in physics. Specialized journals in various areas of the physical sciences began to appear in 1925. The first was the *Journal of Technical Physics*, followed by journals of optics, mechanics, crystallography and a series of other specialities. At present there are no less than 25. There is practically no major branch of physics without its own special journal.

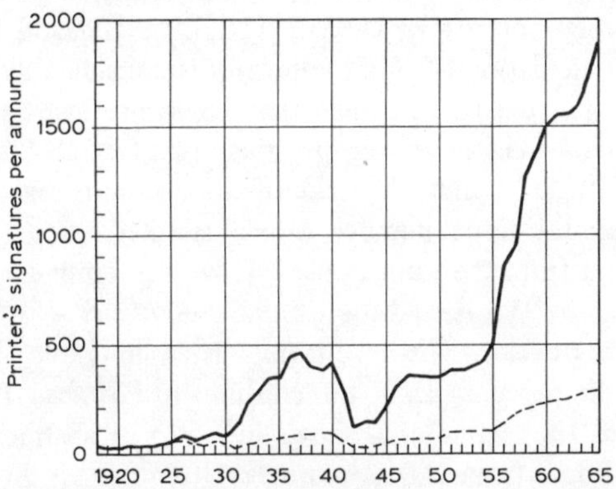

Fig. 4. Printer's signatures from 1910 to 1965 for the basic physical journals (solid line) and for *JETP* (dashed line).

The dashed line of Figure 4 illustrates the growth of *JETP* between 1917 and 1965. The solid line shows the total number of pages of all the journals in the different branches of physics (the 1942–1944 gap in the curve is a consequence of the war). It is evident that in 1925 *JETP* ceased to be the only physical journal, and at present it accounts for only 10–15% of the total volume of the journals.

This raises the natural question whether a general physics journal is needed at present, when there are so many specialized journals. Experience has shown that such a journal appears to be nonetheless needed for the development of science. This is confirmed not only by the fact that it continues to exist, but also because similar journals on general topics exist abroad, for example, *The Physical Review* of the American Physical Society, although a large selection of specialized journals are also published in the U.S.A.

To justify the existence of *JETP* as a general journal of physics at the present stage of its development, let us consider its present role in world science. I shall use for this purpose some data from an original paper by S. Keenan and P. Atherton, 'The Journal Literature of Physics', which was published in the U.S.A. [3]. The method used by the authors is the following: the most important American abstract journal covering all areas of physics is *Physics Abstracts.* In 1961 it printed 20,287 abstracts of articles published in 405 journals of 39 countries. The paper cites numerical statistics on the distribution of these papers by countries, by journals, and by areas of physics. We include here only a few data from this extensive material, which takes up 156 pages.

Table I gives the numbers of abstracts as percentages of the total number for the following countries: U.S.A., U.S.S.R., England, Japan, France, Holland (the participation of other countries is considerably less). It follows from this table that the U.S.S.R. and the U.S.A. produce together 47.5% of the total number of abstracts, i.e., practically half of all the publications in physics. In addition it must be taken into account that only 24 Soviet journals, less than the actual number, were abstracted.

Using the data from the same paper [3], we can compare numerically the contributions of the U.S.A. and the U.S.S.R. in *Physics Abstracts* in different leading areas of physics. The results are listed in Table II. The first two columns show the percentages of the citations in *Physics Abstracts* for given areas of physics. The last column shows the ratio of abstracts from U.S.S.R. journals to abstracts from U.S.A. journals. Of course all these data must be regarded critically since they are based on the material in an American abstract journal, where papers not in English were not readily included, in view of the language barrier. The quality of the scientific work is likewise not considered in such a formal method of comparison. But even under these circumstances it is evident that we in general now occupy the second place in physics, and lead in the study of the solid state. This method of comparison

TABLE I

Number of journals and abstracts in the physical science in the principal countries.

	U.S.A.	U.S.S.R.	England	Japan	France	Germany	Holland
Number of journals	71	24	62	37	14	47	14
Number of abstracts	6316	3317	2729	1560	1268	1240	1043
Percentage of total number of abstracts	31.14	16.37	13.45	7.69	6.32	6.11	5.23

TABLE II

Comparative number of abstracts by fundamental areas of physics for the U.S.A. and the U.S.S.R.

No.	Field of Physics	U.S.A. % of all abstracts	U.S.S.R. % of all abstracts	U.S.S.R. in relation to U.S.A.	
0	All areas in physics	31.14	16.37	0.53	average
1	Electronic properties of solids	25.0	29.7	1.19	
2	Solid state	25.2	26.1	1.04	
3	Optical properties of the solid state	32.1	23.4	0.73	
4	Electricity and magnetism	31.2	22.1	0.71	above
5	Mechanics	27.6	16.6	0.60	average
6	Nuclear physics	25.4	13.8	0.54	
7	Elementary particles	31.8	17.2	0.54	
8	Cosmic rays	24.9	13.1	0.53	
9	Liquids and gases	29.4	13.4	0.46	
10	Acoustics	45.1	19.9	0.44	
11	Technology and materials	40.6	16.9	0.42	
12	Magnetic properties of solids	33.5	13.1	0.39	
13	Heat and thermal phenomena	41.2	15.3	0.37	
14	Optics	35.4	11.9	0.34	
15	Mathematical physics	27.2	8.8	0.32	below
16	General physics	52.6	10.6	0.20	average
17	Astrophysics	35.8	7.3	0.20	
18	Atoms and molecules	45.5	8.6	0.19	
19	Mathematics	35.1	6.7	0.19	
20	Physical chemistry	49.7	6.4	0.13	
21	Geophysics	36.5	4.9	0.13	
22	Biophysics	66.2	2.9	0.04	

shows also that we are behind in certain areas, as for example in geophysics and biophysics. Therefore even such a formal study of scientific periodical literature gives us interesting and useful material for the organization of the science in its planning stage.

The material cited in the paper by S. Keenan and P. Atherton affords also the opportunity to evaluate the significance of individual journals to the development of physics in its international aspect, since they cite the number of abstracts in *Physics Abstracts* from every journal. Thus, *JETP* is in first place among the Soviet journals (all its articles are abstracted), *Solid State Physics* is in second place, and the *Proceedings* (*Doklady*) *of the U.S.S.R. Academy of Sciences* is in third place.

The international significance of *JETP* is furthermore confirmed by the fact that since 1955 it has been translated into English cover to cover, as are a number of our other leading scientific journals now. These translation journals had to be subsidized initially by the U.S. National Science Foundation, but have been self-supporting since 1964.

The content of *JETP* has of course changed over the hundred years of its existence, in accord with the growth and development of physics, although it has preserved its general character. It is therefore natural to pose the following question: which tasks must this non-specialized journal perform, and what should be its content in order to be useful, if separate journals, where any of the articles now published in *JETP* can be printed, now exist for all areas of physics?

Obviously the natural answer to this question is that in physics as well as in other sciences there are scientific papers whose significance goes beyond the boundaries of the interests of specialists in its separate areas. These can be either the discovery of new phenomena, or original methods of research, or broad theoretical information that concerns many areas, etc. All of these papers must be printed in the general journals like *JETP* and *The Physical Review*. Therefore the editing of a general journal has its own peculiarities that are reflected in the correct choice of such papers. Naturally the selection of scientific papers places a greater responsibility on the editors of *JETP* than in a specialized journal. This requires a larger staff of editors who are qualified in a number of areas in physics. And it is impossible not to involve specialists as referees in the evaluation of works. Besides, for an objective evaluation, a system has to be worked out in such a way that the referee can be assured that his opinion is held in full confidence. Experience has shown that in a number of cases the opinions of the referees do not decide the editor's final word. This happens, for example, when the author of the article is himself the most prominent scientist in the given area. But the opinion of the referee is nevertheless important, since it indicates how the article will be received by the scientific community.

It happens often that an article can be very valuable, but not written sufficiently clearly to be understood by a larger circle of physicists. It is generally accepted that any scientific article must satisfy the following demands: if the article is experimental, sufficient data must be communicated in it so that the qualified scientist could himself repeat the experiments; if the article is theoretical, the fundamental premises on which the theoretical structure is founded must be concisely formulated, and the mathematical calculations must be stated sufficiently fully so that the qualified theoretician could verify the results. For *JETP*, however, a journal read by a wide circle of scientists, it is necessary that the article satisfy one more condition: it must be understandable not only to the specialist in a given field, but also to the scientist who is working in different areas. Usually this is achieved by prefacing the article with an introduction that states in more comprehensible form the general significance of the questions dealt with in the paper and of the results. Such demands are frequently resented by authors, because even a prominent scientist is frequently unable to present his thoughts clearly and understandably.

The articles are selected by the editorial panel in accordance with these principles. Those articles which are not of general significance must be rejected, and their number reaches 40% of the submitted articles. The number of rejected articles is determined also by the need to limit the size of the journal (at present this is about 300 printers signatures (of 16 pages each) per year). This guarantees a short enough publication time. For *JETP* as well as for similar journals in other countries this time is about five to six months. This period is determined primarily by the time needed to referee and edit the articles. In those cases where the decision of the editorial panel differs from the opinion of the management of the scientific institution, the editorial board evaluates the articles with special care and makes the final decision. Under present-day circumstances an article rejected by *JETP* may still be published in one of the specialized journals. A physicist's desire to be published in *JETP* is motivated by the fact that the acceptance of a paper by this journal is regarded as a high appraisal by the scientific community.

Experience has shown that at the present rate of scientific work and inasmuch as much important physical research is carried out on an international scale, the six-month publication period is too long. Therefore, as has already been done in the U.S.A., we launched in 1965 *Letters to the JETP*, an offset-printed journal devoted to short articles, 3 to 4 pages. These are little booklets of four printers signatures: they appear twice a month; the publication time is 1–2 months. Up to 400 articles are printed this way annually.

A special editorial board was appointed for these *Letters*, headed by

Academician A. S. Borovik-Romanov. To shorten the publication period, refereeing is dispensed with. What has to be decided, beside the scientific value of the article, is the need for its rapid publication. The *Letters* journal is also very important because it establishes priority of discovery, and therefore enjoys great popularity. For this reason many articles are submitted to the *Letters*, and at the moment 60% of them have to be rejected.

A subject analysis of the articles printed in *JETP* shows that the most deal with solid state physics (46%) followed by articles on plasma (21%), optics, mainly lasers (21%), and finally nuclear physics (9%). In physics, as in other sciences, there is always one field of research which is most intensively pursued at any given time. Thirty years ago this was nuclear physics, right now it is solid-state physics, plasma and lasers. The distribution of articles in *JETP* corresponds to this tendency. Thus there is sufficient reason to think that *JETP* in its present form continues to be needed as a general journal of physics, even after a hundred years of its existence and promotes the development of our science. One cannot be certain that the situation will be the same in the future, since there are many indications of an impending crisis in the methods of scientific information service.

The scale of scientific research has grown to such an extent that the dissemination of scientific information in the basic journals becomes difficult. According to the plot of Figure 1, their number is growing uninterruptedly and will exceed a million by the year 2000. The number of abstract journals is also growing rapidly, and will reach 3000 by the year 2000 (see Figure 1). Obviously, this method of information will likewise exhaust itself. The need to resort to new methods of information makes itself felt more and more. These should be based on contemporary electronic techniques, as for example computers and teletype. Furthermore, it is obvious that with increasing international scientific cooperation, information should be available more and more centrally on a world scale. These questions have to be given full attention because, undoubtedly, a well-established information system is one of the fundamental factors that secure successful and effective development of science.

REFERENCES

[1] de Solla Price, D. J., *Intern. Sci. and Technology* 37 (March 1963).

[2] Tsipenyuk, Yu. M., *Zh. Eksp. Teor. Fiz.* **64**, 3 (1973). [*Sov. Phys.-JETP* **37**, 1 (1973).]

[3] Keenan, S., and Atherton, P., 'The Journal Literature of Physics', N.Y., The American Institute of Physics, 1964.

20. BASIC FACTORS IN THE ORGANIZATION OF SCIENCE AND HOW THEY ARE HANDLED IN THE U.S.S.R.

This, it is said, is the age of the scientific and technological revolution. Science, this implies, with its effect on technological progress, is the principal motive force in contemporary culture and civilization. That is the reason why the development of science is a major concern of nations today, especially of advanced nations. It has become clear that for scientific research to develop effectively in a country, it must have a special very specific organization, different from all other organizations. Questions related to the organization of scientific research are now attracting the widest interest among scientists themselves, for only scientists are capable of fully understanding and evaluating the specific features of organization necessary for the efficient development of scientific work in their fields. Experience has shown that the more involved a country's scientists are in the organization of science, the more successfully science develops in that country.

Organizational measures for the development of scientific research fall roughly into three main groups.

(1) Selection and training of scientific personnel.

(2) Choice of directions of scientific research appropriate to requirements of world science and to the cultural and industrial requirements of a given country.

(3) Provision of the necessary material resources for scientific research, such as research centers, an instrument-manufacturing industry, and information exchange.

These measures, of course, are handled differently for different fields of science. Then, too, the economic and social structure of a country affects the organization of its science.

However, science is indivisible and in the final analysis its achievements are the property of all mankind. The international character of science encourages cooperation among scientists of all countries and establishes a world standard for the evaluation of scientific research. Thus, it is hardly surprising that one factor, perhaps the most significant of them all, is equally important to the healthy development of science in all countries.

From *Daedalus* [J. Amer. Acad. Arts and Sciences] **102,** 167–176 (1973).

This factor relates to the first and most important of the above-mentioned measures for the organization of science: the selection and training of scientific personnel. Experience has shown that, even though it is not yet getting the attention it deserves, it is this factor more than any other that ensures a high level of effectiveness in scientific research.

Science is a field of human activity which must be pursued by creatively talented people if it is to advance. It is common knowledge that only a small number of people have enough creative talent to make a mark in art, literature or music. The same is true of science. Hence, for science to develop successfully, it must, like art, have a supply of gifted people. However, as one can easily see, this supply is harder to come by in science than it is in art, where life itself does the selecting and where assessment is not connected with special organizational difficulties, for a poor novelist's novels will not be read and no one will listen to an untalented composer's music. It is much more difficult to evaluate a person's creative achievements in science, though evaluation here too is made by a kind of public opinion, that of a small group of specialists. Therefore, selection of creative scientific personnel requires the establishment of a special organization which can utilize the evaluation of specialists in the field. This is one of the most difficult organizational problems of scientific research.

The number of scientifically creative people may differ in different countries, but it is always a small proportion of a country's total population. On the other hand, the scientific and technological revolution requires the broadest possible development of scientific research, and this calls for a large number of scientific personnel. It follows that the main objective of the organization of scientific work is to make sure that a country's available talent is used correctly. The first step, then, is to create a situation that will attract gifted people to careers in science. The second is to devise a system for selecting interested people in accordance with their talent. And the third is to create conditions for training talented young people that will enable them to develop their potentialities fully.

Today all advanced countries have specialized secondary and higher schools in art and music – academies of arts, conservatories of music – to which students are admitted at as early an age as possible, as soon, in fact, as they begin to show talent, which talent is thereby given the opportunity for the fullest development. As yet, no such opportunity is available to scientifically gifted young people, though the need for it exists in many countries.

Because selection and training of creative scientific personnel is the basic factor determining the effectiveness of the organization of scientific work in a country, I shall devote most of this article to it and describe what is being done about it in the U.S.S.R.

The question of the selection and training of young scientists has comparatively little to do with the social structure of a country, for in both capitalist and socialist countries education of the young is in the hands of the state. Therefore, the questions under consideration here are of general significance.

Except for slight variations, the system of selecting and training scientists used most widely in the world today is similar to the one used in the Soviet Union. Future scientists attend universities and higher technological schools together with other students, and only afterwards and on the basis of abilities they have shown there, are they admitted to graduate schools to do scientific work. There, for the first time, they engage in scientific research, and only then do they have the opportunity to show their creative abilities. The writing and defense of their theses for the *Kandidat* degree (a Soviet intermediate degree, roughly equivalent to a U.S. doctorate) will be used to judge their suitability for creative scientific work.

This system has several major shortcomings. The principal one is that a person is already between twenty-three and twenty-five years old, a mature adult, before his gifts begin to be discovered and developed, despite the fact that a creative talent for science, though it surfaces at a later age than artistic talent, is usually clearly discernible by the age of eighteen or nineteen, that is, by the time a person completes secondary school. (Pure mathematics is an exception; there, as in music, creative talent may manifest itself at a much younger age.) Consequently, these creative abilities have no opportunity to develop during the four to five years of higher school, precisely those years when the brain is most receptive.

Another shortcoming of this system is that very few higher educational establishments have research scientists on their faculties, and experience has shown that ordinary teachers do not have the qualities needed to train and assess scientific talent. The present system of training scientists may have been acceptable in former times when only a small number of scientists was required and little significance was attached to the effectiveness of scientific research. As a rule, the most talented young people were drawn to science, but had to push their way through to it on their own, irrespective of the routine schooling they received. However, with the increasing scope and demand for scientific research today, it is high time that specialized higher schools were set up to educate and train scientists, like those set up long since for musicians and artists.

In Moscow, the first higher school of this kind was established twenty-five years ago – the Moscow Physics and Technological Institute. It trains scientists in various fields of physics and has a present enrollment of 3,000. It was founded on the initiative of a group of eminent, active physicists, chemists, mathematicians, and other scientists associated with the most intensively developing fields of science and technology and responsible for the scientific and technological revolution.

I participated very closely in the organization of that institute and should like to describe, in some detail, the principles on which we founded it, the problems which confronted us, and how we handled them.

The first, and probably the most difficult problem was the selection of students for admission to the institute. In approaching this selection problem, the important thing was to create a system of entrance examinations, which would disclose a person's creative talents.

Secondary school aims to provide its students with a certain quantity of knowledge according to a specific program; success there is determined by how much of this knowledge the student learns. Secondary school does not develop or assess a student's creative abilities. Therefore, successful completion of secondary school is not a good criterion for selecting young people for scientific careers.

The system of selection we adopted is based on competitive exams, as it is everywhere, but the exams have certain features that enable the examinee to display his creative abilities. Much significance is attached to the choice of examiners. The majority of them are young and all of them are gifted scientists, for we believe that only active scientists are capable of correctly evaluating creative talent. In grading the examinations in basic subjects, we pay special attention to how the student has handled the problems in mathematics and physics. Then comes the crucial test for the applicant who has received at least passing marks in the exams: an interview with a group of two or three examiners. The purpose of the interview is to discover the applicant's level of sophistication, the nature of his talent, and whether he has aptitude for scientific thinking. The interview is held informally so that the examinee may not be prevented by shyness or nervousness from expressing his individual qualities.

The competition for admission to our institute is intense: the 500 available places are sought by eight times that many applicants. Such intense competition gives us a great deal of material from which to select. The quality of our applicants is also affected by the fact that creatively talented young people usually have a natural desire to make a career in science and therefore seek to get into our institute.

Selection of future scientists does not end with admission to our institute, however. Another selection takes place after students complete a two-year study program, consisting mainly of such fundamental disciplines as mathematics, experimental and theoretical physics, mechanics, stresses of materials, and languages. There is a great deal of emphasis on seminars and laboratory work. We try to obtain professors and teachers whose primary occupation is doing research in research centers, to lecture and to conduct seminars in their disciplines, and also to determine the general organization of instruction. Here again we proceed from the assumption that only scientists are capable of creatively instructing and training young people in science.

After he completes the two-year program, a student's selection depends on a state examination in physics, designed not so much to determine how much he has learned, as to find out how well he can use the knowledge he has acquired. The examination is conducted in an unusual manner. It is an oral exam: The student is given his questions and problems ahead of time and is permitted to consult books, journals, notes, even faculty members, before the examination. What is required of him is the kind of understanding of the field of physics that will enable him to use his knowledge in dealing with the problems he is given. Moreover, the problems themselves are posed in unusually general terms. The student is required to indicate the path a solution might follow and to give a quantitative evaluation of this solution. He is given a choice of questions to help the examiners determine both his capacity for independent interest in scientific problems and the depth of his comprehension of them. An examination of this complexity is in itself an improvement over previous methods of selection of young people for scientific careers.

An average of about ten per cent of the students fail at this stage and are given the opportunity to transfer without examinations to the third year of any ordinary college or university of their choice.

The students who pass this examination now continue their training outside the walls of the educational institution. They are sent in small groups to specialized research laboratories of their own choice, usually located at one of several dozen large scientific research centers which we call *bazismy,* or foundational institutes. A chair is created at each of these centers for a supervisor to oversee the further training of students in accordance with a specific program and also to organize the students' participation in research. The only classes students now attend at the academic institution are those in general subjects such as economics and languages. All training in special disciplines takes place at the foundational institute.

There the students are in the hands of young and, for the most part, gifted scientists who lecture and conduct seminars. Usually only five to fifteen students work in a single area of research, first as laboratory assistants, later in more independent work. The student takes his exams and writes his lab reports at the research center during his training period of three years, after which he writes and defends his graduation thesis.

The approach to each student is individual. The full course usually takes six years, but if a student is unusually capable he may, with the approval of the research center executive body, complete and defend his thesis sooner. There have been students who have completed their work a year, or even a year and a half, ahead of schedule.

On graduation, the student is officially designated an engineer-physicist and is usually offered a research position in a scientific institute. By the time a student finishes the course I have described, he has not only acquired scientific research skills, but has also been able to discover the bent of his talents. Consequently, he should be able to decide what kind of work he wishes to do after graduation.

The most gifted students may enter a still higher program for improving their skills in scientific research, known as internship. The foundational institute usually selects a few of its most capable students to keep on as probationers after they have defended their graduation theses. After two more years at the institute, if a probationer shows himself to be an able researcher, he may be offered a permanent position on the staff or he may go on to graduate studies. The graduate school course is three years, after which the student defends a thesis and earns the degree of *Kandidat* of Sciences. Thus, the selection and creative training of scientists are constantly combined with their actual involvement in research, and our experience has shown that this is the best way to train scientists who will go on to do good work in the future.

In the twenty years since our Moscow Physics and Technological Institute was established, our close follow-up on the work of its alumni has confirmed us in our belief that the above system for training talented young people is much superior to the system in operation in other technological institutes and in the universities.

It should be mentioned that our system can still move in a number of directions. Many young people, for example, have talents which extend in several directions, some of which are not being developed at present.

Furthermore, although we generally recognize two basic directions in physics, and accordingly train both theoretical and experimental physicists, given the way physics is developing, this division is no longer sufficient. The present huge scale of experimentation requires the construction of such highly complex apparatus as enormous accelerators, chambers for observing the flow of elementary particles, nuclear reactors, complex electronic devices for space research and superconductor solenoids with cryostats. To produce these it is necessary to be an engineer as well as a physicist. The need has therefore arisen to train physicist-designers.

In the very near future, we will also have to deal with the problem of training scientist-organizers to head research groups. Many extremely important problems in contemporary science can be solved only through research by large groups of scientists. To succeed, these require coordination by a talented organizer – one who is not only a fine scientist, but a fine leader as well, gifted with the kinds of talent required of a manager in industry. The creative talents needed by the head of a scientific collective resemble those of the director in a contemporary film. The director must select and harmonize the performances of a large cast of actors; his organizational talent ensures the success of the film. Nowadays the talent of the director is recognized at least as much as is that of the leading actors. Analogously, there are large scientific research groups, in space research for example, in which the leading scientist-organizer plays a major role, one on which the success of the entire group depends. Yet, as a rule, the success of a scientific achievement is unfairly attributed exclusively to individual scientists. Surely this will have to change in the future, for the leader and organizer deserves high appreciation. Furthermore, such chiefs will have to be selected from among scientists and given special training in scientific management. And this must become a major objective of the organization of science.

The continuing scientific and technological revolution is constantly increasing the scope of scientific research. Statistical data indicate that both the U.S.S.R. and the U.S.A. have close to one million persons working in organizations connected with scientific research, and that this number is increasing. Since the effectiveness of scientific research is determined primarily by the creative abilities of the scientists engaged in it, and since the number of young people with a talent for scientific research is limited, one of the most important organizational tasks of science is to attract more and more gifted young people to careers in science. To do this, we must make the situation of the scientist sufficiently attractive. Not only must he be well provided with material things, but, even more important, he must be assured a high and respected position in society. To one degree or another, of course, his situation is attractive in all advanced countries today.

Much can still be done to improve selection of scientifically talented young people. The examination-interviews of our institute do not, for example, guarantee that the largest possible number of the country's most creative young people will find their way to scientific careers. The selection process could be improved if it were transferred to the last few years of secondary school.

The difficulty here is that the professional teaching staffs of ordinary schools are not adapted to educate their students creatively. Hence, the only way to exert a creative influence on secondary school students is to organize it from the outside. One of the best ways this is done at present is through the organization of what we call *olympiads* for high school students. These are contests in mathematics and physics, instrument design, astronomical observation, and other areas of interest in science and technology, and they are designed to reach as many students as possible. Organized by groups of scientists, they not only spotlight talented young people, but also encourage children to take an interest in scientific creativity from an early age. Contests of this sort are widespread in our country, as they are in the U.S., and this is a good thing.

Another way of encouraging interest in science is through the extra-curricular clubs in various subjects which universities and other higher educational establishments offer secondary school students. Led by young scientists, these clubs familiarize the high school student with the creative scientific process. However, they reach relatively few secondary school students because they are feasible only in cities which have colleges and universities.

Finally, there is a plan afoot to get young scientists, mainly graduate students, to teach part time in the upper grades of secondary schools. In fact, the idea is to make it obligatory for graduate students to teach such subjects as physics, mathematics, biology and chemistry in secondary schools. Not only will this develop the native talents of the secondary school students, but it will help us to discover the best among them early. This would also benefit the graduate students, for it is common knowledge that teaching is the best test of one's own knowledge.

The second organizational problem confronting science is how to choose areas of research appropriate to the basic directions of world science and to the country's cultural needs. The body in charge of this choice must be able to judge the importance of scientific themes, to determine the quality of scientific researchers, and to evaluate scientific results already achieved.

Clearly only a group comprised of scientists is capable of performing this kind of organizational function; in fact, everywhere academies of sciences, scientific associations, academic councils, and similar representative groups are put in charge of this type of work. Everywhere, too, such organizations operate democratically. First they discuss the problems confronting them, then determine their decision by a vote. This is the generally accepted system for handling problems connected with scientific work and is, apparently, the sole correct one. It is especially effective for evaluating scientific personnel. Consequently, all major positions in scientific institutions are filled by secret ballot. Only the placement of rank-and-file scientific personnel is done by one-man decisions at the administrative level. This system does, however, tend to be more centralized in the socialist countries than in capitalist countries. For example, in the U.S.S.R., the Academy of Sciences is a central organization with very broad powers. It appoints by secret ballot the directors and top level scientific personnel for all of the country's leading scientific institutions. Though it is the council of each scientific institution which confers the academic degrees of *Kandidat* and Doctor after passing on dissertations, a special Higher Certification Committee strictly supervises the procedure. This committee is comprised of scientists and it, too, makes its decisions by secret ballot.

In socialist states, the selection of scientific themes to be pursued is also to a certain extent controlled by a centralized body of the scientific community. The Academy of Sciences guides the work of all of its institutes (which are engaged primarily in research on problems of pure science) by means of its power to approve or veto all research plans. Problems connected with technology, agriculture, industrial output, and the like are handled in institutes under the various ministries and in the State Committee of Science and Technology. These institutes are guided by their own scientific councils which also operate on democratic principles.

Obviously, the selection of areas of research is more centralized in socialist states than in capitalist countries where it is determined by executive bodies in the interests of large industrial concerns. The centralization that characterizes the socialist system has a number of major advantages. It is, for example, easier to watch out on a nationwide scale for duplication of research, to make immediate use of scientific achievements on a nationwide basis, and to coordinate large numbers of scientific institutions run by different ministries to do research on pressing problems. But centralization also has its weak points. The slack time involved in implementing plans may become quite lengthy. In addition, centralization is susceptible to bureaucratic stagnation. Of course it is possible to combat these weak points, the most effective weapons against them being community supervision and criticism.

The significance of achievements in science has always and everywhere been assessed through the discussion of scientific papers by colloquies of scientists, not only at the level of the councils of scientific institutions, but also more broadly, at conferences, symposiums, and congresses. International meetings of scientists recognize contributions to world science and make highly objective evaluations which often exert considerable influence on directing bodies within nations.

The efficient exchange of scientific information by journals and correspondence and, most importantly, by personal contacts among scientists and visits to scientific laboratories, is one of the most effective ways of ensuring correct evaluation and direction in research. International cooperation among the scientists and scientific groups of all countries engaged in research not only helps to guide research effectively, but also ensures that science adheres to a world standard.

There are some who believe that the practical implementation resulting from the achievements of basic scientific research can be used as a criterion for evaluating the original basic research. In my view, this is so only in a limited number of instances. The fact of the matter is that it takes such a long time for industry to assimilate a scientific achievement, especially a major one, that research in progress is hardly affected, for the group of scientists involved changes direction before the assimilation is completed.

The third problem pertaining to the organization of science concerns the provision of material resources for scientific research. Money is a prime resource, money for paying scientists and auxiliary personnel, for constructing and equipping research centers, for organizing the exchange of information, and for all kinds of peripheral expenditures. In socialist countries the allocation of financial resources is planned centrally. In capitalist countries, some of the money comes from the government and some from private sources. In both cases a question arises as to how much money should be allocated for science as a whole.

It is known from published data that approximately three percent of gross national income is allocated for scientific work in both the U.S.S.R. and the U.S.A., as well as in a number of other advanced countries. We have no reliable way of determining whether this is indeed an optimal proportion, but it has now been established beyond any doubt that when allocations for science increase beyond it, the number of scientific personnel increases, but the effectiveness of money expended begins to decline. As a matter of fact, some say that the effectiveness of a research institute's work is in inverse proportion to the logarithm of the number of scientific personnel. They argue that science gets the most capable and talented researchers first, and that, inasmuch as the number of creatively gifted people is limited, quality must decline when too many researchers become involved. In my view, a country's success in science is dependent fundamentally on the number of creatively talented people it can tap.

For a scientist to perform efficiently, a certain average sum of money must be spent on him annually (for a physicist, this sum is about $30,000.) Therefore, since we know the number of persons with talent for scientific research and we know the sum that has to be spent on each of them, it is possible to determine the amount of money a country can productively spend on science. It may very well be that three percent is correct. However, should our efforts to improve the selection and education of talented young people succeed, eventually, in increasing the total number of competent scientific researchers, that sum would also have to be increased.

If the effectiveness of scientific research could be determined as definitively as that of industrial production, we could determine how much money should be invested in it merely by solving a problem in arithmetic. Calculations of this kind have, in fact, been attempted when research was closely linked to practical industrial problems, but they have never been very convincing.

The quantitative effectiveness of a scientific institution's work might, indeed, be determined as follows. Assuming that scientific research is creative exploration, and that unsuccessful attempts usually precede the solution of a problem, we could take a ratio of the number of successfully solved problems to the number of total attempts to solve them as the coefficient of the effectiveness of scientific research. Since this coefficient will inevitably be small for difficult major problems, it cannot be regarded as the sole indicator of the quantitative effectiveness of a given scientific group's work. However, when the coefficient remains at zero, the scientific institution concerned should be closed down.

I doubt that it is possible to devise a method for the quantitative evaluation of scientific creativity, any more than it is possible to do so for creative art. We have not been able to solve much simpler problems. We cannot quantitatively compare, for example, the achievements of major individual scientists whose work is well-known and universally recognized, say, of Newton and Einstein. It is as impossible to do this as it is quantitatively to compare the work of two great artists like Rembrandt and Picasso.

Nevertheless, present efforts to study the processes of scientific creation and to evaluate specific scientific achievements may be very useful with regard to the organization of science. It will no doubt emerge that the most important factor ensuring the effectiveness of scientific development in all fields of knowledge is the ability to attract the most gifted young people to science and to select and train them correctly.

PART FOUR

21. PHYSICAL EXPERIMENTATION AT SCHOOL

To be able to understand a physical experiment, pupils have to perform it by themselves. But the understanding becomes even better if the pupil personally makes the necessary devices for experimentation. Therefore the participation of pupils in the preparation of such devices must be encouraged by all means.

In this connection I would like to point out the poor status of physics demonstrations in our secondary schools. There are some entrenched types of demonstration devices which were described in physics courses since the time of Kraevich and are still being made. But physical phenomena must be presented with emphasis on their links with real life. In this manner the pupil will be able to understand the connection between theory and practice and do away with that severe affliction of our schooling system – its abstract nature whereby knowledge and real life never touch one another. Many achievements of physics were actually derived from everyday life. That Archimedes got an insight into the law of buoyancy while taking a bath is far more than a mere anecdote. As a matter of fact this is one of the simplest and most eloquent examples of how an observation of nature can reveal the fundamental laws of hydrostatics. The laws of physics can be effectively demonstrated on such commonly used machines as automobiles, tractors, radio receivers, telephones, etc. For example, the laws of electric induction can be demonstrated just as instructively by means of a modern car speedometer as on the disk of Arago. In selecting a suitable demonstration for a given physical phenomenon the teacher can show a great deal of creativity and inventiveness.

In designing such devices efforts must be made to stimulate the creative abilities of the children and provide them with everything necessary for revealing their inventive talents, even in trifles. A hand-made device built with the simplest of means but in an intelligent manner is far better than an accurate copy from a physics course, even if made by the pupil himself.

From a review of the exposition 'Young Technicians Help the School', 1940.

22. PROBLEMS IN PHYSICS

The problems published in this collection were compiled by me for students of the Moscow Physical-Technical Institute, where I taught a course in general physics in 1947–1949. The collection also includes problems given at examinations for postgraduate studies at the Institute of Physical Problems at the U.S.S.R. Academy of Sciences. The problems were selected and prepared for printing by I. Sh. Slobodetskii and L. G. Aslamazov, who recently graduated from the Physical-Technical Institute.

In compiling these problems I had a certain objective in mind, which explains their unconventional formulation. The following explanation will make their solution of greater interest to the reader.

Problem solving is of particular importance in the study of the exact sciences such as mathematics, mechanics, physics, etc. It enables the student to apply his own knowledge to the solution of practical questions. Moreover, for the teacher it represents an excellent means for finding out whether the student really understands the subject and whether his knowledge is a mere accumulation of facts learned by heart. Teaching by means of problem solving also helps in revealing and cultivating the creative scientific thinking of the young generation. The need for this was quite evident 25 years ago when the Physical-Technical Institute was founded as an institution of higher learning intended specifically for the selection and training of research workers. It is well known that fruitful scientific work requires not only knowledge and understanding but also a capacity for independent analytical and creative thinking. In effect, these problems were compiled as a useful means for the discovery, evaluation and cultivation of these qualities during the teaching process.

I strove to achieve this end by formulating the majority of questions in the following manner. A small problem is presented, and the student, using the known laws of physics, must analyze and describe quantitatively the natural phenomenon involved. These natural phenomena were selected in terms of their scientific or practical interest within the scope of the students' level of knowledge.

From the book *Problems in Physics* [Fizicheskie zadachi], 1972.

Most of the problems allow a number of approaches to their solution in order to reveal the student's individuality. One example is the problem of the trajectory of an airplane with a state of weightlessness in its cabin. Here the standard approach would be to write the equation of the plane's motion in the gravitational field of the Earth, taking the resultant of forces acting on a point inside the plane as equal to zero. There is, however, another, simpler solution. If the plane follows the trajectory of a freely flying body, which is close to a parabola in the Earth's field, then a body located inside the plane can be in a state of weightlessness. The curious student can pursue the matter further and find out what is required during the flight in order simultaneously to achieve a state of weightlessness at all points of the plane's cabin. Another question that can be considered is what navigational instruments are necessary for the pilot to fly the plane along the trajectory required for weightlessness, etc.

A characteristic feature of our problems is that they have no definite answer because the student is allowed to proceed further and further with the analysis of the problem posed, depending on his own abilities and inclinations.

The student's answers provide a clue to the trend and nature of his scientific thinking, which is of particular importance in the selection of post-graduate workers. The independent solution of such problems stimulates the capacity for scientific thinking and cultivates an attraction to scientific problems.

Most of the problems have another distinctive trait. They do not contain numerical values of physical constants or parameters, and the student has to choose them personally. Thus, in the problem of weightlessness in an airplane the student is asked to determine the duration of weightlessness, though it is mentioned that the airplane is of a modern type. Here the student has to pick out the airplane's ceiling and maximum speed for himself. We did so because few people in our country appear to care whether the future scientist or engineer is taught to imagine the actual magnitudes of such common physical values as velocity, current, voltage, strength, temperature, etc. in concrete terms.

In solving a scientific problem the research scientist always has to visualize clearly the magnitude and relative significance of the parameters used for describing the phenomenon examined. Thus he will be able to pick out the crucial parameters for the study of the given phenomenon. Hence the importance of teaching the young generation that the symbols which represent physical magnitudes in the formulae should always be visualized in terms of concrete quantitative values. In physics, contrary to mathematics, both the parameters and the variables in a mathematical equation necessarily represent concrete quantities. In our problems we therefore train the students to seek out the required magnitudes for themselves in the literature.

The students of the Physical-Technical Institute showed interest in these problems and frequently subjected them to collective discussion. On the examinations, the students were always given complete freedom to use literature for solving the problems. Usually a few (up to 5) problems were given per examination, so as to enable the students to choose 2–3 of them. Thus, the inclinations of a student could be gauged from his selection of problems. For postgraduate examinations, new and more complex problems were prepared; in these cases, however, the student was allowed not only use of literature but also freedom to seek advice. Indeed, the scientist must cultivate the skill of using the advice of others, apart from learning the use of literature. In scientific work, discussions and consultations with colleagues and instructors are essential for success; this, however, requires a proper training from the very beginning of the studies.

We usually allowed about one hour for the solution of each problem. All problems have to be solved in writing, but the capabilities and character of the student become evident mostly in the course of a verbal discussion of the written text. The greater the talent of the young scientist, the sooner it becomes evident. The discussion of all these problems rarely took us more than an hour.

Today there is no doubt as to the enormous importance of science for the development of the culture and economy of a modern state. The number of Soviet scientists and research workers is constantly rising, having already exceeded half a million. For this reason the training and education of young scientists is nothing less than a major, independent objective of the state.

Apart from the Moscow Physical-Technical Institute, there are a number of other institutions of higher education in the U.S.S.R. intended specifically for the training of research cadres. Teaching at these institutions undoubtedly has specific traits of its own, as distinct from the universities which produce cadres for industry and the national economy. I believe that problems comparable to those compiled by us can be used not only in the teaching of physics but also in other exact sciences such as mathematics, mechanics, chemistry, etc. The solution of problems comparable to those given here will therefore improve the training of future researchers.

(1) What should the trajectory of a modern airplane be so as to create a state of weightlessness? How long can the weightlessness be maintained?

(2) The tire of a car suddenly bursts during a race. What must the speed of the car be so as to prevent the tire from falling off?

(3) How many times can the height of an acrobat's jump be increased by a one-time application of a springboard?

(4) An equilibrist having a weight P is standing on a sphere of radius R and mass M. The sphere is located on a horizontal plane where it can roll without sliding. Find out how should the equilibrist should walk on the sphere to make it roll, and what is the relationship between the coefficient of friction of his soles and the acceleration of rolling.

(5) Explain why a man can run on a very thin sheet of ice but not stand on it without falling through.

(6) Estimate the order of magnitude of the speed at which a man has to run on water to keep from sinking.

(7) A spaceship is flying from Earth to Mars. One half of its surface is blackened so as to absorb the entire solar radiation, while the other half consists of polished metal which fully reflects the solar radiation. Determine the influence which the light pressure exerts on the translational and rotational motions of the spaceship. Determine quantitatively the magnitude of this effect with respect to a spherical spaceship having a weight of 5 tons and a diameter of 300 cm.

(8) Determine the distortion of the surface of a fluid, caused by the gravitational force of a sphere. Discuss the prospects for an experimental observation of this effect for determining the gravitational constant.

(9) When a pebble or a rain drop falls on water, a water spray flies upward. Explain why. What exerts a greater influence on the height reached by the spray: the pebble's size or the velocity of its fall? What is the maximum height of the spray?

(10) Why can liquid nitrogen be poured on the hand without risking any 'burns'?

(11) Determine the limit of the audibility radius of a conversation in the open air.

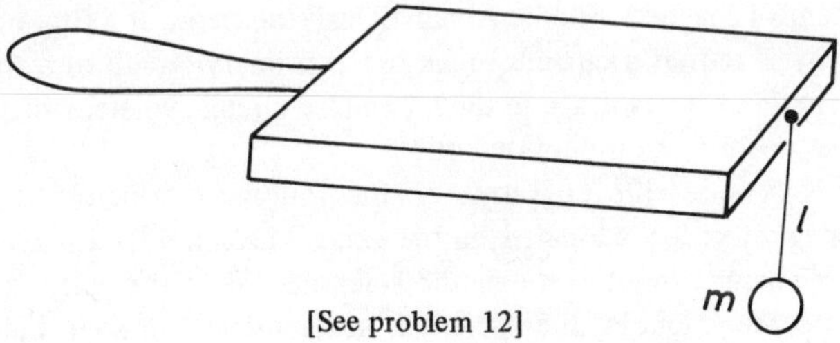

[See problem 12]

(12) The watchman's rattle of times past, which was used to show potential malefactors that the guard is not asleep, consisted of a wooden board with a handle at one end and a ball of mass m suspended on a string of length l at the other end (see drawing). Find out the motion of the handle at which the ball will knock at a period T.

(13) Explain why there have been cases where the front end of a cannon has flown off during a shot.

(14) What motions must a man perform in order to rotate a hoop around himself?

(15) List the factors which affect the accuracy of a pocket watch. Evaluate the relative significance of these factors.

(16) A river surface forms a sloped plane. Can a body float freely on the river at a velocity exceeding the maximal velocity of the current?

(17) An airplane is flying at a velocity close to that of sound. Its fuselage is heating up because of friction from the air. Determine the maximum temperature to which the airplane surface can possibly be heated.

(18) An insulated copper ball of a given radius, coated with a known amount of polonium, is placed in a vacuum. It develops a charge owing to the escape of alpha particles. Determine the increase of potential with time and its maximum value.

(19) Explain the use of electromagnetic machines rather than the electrophorus device for electric power production.

(20) A pulse of current of intensity I is let through a thin wire of diameter d. The wire will break down after an interval t. Calculate the magnetic field; determine the magnitude of the maximal magnetic field that can be obtained in this manner and the duration of its existence.

(21) A lightning rod is grounded by means of a circular copper tube of a diameter of 2 cm and wall thickness 2 mm. Lightning strikes, converting the tube into a round rod. Explain this phenomenon and evaluate the force of the lightning discharge.

(22) Explain which of the following devices allows a more accurate measurement of magnetic fields: a ballistic galvanometer or a fluxmeter?

(23) Why is it that a current break in the primary circuit of a transformer does not result in overvoltage in the secondary circuit, whereas such a phenomenon does occur in an induction coil?

(24) A magnetic rifle operating on the following principle is proposed. A cylinder (projectile) is located on the axis of a solenoid, close to the latter. Abruptly a current is let through the solenoid. When the cylinder is drawn halfway into the solenoid, the current is automatically broken. Calculate the

initial velocity of the projectile that can be achieved in practice in such a rifle. Calculate the necessary output of the generator.

(25) Describe the reflection of white light from the side of a soap bubble depending on the size of the bubble and its wall thickness.

23. SOME PRINCIPLES OF THE CREATIVE UPBRINGING AND EDUCATION OF TODAY'S YOUTH

It is generally recognized that the achievements of science affect the general level of human culture. In the twentieth century, however, these achievements have become so great that their application is exerting a global influence on the structure of society. Because of this process, which has assumed the proportions of a scientific-technological revolution, the education of youth can no longer be considered separately from the social implications of this revolution.

In my opinion there are two phenomena which result from the present scientific-technological revolution and exert a profound influence on the whole structure of education.

One of the most significant consequences of the application of modern science and technology in industry is the high productivity of labor. This can be attributed mainly to the increasing use of machines instead of human physical labor – a change made possible largely by the extensive application of electric power. Automation is ever gaining new grounds, and human tasks are being relegated to a push-button control of machines, lathes, cranes, etc. Hence the impressive increase of the productivity of labor in the advanced countries by a factor of several times in comparison with the past century, in agriculture and industry alike.

In the nineteenth century, nearly 80–90% of the population lived in villages and barely managed to produce enough food for its own needs and those of the urban centers of the country. Today, on the other hand, there are countries where only about 10% of the population are engaged in farming but their output exceeds by far the national demand. The high level of productivity attained by modern industry is evident from the following example. If the number of automobiles manufactured by a modern company is divided by the number of employees, it can be found that each of the latter makes more than one car per month.

Address at the International Congress on Problems of the Training of Physics Teachers at Secondary Schools, Eger, Hungary, September 11, 1970. [Slightly different English text appears in *Scientific World* **15** 12–15 (1971) under the title, 'Science Teaching and Scientific Method'.]

Economists believe that at the present productivity of labor between one third and one fourth of the work force of an industrialized country can provide everything necessary for the entire population: food, clothing, housing, transportation, etc. As for the rest of the present industrial work force, it is engaged mainly with military production, economic aid to less developed countries, research, various services to the population, tourism, radio, television, cinema, sports, press, etc. Employment in these branches is practically unlimited, being apparently a function of the available manpower.

Because of the vastly increased productivity of labor and the resulting reduction of the work load on the population, education of the young generation can now go on much longer.

In the nineteenth century, for example in England, the most industrialized country at the time, only a small fraction of the population could afford to send their children to school until the age of 20–23 years. The majority of youths were already working in factories or farms from the early age of 14. Such was the fate of Faraday, who at the age of 14 was an apprentice in a bookbinder's shop. In those years, the working day was often as long as 12–14 hours.

Today there are no economic reasons which could prevent an industrialized country from granting all its youths not only a full secondary education to the age of 16–18 but also a higher education – to the age of 20–23 years.

The considerable increase in the number of students observed today in the more developed countries is largely due to the high productivity of labor. In effect, enrollment in the universities of these countries has doubled during the last decade. Extrapolating this growth, we conclude that within a few decades higher education will perhaps cover the entire population of these countries. This trend will obviously affect the structure of education as a whole and the secondary school in particular.

Even the leading economists and sociologists of the last century did not foresee the present enrichment of states due to the high productivity of labor. On the contrary, at that time the growth of industrial productivity was indissolubly accompanied by the process of the 'impoverishment of the proletariat'. Today the reverse phenomenon is taking place.

The present growth of public wealth as a result of the increased productivity of labor and the development of manufacture for mass consumption will lead to an impressive rise of the per-capita income. The welfare of the working class is constantly rising. Any unemployment and poverty existing in some countries are due entirely to defects of the social structure; they do not in any way reflect the economic potential of the country.

A new social problem is emerging with the rise of the living standards of the population. This is the problems of 'leisure'. Though widely discussed nowadays, the leisure problem still defies solution beyond the recognition of its close ties with the upbringing and education of the young generation.

Briefly, this problem can be outlined as follows. Today the average man works for about 7–8 hours a day. Allowing him another 7–8 hours for sleep and 2 hours for eating, transportation etc. we are left with a leisure of 7 hours or so. As in the past, Sunday serves as a day of rest. But leisure will continue its growth in parallel with the steady increase in the productivity of labor. Its present growth, for example, results largely from the introduction of computers. In this connection many sociologists and economists foresee a new revolutionary increase in the productivity of labor not only in industry but also in various services.

In view of the expected further decrease of working time, it is evident that leisure hours will soon exceed those spent on the job.

Now we are confronted with the social task of providing humanity with suitable conditions for a rational use of leisure.

Aldous Huxley emphasized this problem in a very eloquent form. Those who have read his book, *Brave New World* [1] remember that the leisure problem of the population of the 'Brave World' was solved by such occupations as sports, various primitive spectacles and sex, not forgetting the extensive use of drugs. The main concern of the leaders of Huxley's 'Brave New World' was to steer the population away from social problems. For this purpose, people were discouraged from independent and critical thinking from early childhood.

Huxley's forecast of the use of leisure is now materializing in the most industrialized capitalist countries [2], where the rapid rise of living standards is accompanied by moral and social degradation and an ever-increasing use of drugs. Because of a lack of cultural interests, many youths cannot make a proper use of their leisure or material wealth. Adolescents rapidly become satiated with sports and other spectacles. Sex likewise knows no boundaries. The material wealth brings an abundance of various 'gadgets' – radio, cameras, cinema, cars, etc., but the pleasure from their primitive utilization rapidly wanes. Feeling themselves well protected by their parents, the young people show little concern for their future and feel no need to struggle for their existence. Under such circumstances, the young generation lacks any tasks the solution of which could stimulate its will and its energy.

All this drains any inner content from the life of young people. Moreover, the traditional principles of the capitalist society place particular emphasis on individualism in the process of upbringing at home and at school alike. Thus the young people grow in an atmosphere empty of any broad social ideals such as service to humanity, to science, to the arts; they face a bleak, meaningless life. The expanding use of drugs among youths provides brief ephemeral escape from reality, but at the price of a destruction of the nervous system which in turn makes the victim plunge into an even deeper mental depression. Criminality is on the rise among the youths.

Not surprisingly, voices of protest against this reality are rising among young people. The first symptoms of the protest of the young generation against the existing social order became evident long ago in the form of beatniks, hippies, etc. Though not of mass proportions, this phenomenon can only take place in a society which provides a surplus of wealth and leisure. Manifestations of this kind obviously reflect the disgust of today's youth with the bourgeois nature of modern civilization.

Student unrest is becoming much more ominous. Today it has assumed the proportions of a major social phenomenon with which governments have to cope. Statistics from the U.S.A. show that in 1968–1969, 55% of all school graduates became university students. Present enrollment in American institutions of higher education totals 7.5 million people [3]. Students clearly have the magnitude of a major political force.

Analysis of the student unrest which has become so prevalent in all developed capitalist countries reveals the active part taken by young people from wealthy families [4].

Obviously the unrest is not due to economic causes. It reflects a dissatisfaction with the ideology of the existing social order. The social environment in which the young people grow cannot possibly give them the ideals they need, because the individualism typical of the capitalist society cultivates a striving for material riches rather than broad social ideals.

There was a time when religion provided an ideological basis for the social activities of mankind. Today, largely because of the achievements of science, most people have realized the primitive character of doctrines forming the foundation of beliefs, which can now satisfy only a negligible fraction of humanity.

The present student movement is in many ways a rebellion because the young people are still looking for ideals and social orders worth fighting for. The unrest is just beginning to assume a meaning of its own; this process will probably continue for some years to come.

As it turns out, modern society is still unfit to make sensible use of the material riches and free time which the scientific-technological revolution has placed at its doorstep. Some bourgeois sociologists point to certain signs of social degeneration in the most developed capitalist countries. Sociological studies of prosperity and leisure among the masses are published at an increasing rate in recent years. The authors of these works stress the inherent dangers of the imminent further rise of living standards and leisure time, if left uncontrolled. Some of them see no way out of this predicament and conclude that this process perhaps represents the final episode in a cycle leading toward the doom of modern civilization [5]. It has been said that the inability of people to make proper use of their prosperity and free time may be no less fateful than the disaster of a general nuclear war.

Naturally, such forecasts are inadequate and premature. A solution to the problem can be sought in two opposite directions. One of them was vividly described by Huxley in his Utopian novel: filling the leisure time with occupations which satisfy only the most primitive animal needs, and bringing up children to be indifferent toward any spiritual or social problems. The other way out is diametrically opposite: cultivation of exalted spiritual requirements in the young generation, which will thus be able to use its leisure and prosperity in an interesting and socially beneficial manner. For this purpose, people generally and youths in particular must be taught to understand a purpose of life, to develop an interest in the solution of social problems, to cultivate spiritual values essential for the perception of science and arts. Humanity will undoubtedly choose this way. Since education represents a major factor in the upbringing of man and in the development of his moral qualities, this is indeed the new task which the scientific-technological revolution poses before our schools and universities.

So far the approach to education has been largely utilitarian. People have been taught to perform well on a professional level – as engineers, physicians, lawyers, etc. The objective has been to make people more efficient and conscientious in their work. Now the time has come when higher education is essential for every individual in order to teach him how to make proper use of his prosperity and leisure in a manner interesting to himself and beneficial to society.

What should be the nature of this education? A definite answer escapes us so far, but some general aspects of the solution are already emerging.

Experience shows that the people most satisfied with their work are those engaged in creative occupations: scientists, writers, artists, painters, theatrical directors, etc. People in these professions usually do not divide their time between work and nonwork, between work and leisure. They live in the world of their own activity and see no purpose in life except their work. Any work can be made attractive and interesting if it contains an element of creativity. Naturally, the process of creativity is to be understood broadly in this connection; it emerges in any form of human activity where man has to decide for himself what to do, rather than follow instructions.

In modern mass production everything has to be done strictly according to instructions; the individual worker becomes part of a smoothly operating mechanism and is given no room for creative initiative. Mass production becomes boring to the individual, as Chaplin has shown in his film *Modern Times.*

Some Utopians predicted long ago that one day man will work only part-time in manufacture and devote the rest of his time to some interesting, creative work in science or art. This solution of the problem is unrealistic. Experience shows that fruitful work in science or the arts requires talent, and it seems that only a small percentage of the population are talented enough to be successfully occupied as professional scientists, writers, designers, artists, etc. Therefore the problem today consists in finding ways for imparting an element of creativity to the free time of the ordinary person, so as to make it attractive and sensible.

Life teaches us that most people indeed can find a creative occupation in their free time. Such occupations can be of a humanitarian, social, scientific or technological nature. Many people are already devoting their leisure time to such pursuits. But we also know that such endeavors require a certain level of education and most importantly, the ability to introduce a creative element.

I shall give a simple example to illustrate this point. Many people spend their free time in travels. But for a truly interesting and meaningful visit to a city one has to know something about its history, for instance. The visitor will feel more satisfied if he can truly understand the sights and compare them with the history of other countries or with modern times. This, however, requires some training in accordance with the creative capacity of the individual.

Thus, education must not only impart the knowledge necessary for the creation of worthy citizens but also cultivate the capacity for independent thinking which is so essential for creative perception of the surrounding world.

The creative capacities of the human mind usually emerge at an early age and can be cultivated in the secondary school. Their nature and trend, however, normally crystallize at the age of 18 or so. The higher education, which begins at that age, must therefore be specialized and fitted to the individual capacity of the student. But in order to cultivate in all people the ability for meaningful leisure occupations the government must obviously make higher education available to the entire population regardless of whether this is necessary for the occupation of the individual, or not.

Leaving aside the general subject of the great social importance of the creative education of youth, I would like to tell you about the experience I have gained after many years of scientific and administrative work in connection with the manner of teaching for the cultivation of the creative capacities of the students rather than forcing them to absorb a multitude of facts and remember the laws of nature.

I developed an interest in this problem long ago, independent of the above-mentioned considerations of human education for a meaningful use of prosperity and leisure time.

The selection and education of young people for a creative scientific work is a crucial matter for the successful development of science.

Since the education of man begins essentially in the secondary school, I shall examine in a general manner the changes to be made there in order to cultivate the capacity for independent thinking.

So far the main task of secondary education has been the accumulation of a given amount of information in various fields of knowledge in order to convert every youth into a full-fledged citizen of his country. But the training of the creative capacity of the student requires an individual approach, which greatly complicates the task of teaching.

The creative capacities of young boys and girls become evident at a relatively early age, whether in the exact sciences, the arts, or literature. The school must naturally be aware of these differences in the inclinations of youth and avoid any forced intervention in this respect. I have always stressed the extreme importance of an early detection of the creative capacity of the future scientist during his upbringing. This objective must be pursued during the school years, and the sooner, the better.

The development of independent thinking is a major factor in the cultivation of the creative capacity of man. In my view this can be achieved in the following principal directions: a capacity for scientific generalization, i.e. induction; an ability to apply theoretical conclusions for forecasting the course of actual processes – deduction; finally, the detection of contradictions between theoretical generalizations and processes occurring in nature – that is, dialectics.

Of the natural sciences, mathematics and physics are obviously the most suitable disciplines for cultivating the capacity for general scientific creative thinking, because independent thinking can be trained there from an early age, mainly by the solution of problems and examples. If we compare the development of creative thinking in young people who have dedicated their lives to physics and to mathematics, we will probably conclude that physics is much closer to real life and has a much more direct approach to the study of natural processes, especially in view of the fact that laboratory exercises teach the youth from an early age how to make theoretical generalizations from his observations, (the inductive approach to the study of nature). By solving problems, the pupil develops the capacity for deductive thinking. As for dialectical thinking, the teacher can stimulate it by citing examples in which discrepancies between theoretical concepts and experimental data have led to major discoveries in physics.

Physics is a highly appropriate subject for the cultivation of creative thinking in natural sciences from an early youth. Consequently, the organization of physics instruction in schools becomes a responsible task.

Practical work and seminars contribute a great deal to the development of creative thinking in physics. Other effective approaches are problem solving and participation in olympics, which truly reveal the creative abilities of youths.

Our experience shows that the problems commonly included in the collections do not always possess the ingredients necessary for the stimulation of creative thinking. To solve these problems, the student often has merely to substitute the given data into the proper formulae, and the answer is ready. Here the only independent act of the student is the selection of the necessary formulae into which the given data are to be introduced.

In my opinion the problems must be made less rigid in order to give students more freedom in selecting the necessary magnitudes from experience. Here are a few examples of such simple problems. Find out the output of the pump motor necessary for maintaining a water jet high enough to extinguish a fire on the sixth floor. Or, what must be the size of a lens if the solar rays collected in its focus are to make an iron wire glow? Obviously the pupil has to pick out the pertinent data from his own experience or from a handbook. I gave my students problems of this kind, though naturally of somewhat greater complexity. They collected them over a number of years and published them in booklet form[6]. These problems fascinate the students; the fact that they lack a definite solution triggers a lively discussion. A similar book of problems can be compiled for the secondary school.

Nowadays special schools for gifted children are being set up in the Soviet Union and elsewhere in order to prepare them for scientific work.

With respect to arts this may well be a good idea because the creative capacity in such areas as music or fine arts usually become manifest at a much earlier age than the ability of creative thinking in sciences.

But the schools set up for children gifted in such fields as mathematics, physics, chemistry, biology etc. are even harmful. This is so for the following reasons. For a school the loss of a gifted pupil means a heavy bloodletting and a decline of the general vitality and level of the school. Indeed, the gifted pupil can spend much more time with his classmates than does his teacher, and mutual assistance between them becomes a powerful stimulus. Gifted pupils often contribute more than their teachers in the education of their comrades. But this is not all.

It is well known that the teacher himself learns in the process of instruction. To be able to explain a theorem the teacher must first understand it thoroughly, and any gaps in his understanding become apparent during the explanation. For their own intellectual growth the gifted pupils need the company of others. Such mutual instruction does not exist in schools for gifted children, and this has an adverse effect on the development of human abilities. Naturally, there are also other well-known reasons against such selective elite education – notably self-conceit and presumptuousness, which have an adverse effect on the normal development of the youths.

Following the publication in *Komsomol'skaya Pravda* of part of my report on instruction in secondary schools, I received many letters on the matter, from which it became evident that I did not express my ideas with a sufficient degree of clarity. I am not opposed to the special schools; the truth is that I have a different concept of their objectives.

In my opinion the task of the special schools is analogous to that of [specialist] clinics in comparison with hospitals.

Clinics are engaged in the study and development of new diagnostic and therapeutic methods. For this purpose they are staffed with the most highly qualified personnel, their task being to introduce advanced methods into daily practice and thus raise the level of medical care given to patients in ordinary hospitals. For this purpose clinics must naturally be specialized in specific categories of disease. The usefulness and necessity of such organizations in the medical care system are generally recognized and accepted. The same applies to the advancement of instruction.

The objective of the special schools is to study and work out advanced methods of education and upbringing. The special schools must possess carefully selected cadres of teachers and an exemplary organization. Naturally, such schools cannot cover all fields of knowledge and have to be specialized in such disciplines as mathematics, physics, biology, etc. In my view the main task of the special schools is to raise the level of teaching on a large scale in the country. If this is so, then the organization of these schools and the selection of their teachers and pupils must conform to the stated purpose. There will always be a need for special schools in the major branches of knowledge, the purpose of which is to work out and introduce advanced teaching methods on a national scale.

Teachers play a major role in the cultivation of creative capacities in their pupils. Here we face great difficulties as it is practically impossible to provide the secondary schools with the sufficient number of gifted teachers capable of approaching the youths in an individual manner and cultivating their independent thinking.

Most teachers strive to give their students a certain amount of knowledge and assess their success in terms of how well they have managed to assimilate it. Moreover, the school has no criteria for evaluating independence of thinking. For this particular purpose the selection of suitable teachers constitutes the most difficult task. I think the problem can be solved, though not easily. The solution resembles the one now being applied on a large scale in a Moscow institution of higher education established especially for the training of research workers for leading scientific institutes, most of them belonging to the U.S.S.R. Academy of Sciences.

Our basic idea can be outlined as follows. The history of science shows that scientists who have students of their own and work together with them are the most prolific of all. Many major scientists can be mentioned to prove this point. Thus, Mendeleev discovered the periodic system of elements while looking for a way to explain the properties of chemical elements in a manner best remembered by his students. The young Lobachevski, while teaching geometry in a school for adult secondary school students, found no satisfactory way of explaining the obviousness of the postulate stating that parallel lines never intersect; thus he discovered the non-Euclidean geometry. Stokes, while formulating problems for mathematics students, asked them in one of these problems to prove that the surface integral is directly related to the magnitude of flux or current passing through the contour. This is now known as the Stokes theorem, though in fact he never published its proof, leaving this to the students themselves. This theorem became a fundamental one, being the basis of Maxwell's equations. In deriving his equations, Maxwell, in his famous treatise, refers to the collection of problems compiled by Stokes. The list of such examples can be continued to our own time. Thus, Schrödinger found his well-known equations while explaining de Broglie's work to a group of postgraduate students at University of Zürich, where he did this at the request of Debye, who told me personally of the manner in which the basic equations of quantum mechanics were found.

On these grounds, young research workers at many scientific institutes are encouraged to give small courses of lectures to the students and conduct seminars, usually on special subjects. This takes them not more than one working day per week, and their work is suitably remunerated. We think that the young research workers gain from this as much as the students themselves. Some young research workers even go to secondary schools on their own initiative to teach physics in the senior classes; this likewise proved beneficial.

In my opinion it is quite possible to organize the teaching of physics in the senior classes of secondary schools on the same principles with the participation of young scientists from research institutes. This will be beneficial to them as well as to the pupils; the difficulty lies in the organization. The research workers should be spared any excessive burden; they should not devote more than one working day per week to this purpose. But in the secondary schools such an initiative causes various organizational difficulties in the distribution of work. A large number of teachers have to be employed, as every one of them can only spare a limited time, and this in turn complicates the job of the administrative apparatus.

To sum up I would like to point out once again: the correct education of modern youth must involve cultivation of their creative capacities with particular reference to the individual inclinations and abilities of every person, starting from the school bench and proceeding in the universities. This is a basic task, the solution of which might determine the future of our civilization on a global scale, a task no less important than that of maintaining peace and preventing an atomic war.

If humanity is to develop further along the path of humanism, culture and social progress, we scientists and intellectual workers must all participate actively in the solution of problems involving the healthy and progressive education of the next generation.

REFERENCES

[1] Huxley, A., *Brave New World*, New York, 1932.
[2] Huxley, A., *Brave New World Revisited*, London, 1959.
[3] *Projections of Educational Statistics to 1977–1978*, Washington, 1969, pp. 12, 13, 30.
[4] *Youth in Turmoil*, New York, 1969.
[5] Stent, G., *The Coming of the Golden Age: A View of the End of Progress*, Garden City, New York, 1969.
[6] Kapitza, P. L., *Fizicheskie zadachi* [Problems in Physics], Znanie Publishing House, 1972.

24. PROFESSOR AND STUDENT

Our traditional evening for graduates and senior students of the Moscow Physical-Technical Institute consists of two parts. The first part is of a solemn nature, and dedicated to the life and activity of our institute. During the second part we review our own activities, meet with old friends, and have a good time. I have to participate in the ceremonial part of the program, which is less attractive than the second part. But we must remember, dear friends, that in a good meal the nourishing dish always precedes the much-awaited dessert. As a cook I find it rather difficult to prepare a tasty dish for you within the 15–20 minutes placed at my disposal, and to discuss our matters in a serious but not soporific way. For we have a lot of matters to consider in connection with our institute.

As you all know, the Physical-Technical Institute was established about 20 years ago. The basic idea behind the creation of this institute was quite simple and obvious. Science was developing at a fast pace in the U.S.S.R. and abroad. This led to the establishment of many research institutes which began to attract the best scientists and eventually covered the whole of pure science. This took place at the expense of the universities, which lost much of their teaching staff, many professors and even some of the equipment used for instructional purposes. This deprived the students from participating in research while still in the universities, and they had to train in other institutes before starting scientific work of their own.

This gap between the universities and research institutes proved extremely harmful to the training of young cadres and therefore had to be eliminated. This was the objective of the Moscow Physical-Technical Institute, where instruction is closely related to research, and students learn with modern equipment under the guidance of young scientists actively involved in research in which the students themselves can take part from the second and third year onward. This approach did away with many of the shortcomings in the organization of our science, which resulted from its rapid growth.

The success of this system is quite obvious. This can be seen from the

---

Address at a meeting of graduates of the Moscow Physical-Technical Institute, 1963.

proliferation of MFTI[1]-type institutions for the training of young scientists in the U.S.S.R. Despite its achievements over the years, however, the system is still afflicted with major defects which have to be eliminated. Our objective is to establish a diagnosis of these shortcomings and find suitable ways for their elimination.

One of the first graduates of our institute, Comrade Belotserkovskii, is its present rector. Apart from having successfully completed his studies with us, he understands thoroughly the spirit and significance of the instruction and training system applied at MFTI; working with him has become an easy and pleasant task, and we often discuss together the measures to be taken for improving our work.

We still have many vacancies in our professorial body, as our attempts to attract top-level professors are not always successful. And there is yet another shortcoming. The institute does not perform all functions within its power. Let me dwell on these functions at some length. As for the selection of professors, you know all too well that there are good ones, average ones, and others who are less than average. Nothing can be done in this respect.

The worst part is that the teaching of the basic disciplines is not always placed in good hands in our country. There was a time when the fundamental subjects in the universities – general physics, chemistry, mathematics, mechanics – were entrusted to the most prominent scientists, and the teaching of such disciplines was regarded as a matter of great honor. This is no longer so, though I find it difficult to say why. For a correct education of the young generation, it is indeed crucial that the fundamentals of knowledge be imparted by major scientists capable of giving the students the necessary tools for completing the construction. Indeed, nothing can be built on a poor foundation.

How to correct this state of affairs? How to attract top-level professors, teachers and scientists to the universities? Perhaps this could be done with the assistance of modern technology – for example by means of educational films in which a scientist of top renown in the field (or even a group of scientists) could teach physics, chemistry, or mathematics.

Naturally, the participation of our best professors can be assured in this manner. But let us consider the actual consequences of such a step. Perhaps the administration of the institute would welcome the idea, which means a cut in personnel and no more efforts necessary for seeking and attracting new teachers. From the Ministry's viewpoint these advantages are very convincing. By producing such films it would be possible to save manpower and other

[1] MFTI – Moscow Physical-Technical Institute

expenditure at the universities. Some students would be only too glad as sleep comes easier in a darkened lectured hall.

All this is of course absurd. Just imagine our institute with a projection apparatus replacing the professor, only students and projectionists. Such a dark, boring place would hardly deserve the name of *alma mater.* But this is not the point. Perhaps the students will sooner or later get used to this state of affairs; at any rate, they will survive. But the teachers themselves would never adapt to it. The point is that the instruction of students is not the only function of a university, since the professors and other teachers need teaching too.

A good scientist always learns while teaching others. To begin with, he tests his own knowledge because only by being able to explain something to others can one be sure of having clearly understood it himself. Second, fresh ideas often come to the mind while looking for a suitable way of explaining something to others. Third, questions, even absurd ones, asked by students after a lecture often stimulate the mind and provide a totally new viewpoint of a phenomenon approached conventionally so far, and all this spurs creative thinking.

Finally, students have a better and wider knowledge of physics than their teacher does. Being a specialist in his field, the teacher has a narrow approach, in contrast to the broader viewpoint of his students. And the teacher can learn a great deal from his student just by talking with him.

This is why the young scientist must engage in teaching. A good university is one which allows ample room for the development of teachers and students alike.

To show that this is no mere phraseology I shall cite a number of cases in which major discoveries came as a result of teaching work. These examples are eloquent enough to prove my point.

One of the most classic examples concerns Mendeleev and his periodic system. Mendeleev sought a way of explaining the properties of the elements to his students in a more systematic manner. He distributed the elements on cards which he arranged in various sequences; eventually he found that the cards formed an integral system when arranged as a periodic table. On March 1, 1869, the table was printed as a separate edition, and some time later it was published as a supplement to the second edition of the *Fundamentals of Chemistry.* Thus the periodic system of the elements arose from Mendeleev's pedagogical activity as professor of the St. Petersburg University.

Another case comes from the realm of mathematics. At the beginning of the 19th century, the Russian government decided that all clerks must have a secondary education. Those who had no school-leaving certificate had to obtain it. To make things easier for them, suitable courses were established leading toward the required graduation exams. Lobachevski was a geometry teacher in one of these courses. He was 24–25 years old at the at the time. At this young age he had to explain the principles of Euclidean geometry to elderly clerks who just could not understand the origin of the axiom which states that two parallel lines never intersect.

For a long time Lobachevski sought a suitable explanation; eventually he concluded that such an explanation simply does not exist. He realized the possibility of devising a geometry where the lines always intersect. Thus he created his non-Euclidean geometry – a new branch of mathematics which, as you know, is of fundamental importance to modern physics.

The well-known physicist Debye told me of another case in point. At that time Debye was a professor in Zürich. One of his pupils, Schrödinger, was still a totally unknown young man at the time. Debye was acquainted with the work of de Broglie, who had proposed the well-known hypothesis of the wave structure of the electron and had shown that under certain conditions of interference the motion of the electron can be replaced by a wave motion. Many physicists rejected this idea of an equivalence between wave motions and quantum processes, between wave and corpuscular motions. Schrödinger was among the opponents of this view. Asked by Debye to tell the students about de Broglie's work, Schrödinger declined at first. But Debye applied his leverage as a professor, and Schrödinger eventually consented. He started to look for a precise and comprehensive formulation of de Broglie's ideas in a mathematical form. And when he presented de Broglie's work in this manner, which he considered most accurate, Debye exclaimed: "Listen! You have just found a remarkable new equation of fundamental importance in modern physics!" Thus the wave equation – one of basic significance in modern physics – emerged in the course of a pedagogical endeavor.

Let me give you yet another example. Stokes taught theoretical physics in Cambridge during the second half of the last century. A young man came to him for a postgraduate examination. Such examinations were very difficult at the time because of the scarcity of postgraduate positions – only two or three; hence the fierce competition. Stokes used to give about ten problems among which the student had to make his own choice. The students were given a specified number of hours. Stokes often gave insoluble problems in order to find out whether the student could recognize it as such. One of these problems (those were pre-Maxwellian times) was to find the distribution of velocities in a gas. This distribution of velocities was unknown at the time. Bernoulli and all the others believed that these velocities were roughly equal.

To Stokes' surprise, one young man solved this problem quite correctly. Perhaps you have guessed that his name was Maxwell.

Thus the law of the distribution of molecular velocities in a gas was discovered by Maxwell during an examination.

Many similar examples can be cited. It seems quite obvious to me, however, that if the teaching activity can be so fruitful in such major, fundamental matters it can certainly be of benefit on a more elementary plane as it often exerts a positive effect on the science and scientists of our time. For this reason the universities should not be regarded merely as training grounds for young scientists as in reality they stimulate the scientific genius of renowned professors as well. Therefore, the universities must be so organized as to provide ample opportunities to their teaching staff in this direction.

This view is not generally recognized in our country. Universities and other institutions of higher education still adhere to the principle that part of the staff should be engaged in research and the rest – as teachers. Instead of the present state of affairs, a system should be adopted whereby the scientists devote only a small fraction of their time to pedagogical activity. Only then will the universities be able to perform all their functions – the instruction of students and teachers alike.

The replacement of the professor by a projection apparatus is therefore quite absurd as it precludes the second function of any institution of higher education, which will undoubtedly develop in the near future. To this end we must devote our efforts.

I am telling you all this because soon you will leave the Physical-Technical Institute and start work in one form or another at a research institute. If you wish to advance as scientists without getting old, you must develop your knowledge without losing contact with the next generation, you must teach the young generation and at the same time learn from it. Once severed from the instruction of the young, you will start growing old in age and backward in science.

I wanted to leave you this small heritage, which I regard as particularly important.

25. REMARKS ON THE ANNIVERSARY OF THE PHYSICO-TECHNICAL INSTITUTE

Usually I avoid speaking at anniversary sessions because I do not know what to say. But when asked to speak at the 50th anniversary of the Physical-Technical Institute I could hardly refuse.

Much can be said here of the extraordinary achievements of the FTI and and its remarkable founder, Abram Fedorovich Ioffe. This institute is a leading research center, and not only in the U.S.S.R. It has educated the pioneering cadres in Soviet physics. Abram Fedorovich has indeed contributed a great deal to the promotion of physics in our country. We, his assistants in those times, are proud and satisfied with the results obtained. I do not intend to speak of all this, but in my view it would be worthwhile to consider the reasons for the achievements of the Physical-Technical Institute as one of the leading research centers of the country.

There is in my view a somewhat neglected aspect of FTI's life which has contributed much to its achievements. What I intend to say here is not a trivial matter, and therefore I shall begin with an introduction.

An anniversary obviously means that the institute has reached a certain age. If so, it must have been born. But everything born has to die – such is the dialectics of nature. Death spares no one – infusoria, men, institutions, states. Eventually our entire planetary system is doomed to die. What is born must die some time, which means that it has reached a certain age.

But what do 50 years represent for a research institute? To what age does this interval correspond? Is this Youth? Maturity? Senility? How can one measure the age of an institute? I shall start with a discussion of this topic.

Human age is determined on the basis of everyday experience. We know that from the age of 50 years onward, man begins to suffer from sclerotic phenomena, signs of an approaching old age. But what are the corresponding changes in research institutions? What are the manifestations of sclerotic phenomena there? These are not difficult to define. Let us mention some of them.

Senile gluttony develops in man. That is, the elderly tend to eat more

---

Address at the grand assembly to celebrate the 50th anniversary of the A. F. Ioffe Physical-Technical Institute, 1968.

than they can usefully digest. Gluttony afflicts also institutes, which accordingly begin to consume more resources than they can rationally use.

Old age has yet another manifestation in man – the appearance of cells which take no active part in the activity of the organism but rather burden it and bring about a senile obesity. Similar phenomena take place in institutions. With time they accumulate cadres which take no active part in any useful work. Institutes, as we know, find it almost impossible to free themselves from this burden.

Another noteworthy sclerotic phenomenon is senile garrulity. With age, man begins to talk too much without paying much attention to the response of his audience. In the same way, institutes begin to publish a lot of papers, necessary or not. This is indeed a sclerotic phenomenon.

With age, as we know, man loses his capacity for procreation, and the same applies to institutes. Life shows that a healthy, vigorous research institute educates young people and trains new cadres who bud off to form new institutes and laboratories where they pursue fresh trends of research. An institute that no longer does this is indeed afflicted with senile sclerosis.

But one specific aspect of senile sclerosis has a most adverse effect on the healthy activity of both the individual scientist and the research institute. A young scientist at the peak of his powers works most actively and assimilates advanced research methods, thus forming thinking patterns of his own which he usually preserves for life. In the second half of his life, the mature scientist works on the basis of experience and methods dating back to the first stage of his scientific activity, and shows little readiness to accept new developments and keep abreast with the advances of science. This is one of the most unfortunate manifestations of old age.

No less a genius than Einstein himself during the last decades of his life rejected such new ideas of quantum theory as the uncertainty principle; many have heard of his disagreement with Bohr on this matter. Such age-linked phenomena are common among scientists.

To characterize the corresponding processes in research institutes I shall use the popular analogy which regards scientists as soldiers of an army engaged in the conquest of nature for the benefit of mankind. This army forms a continuous front which comprises all branches of science and extends in all directions. The war being waged on a global scale, scientists indeed form units of an international army with a single objective – victory over nature.

Proceeding with this analogy, we see active fighting in some sectors of the front and what can be characterized as trench warfare in others. In the areas of intensive military activity there are breakthroughs and advances into new fields of science. These breakthroughs are usually associated with the discovery of new natural phenomena, the development of new research procedures, or the creation of a new theory. All these are major victories over nature. Life shows that intensive activities in a given field of science last not more than 5–10 years, depending on the magnitude of the breakthrough. Once the intensive activity has subsided, the front reverts to a state of trench warfare, when preparations are being made for the next breakthrough.

Many research institutes fail to notice the ending of active fighting in such sectors of the front and continue to keep major forces there instead of transferring them to the area where the next breakthrough will be made. This means a big waste of senior research forces and a deterioration of the scientific work of the institute. An institute has indeed entered a stage of senile sclerosis if its research work is no longer focused on new frontiers in science.

Can these manifestations of senility be overcome? I believe they can – on the human and institutional plane alike.

What means do we have at our disposal against the onset of senility? Experience points to one powerful means, perhaps the only one, for fighting old age. I can best define it by citing the words of Rutherford. He said to me: "Kapitza, I feel young because I work with young people."

Having students, and working with the young generation, is the best way in which a scientist can retain the vigor of youth and keep pace with the advances of science.

The same is true of research institutes. A research institute employing a large, select young force and actively involved in the education of young scientists entrusted with responsible tasks is indeed immune from senile sclerosis. Such close links with the young generation never ceased to exist in the FTI, which accordingly kept its position as a vigorous institute of advanced research. The living ties with the young generation developed quite naturally at the FTI, which was actually formed at the Polytechnical Institute by a group of young teacher-researchers headed by Abram Fedorovich. Close relations have always existed between these two research insitutions. Selecting the best young people from the Polytechnical Institute, the FTI went on teaching and training them in the course of research work. This is FTI's fountain of youth and guarantee of prominence.

A review of the problems studied at the FTI for the 50 years of its existence shows that they always belonged to the active frontiers of science. In its early days the institute began work on the structure of the atom – the most topical subject of the time. Later the quantum theory began to emerge, and the institute soon took it up as a major field of research. Next came nuclear science, then semiconductors, plasma research – always keeping pace with the most advanced trends of world science. The institute was never afraid of switching from one subject to another. On the other hand, many of the major research institutes of the country recoil from any change of direction, thus gradually giving way to stagnation.

We saw long ago that the recruitment of carefully selected and well-trained young personnel from the Polytechnical Institute constitutes a major advantage for the FTI over research institutes which keep apart from universities. When science began its rapid growth in our country after the war, a group of Moscow academicians undertook to promote this fruitful connection between universities and research institutes. For this purpose a specialized school for the training of young scientists was established in cooperation with various research institutes.

Thus the Moscow Physical-Technical Institute arose, which in cooperation with what we refer to as basic institutes is engaged in the training of young scientists for these same institutes. During the first two-three years all students are taught such general disciplines as physics, mathematics, mechanics, etc. at the Moscow Physical-Technical Institute. The subsequent specialized training takes place at the basic institutes, where the students immediately take part in research. Under this educational system, after the defense of the thesis the basic institute chooses the most capable recruits, trained on the spot. Now, 20 years after the establishment of the Moscow Physical-Technical Institute, the accumulated experience has shown the great advantages of such ties between universities and research institutes. It has become quite clear that young people educated and selected in this manner indeed become leading scientists at the research institutes.

Thus, the new ideas which led Abram Fedorovich Ioffe to the foundation of the Physical-Technical Institute were adopted on a large scale and exerted a favorable influence on the organization of research in the U.S.S.R.

I wish the Physical-Technical Institute many more years of inexhaustible youth and unabated vigor in advanced research.

26. FOR THE GOOD OF THE PEOPLE

Sowing in science brings a rich harvest for the people.

D. Mendeleev

As in any other sphere of human activity, the growth of science closely reflects its environment. The socialist way of life undoubtedly exerts an influence on the development of Soviet science. We have moral ideals of our own, as befits a socialist country. We have our own outlook and evaluation of events. Thus, the ideology of asceticism, of the life of a recluse, or extreme individualism is totally alien to Soviet people. In our country, man can only live well and happily when his own interests closely coincide with those of the whole country and people. We do not strive for the accumulation of personal wealth. The Soviet citizen finds satisfaction in the benefit he personally can bring to his own country.

Soviet scientists are first and foremost working citizens of their country. Consequently, their ideals and aspirations coincide with those of the people as a whole. Their scientific work follows the trends which guide the entire country.

On these grounds we can speak of Soviet science as an entity distinct from its capitalist counterpart.

This difference does not apply so much to the contents of research, since the discovery of a natural phenomenon or its correct theoretical explanation are unrelated to the place or time of their occurrence. The main distinction of our science lies in its purposefulness and in the evaluation of its achievements.

The principal objective of Soviet science is to place a powerful tool in the hands of the working people for the fruitful development of their spiritual and material culture. To achieve this end, every scientist must keep the objective of Soviet science in mind from the very beginning of his work. Regrettably, many of our scientists tend to forget this. Sitting in his study or labora-

---

Article in the newspaper *Komsomol'skaya Pravda* 1940. [This article has also been titled 'Science for the People'.]

tory, in isolation from the external world, the scientist may begin to regard his own work as an end in itself. Such detachment from life is extremely dangerous for a scientist. He must feel the needs of his people and country more acutely than any other worker. He must remember that for a Soviet scientist, detachment from life is tantamount to premature death. Conversely, a scientist who can feel the pulse of everyday life around him can be assured of success. Such scientists can easily contribute to the country, which will reciprocate by giving them everything necessary for their research.

Naturally, such contributions should not be interpreted as a narrow-minded practicalness in science whereby every achievement is immediately converted into an invention or process which would save a given amount of money. This is merely one of the simplest ties between science and life.

But other, more fundamental and valuable ties also exist. A major scientific achievement may open new horizons or provide a better understanding of a natural phenomenon and yet be devoid of direct practical applications. The latter fact, however, does not in any way detract from its great and diverse influence on our material or spiritual culture, an influence which simply cannot be rendered in coarse quantitative terms. For example, the significance of Darwin's work on the origin of species does not lie in his contribution to the raising of pedigreed cattle but rather in the fact that it has provided a radically new concept of evolution and a powerful weapon in the struggle toward a progressive culture.

Every young, talented scientist after one or two years of work on a given subject inevitably faces a difficult and responsible dilemma: what to work on next? He can easily see that the number of problems and subjects in science is far greater than that of scientists all over the world. For a Soviet scientist the only correct choice is one based on the needs of life and the aspirations of the country. A choice based on these criteria comes quite easily; moreover, it keeps disappointments away and gives the scientist the feeling of being a full-fledged Soviet citizen.

Above all, the young Soviet scientist must remember that science in the U.S.S.R. is inseparable from the life and development of the country as a whole.

PART FIVE

27. IN MEMORY OF ERNEST RUTHERFORD

In Ernest Rutherford world science has lost the greatest and most daring experimental physicist of our time. I have no doubt that his name in the history of physics will be on a level with that of Faraday.

Like Faraday, Rutherford was essentially an experimentalist endowed with an extraordinary intuition. His intuition led him to the experiments which enabled him to find simple and clear solutions for the most difficult and fundamental problems of science. In physics, as in any other science, there are a number of basic problems the solutions of which point to the future trends of research. Not many scientists have managed to set more than one such trend; Rutherford, like Faraday, set quite a few.

In 1903, while working alone in the small provincial university at Montreal, Canada, Rutherford demonstrated that radioactivity represents a spontaneous decay of the chemical element radium, discovered by the Curies. He proved this in a brilliant and quite convincing manner, having obtained an emanation and helium from radium. The daring idea which guided his work ran contrary to the long-established concept of atomic immutability. His research provided a total new foundation for our concepts of matter and is of crucial significance to our cosmological outlook.

In 1911, Rutherford established the atomic model. He was able to show experimentally that the atom of any substance resembles the solar system, being composed of a positively charged central heavy nucleus around which negatively charged electrons move. In 1913 this model provided the basis of Bohr's theory of atomic structure and spectra. Today, these concepts are of fundamental importance in physics and chemistry alike.

In 1919, Rutherford proved experimentally the possibility of an artificial breakdown of matter. He broke down the nucleus of nitrogen, thereby establishing the trend toward modern nuclear physics.

His death came as a complete surprise to all those who knew him. All the time he instructed his students at the Cavendish Laboratory, where he established the greatest school of English physicists. His creative mind was

---

Article in the newspaper *Izvestiya*, October 1937.

incessantly active, and he took a prominent part in the collective work of scientists from all countries.

Rutherford was not only a scientific genius but also a great teacher. He left a large number of students who are now scattered all over the world.

Rutherford's work has received world recognition. As long ago as 1908, he was awarded the Nobel prize, in addition to all possible international distinctions. He was honorary member of academies all over the world, including the Academy of Sciences of the U.S.S.R.

28. THE SCIENTIFIC WORK OF RUTHERFORD

Ernest Rutherford, known all over the world as the greatest scientist of our time, was born in 1871 in the village of Brightwater near the town of Nelson in New Zealand. A man who received all the international prizes and distinctions that a scientist could ever gain, Rutherford began his life in a very modest way. He was the fourth child of a small farmer who had eight more children after him. His father, who was a modest flax grower, could not afford to send his 12 children to school, and Rutherford depended on scholarships from primary school to the university.

He was an active, lively, joyful boy who loved hunting and sports. While at school and at the university he was a forward in a soccer team. But he also loved reading, model making, and working with mechanical things. As a little boy he made a photographic camera, which was a rather difficult task at the time. On graduation from secondary school in 1890, he enrolled in Canterbury University College in Christchurch. This was a small provincial university with only 150 students and seven professors. From the first day he felt an attraction to science and started research work.

During his student years Rutherford showed great interest in radio waves, discovered by Hertz. He was fascinated by the idea of the wireless telegraph, but the problem at the time was to devise a detector for the electric oscillations excited by the incoming waves. Rutherford found that high-frequency oscillations demagnetize iron. This phenomenon can be easily demonstrated by placing a magnetic needle beside a magnetized bundle of iron wires located in an oscillatory circuit. The reception of radio waves will cause a marked deflection of the needle. Rutherford published this discovery, which quickly made an impression in the small university and a reputation for himself.

In 1891, the students organized a small scientific society where Rutherford, still a very young man, delivered a report on the evolution of matter. In this report he expressed ideas quite revolutionary for the time: he maintained, for example, that all atoms consist of the same components. This report met with a negative response, and he was forced to apologize before the society.

From a report read at the N.D. Zelinski University of Physical Chemistry, November 14, 1937.

At that time, in 1891, Rutherford could not possibly have had any data for such an assertion. Radioactivity was discovered only in 1896. The atom had been regarded as an indestructible entity ever since the days of Dalton, and we can only admire Rutherford's daring in expressing his idea, which he confirmed experimentally 12 years later.

In 1894, he graduated from the university and, having won the so-called 1851 Scholarship, he went to Cambridge, England. The 1851 Scholarship is the largest award that can be won in Britain by a young scientist; it paid Rutherford's entire scientific work for the next 2–3 years.

The year 1895 was a year of reforms at Cambridge University. Until that time, students graduating from other universities were not allowed to work in the Cambridge research laboratories. But, on the initiative of J. J. Thomson, this restriction was lifted and graduates of other universities were permitted to work in the Cambridge laboratories.

Rutherford was one of the first young scientists to benefit from these changes. He entered the Cavendish Laboratory, headed by J. J. Thomson. McLennan, Townsend and Langevin were admitted together with him. Throughout his stay at the Cavendish Laboratory Rutherford worked in the same room with Langevin and they became close friends. This friendship between the two scientists, who had started their careers together, continued until Rutherford's death.

In Cambridge Rutherford continued his work on radio transmission. He established a radio link between the laboratory and the observatory, located more than two kilometers away. Rutherford was the first to transmit radio signals over such a long distance at the time. He would have probably gone far had he pursued this line of work, but the practical solution of the problem did not attract him. What interested him far more was the ionization of gases by Röntgen rays, the nature of which was then unknown. He started work together with J. J. Thomson. Rutherford established the phenomenon of the saturation current in ionization. This work, published in 1896, can be regarded as fundamental in this field.

During the same period, in 1896, Becquerel discovered radioactivity. Fascinated with this phenomenon, Rutherford began to study it. He was the first to show that radium generates two kinds of rays, which he named alpha and beta rays, and which differ in their capacity to penetrate matter. Rutherford demonstrated that these rays differ from ordinary radiation.

By 1897, Rutherford, though still a young scientist, had a well-established reputation. In the same year he was invited to take up the physics chair at McGill University in Montreal, Canada, which he accepted. Thus for the next ten years (1897–1907) he worked in Canada. Those years, spent in a small provincial university, were the most prolific ones of his career. I find this fact particularly instructive for young scientists. We often hear young scientists, just beginning their careers, complain of poor working conditions, lack of laboratory equipment, and this and that. But imagine a young scientist living halfway across the world from his native country in total isolation from the entire scientific world, in a remote place where all journals arrived over a month late. This man, however, is full of ideas and vibrates with enthusiasm, and in that distant corner of the globe he creates the most advanced and revolutionary concepts of contemporary science. Rutherford began to attract young scientists from all over the world.

Rutherford's work in Canada culminated in a series of major discoveries. One of these was the thorium emanation. The young chemist, Soddy, joined Rutherford in a study of the chemical nature of the elements obtained in the course of radioactive decay, in view of the need to determine the chemical aspects of radioactivity, as well as the physical ones. Radioactivity was still a mystery at the time, and Rutherford, together with Soddy, was the first to show that it represents a spontaneous transition of one element into another – a process now named radioactive decay. This process yields either alpha rays, which consist of rapidly moving, positively charged helium atoms, or beta rays, which are fast electrons. Rutherford assumed on these grounds that the thorium emanation represents an element distinct from the thorium itself. Together with Soddy, he determined the atomic weight of the emanation by the diffusion method. It became clear that the emanation consists of a noble gas.

The theory of radioactive decomposition, formulated by Rutherford and Soddy in 1903, constituted a veritable revolution. When Rutherford, as a young student, spoke of the evolution of matter in 1891 before his fellow students he had no evidence at all to prove his point; but when he demonstrated this with purely experimental data the impression produced was colossal indeed throughout the world of science. Yet even some major scientists could not assimilate such a revolutionary concept. One of these was Kelvin, who to his last day rejected the idea of radioactivity as a decay of chemical elements, which he considered as indivisible components of matter.

In the same year, 1903, at the age of 32, Rutherford was elected to the Royal Society (a scientific society equivalent to the Academy of Sciences of the U.S.S.R.). But this is no rare exception in the British academy. Scientists there are usually awarded this honor immediately after having made a major achievement in research; hence the numerous cases of young scientists only

25–28 years old being elected. This is a major advantage of the British academy over its counterparts in other countries; not surprisingly, it represents an active research center.

In 1907 the physics chair in Manchester – one of the main English universities – was vacated. During the nineteenth century this chair was held by such scientists as Dalton, Joule, and others. Rutherford was offered, and accepted, the professorship. Between 1907 and 1919, at Manchester, Rutherford undertook several fundamental works comparable in importance to those of Montreal. One of them concerned the scattering of alpha particles during their passage through matter. On the basis of the latter work Rutherford established a new atomic model which is accepted to the present day.

In 1908, Rutherford was awarded the Nobel Prize in chemistry. In 1919, he discovered the artificial disintegration of matter and showed that apart from the spontaneous decay of radioactive elements the nucleus can be broken down artificially by bombardment with alpha rays. This discovery, first made with nitrogen, was later confirmed with other light elements. Thus Rutherford founded a totally new field of nuclear physics – the artificial disintegration of the nucleus.

In Manchester, Rutherford attracted a multitude of young scientists as he did in Canada. Among those who worked with him were also foreigners such as Geiger of Germany, Bohr of Denmark, etc.; his students achieved many outstanding results in his laboratory.

In 1919, Rutherford took up a chair of experimental physics in Cambridge, where he spent the rest of his life as director of the Cavendish Laboratory after the retirement of J. J. Thomson. Here Rutherford continued his work on the artificial disintegration of the elements. He guided the work of his students, and two of the most important discoveries of nuclear physics during the last decade came from his laboratory: Chadwick's discovery of the neutron, and the work of Cockcroft and Walton on the artificial disintegration of matter by bombardment with artificially accelerated proton beams.

We see that having started his experimental work on radioactivity in 1896, Rutherford steadfastly pursued this trend of research until this branch of knowledge developed by the end of his life into a separate science – nuclear physics.

To understand the significance of each of Rutherford's discoveries one must consider their historical background. This task, however, is well beyond the scope of my present report. But it would be highly instructive to dwell on a number of examples which illustrate Rutherford's methods of work and his ways to success.

Rutherford was an experimentalist in much the same way as Faraday. He had little use for formulas and only rarely resorted to mathematics. Once, while trying unsuccessfully to derive a formula in one of his lectures, he dismissed the difficulty by merely writing the final result and noting: "This will be obtained if everything is deduced correctly".

But of experimentation he was a master. It seemed that he could actually 'see' the phenomenon he was working on, even if it took place in the infinitesimal nucleus of the atom.

Broadly speaking, researchers in physics can be divided in two categories. One of these can be named the German school; here the experimentalist starts from certain theoretical assumptions and tries to test them experimentally. By contrast, the physicist of the English school starts not from theory but from the phenomenon itself, which he strives to examine in order to find out whether it can be explained by existing theories. Here the experimentalist is concerned mainly with the phenomenon itself, its analysis and explanation. And if such a classification of physicists is at all possible, then Rutherford is a prominent example of the latter trend in experimental physics. Rutherford's main objective was to understand the phenomenon. The experiment must be set up so as to show quite clearly the nature of the phenomenon. In view of this objective the accuracy and complexity of measurements must be such as to provide an insight into the phenomenon examined.

One example is the study of alpha particles. Radium emits alpha particles which, as Rutherford was able to show at the very beginning of his experiments, are distinct from ordinary radiation. But what is their nature?

Rutherford decided that, being emanated from radium, these particles must represent some existing element. In order to determine the element in

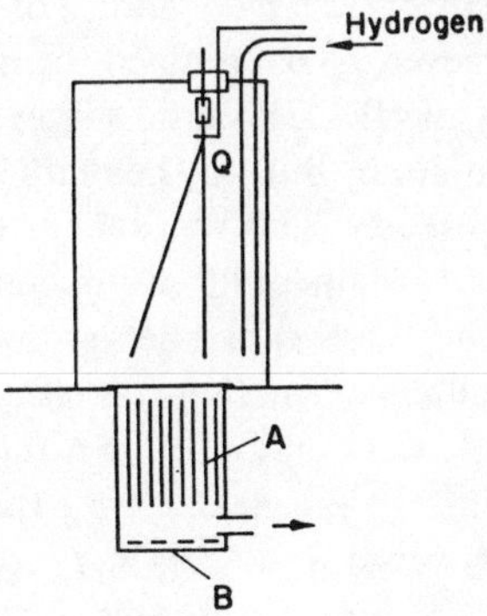

Fig. 1.

question, one merely has to determine their mass, and just to a level of accuracy necessary for identifying the element.

For this purpose Rutherford set up an experiment highly characteristic of him. I shall describe it here although its value is strictly historical now that we possess more sophisticated and accurate methods for determining the mass of alpha particles. But Rutherford's method is strikingly simple and straightforward.

Figure 1 shows the device used for these experiments. The simple electroscope *Q*, made of gold foil, is located over 20 parallel metal plates *A*. The gap between the plates is only 1 mm, so that the alpha rays emitted by the radioactive salt *B* (placed on the bottom) reach the electroscope chamber as a parallel beam. Hydrogen is let through the device in order to remove the emanation and increase the path of the alpha rays.

Ionization inside the electroscope chamber could be stopped almost completely by the application of a strong magnetic field directed in parallel to the planes of the plates *A*. By this simple procedure Rutherford was able to show that alpha rays are composed of rapidly moving, charged particles. If one half of the gaps between the plates are covered up on the electroscope side, it can be demonstrated that at a given orientation of the magnetic field the ionization stops at a lower field strength than at the opposite orientation. Thus Rutherford established the sense of alpha ray deflection by a magnetic field and concluded that alpha particles bear a positive charge. By creating an electric field between the plates *A* after having connected them alternately with opposite poles of a battery Rutherford stopped the ionization and obtained a deflection of the alpha rays by the electric field. From these data he determined the velocity of alpha rays and demonstrated that they represent a current of positively charged atoms of a mass greater than that of hydrogen atoms; moreover, he determined the ratio between their charge and mass with an accuracy of 10%. This ratio suggested that the alpha particle is a doubly ionized helium atom. But this helium identity of alpha particles had to be proven more rigorously. This was done later, in 1909, when Rutherford had greater quantities of radium at his disposal in Manchester.

The device used for these experiments was extremely simple too. It is shown in Figure 2. Radium emanation was placed in a small, thin-walled glass tube *A*. The tube walls were only 0.01 mm thick, and the high-speed alpha particles could pass through the glass, leaving the emanation inside. This tube was placed in the glass vessel *B*, ending in a capillary discharge tube with the electrodes *C* and *D*. Vacuum was created in the space around the tube *1* by the raising and lowering of mercury located in the vessel *B*. The tube containing the emanation was left in the device for two days, after which the gas formed by the passing alpha particles was compressed by raising the mercury in the discharge tube. The gas glow in the tube showed yellow lines typical of helium. That this helium did not diffuse from the tube containing the emanation could be easily shown by a control experiment in which this tube was filled with helium. In the latter case the spectrum obtained contained no helium lines. Thus it was demonstrated that the alpha rays are helium atoms.

The two experiments just described are extremely simple and can easily be performed by any student. But at that time these experiments, so straightforward and correctly designed, solved a problem of fundamental importance and brought about a revolution in the concepts of matter.

Rutherford was not satisfied with a study of alpha-ray beams by an observation of the ionization they produce; he sought a procedure for the detection of individual alpha particles. His first solution of this problem was based on the phenomenon of scintillation.

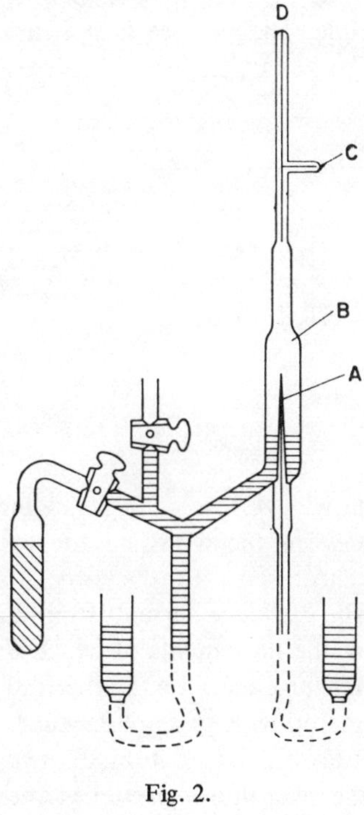

Fig. 2.

Crookes had found that some substances emit light, or luminesce, when bombarded with positive rays. This phenomenon was most pronounced with zinc blende. When Rutherford together with Geiger placed zinc blende under the microscope and directed a beam of alpha rays toward it, they saw discrete flash points in the microscope's field of vision, instead of a uniform glowing background. They concluded that the flashes occur at the sites where the alpha particles strike against the zinc blende. Thus they were able to determine the number of the emitted alpha particles by counting the scintillation points in the zinc blende.

Another method devised by Rutherford for the determination of alpha particles became even more valuable than the scintillation counter owing to the invention of the amplifier tube. This is the counter method based on a phenomenon discovered by Townsend. If a point is placed in a gas under reduced pressure, a voltage can be selected at which a discharge does not appear yet. Now if a very slight ionization is produced in the surrounding gas, even by a single alpha particle, there will be an immediate discharge for some interval of time. In 1908, Rutherford and Geiger built the first counter operating on this principle. Their device is shown in Figure 3. Instead of a

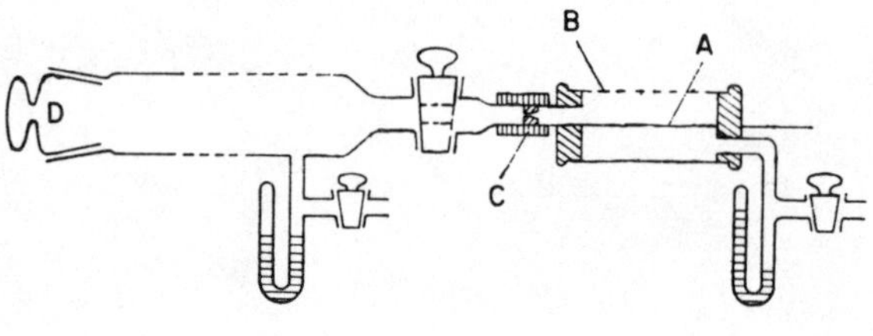

Fig. 3

point they took the thin wire A, located in the cylindrical vessel B. A critical voltage was created between the wire and the cylinder. The opening C, covered by a thin mica sheet, serves for the entry of alpha rays from a source placed in the vessel D. The discharge currents from the wire are recorded by a string galvanometer from the movements of which the alpha particles can be counted. In the counter proposed by Rutherford and Geiger the string galvanometer is now replaced with an amplifier tube, which makes the device extremely sensitve. This instrument, in its modern version, is one of the main tools currently used in the most thorough studies of cosmic radiation.

Having a procedure for the counting of alpha particles at his disposal, Rutherford embarked upon an analysis of a whole series of previously unapproachable phenomena.

A young scientist named Marsden started work in his laboratory in 1910. He asked Rutherford to give him any simple task. Rutherford asked him to count the alpha particles penetrating through matter and determine their scattering, pointing out at the same time that in his view Marsden would not find anything of particular importance. Rutherford's considerations were based on J. J. Thomson's atomic model, accepted at the time. According to this model the atom was visualized as a sphere 10^{-8} cm in diameter with an evently distributed positive charge, and sprinkled with electrons. The radiation spectra were regarded as the result of harmonic oscillations of the electrons. Alpha particles should easily pass through such a sphere, and there should be no appreciable scattering. The entire energy lost by the alpha particles on their way is spent in knocking out electrons, i.e. for ionizing the surrounding atoms.

Marsden began his observations under the direction of Geiger. Soon he found that the majority of alpha particles passed through matter but some scattering nevertheless existed; moreover, it appeared that some of the particles jumped backwards. Hearing of this, Rutherford dismissed it as impossible saying that this would be "just as surprising as if a gunner fired a shell at a single sheet of paper and for some reason or other the projectile bounced back again".[1]

This expression of Rutherford's shows the concrete and figurative manner in which he visualized a phenomenon.

Marsden and Geiger published their work. As for Rutherford, he realized that the prevailing concept of the atom was incorrect and had to be radically revised.

In his studies of the law which governs the distribution of the reflected alpha particles, Rutherford tried to determine the distribution of the field inside the atom, necessary for explaining a scattering by which some alpha particles could even be turned backward. He concluded that this could only take place when the whole charge is concentrated at the center, rather than spread throughout the volume of the atom. This center, which is named the 'nucleus', is very small – only 10^{-12} – 10^{-13} cm in diameter. But where would the electrons be placed? Rutherford concluded that the negatively

[1] Quoted in *Rutherford* by A. S. Eve, New York, Macmillan; Cambridge, Eng., Cambridge University Press, 1939, p. 311.

charged electrons form a circle, i.e. they rotate in such a way that the centrifugal force produced by this gyration balanced the attractive force exerted by the positive nucleus of the atom. Thus the atomic model was found to be something akin to a solar system where the nucleus is equivalent to the sun and electrons rotate like planets. Thus Rutherford created his atomic model.

This model caused a general perplexity as it ran contrary to some principles regarded as indisputable by the physicists of the time. Rutherford realized, of course, that according to Maxwell's theory, any electrons rotating around a center necessarily emit light, lose kinetic energy and eventually must fall into the nucleus. To shake the very foundations of Maxwell's theory was an extremely difficult task in those days. Not surprisingly, Rutherford's atomic model was not accepted at first.

This state of affairs continued for two years. At that time the young Danish scientist, Niels Bohr, came to work with Rutherford. They often discussed this atomic model. Bohr also realized that the model contradicted the basic laws of physics of the time. Consequently he started working on this paradox. Convinced of the experimental soundness of Rutherford's model, Bohr strove for its theoretical explanation. He had the brilliant idea of applying the recently published quantum theory of radiation for this purpose. The latter theory was first proposed by Planck and extensively generalized by Einstein.

In 1913, Bohr published a theoretical explanation of Rutherford's atomic model, which now bears the name of the Bohr–Rutherford model and serves as the basis of modern atomic physics.

One of the most important traits of Rutherford as an experimentalist was his extraordinary capacity for observation, his ability to generalize an observed phenomenon and detect its most essential aspect. Many examples can be cited to proved this point. When Rutherford discovered the thorium emanation, he started from the discovery of a difference between the ionizations caused by thorium with a closed and open electroscope gate. It seemed that the air current passing through the preparation altered the radioactivity of the thorium itself. Rutherford began to collect this air and soon found that it was radioactive. Thus he discovered the emanation. Most scientists, faced with such a discrepancy, would begin a study of the phenomenon with the gate either open or closed. But Rutherford immediately asked himself why does the phenomenon take place precisely in this manner rather than otherwise, whereupon he began to investigate the problem. Why, always why – this little question holds the key to great discoveries.

Here is another example. Rutherford's extraordinary perceptiveness became evident once again in the discovery of the artificial decomposition of matter. Observations of scintillation often revealed that rays of very long path – much longer than that of the bombarding alpha particles – escaped from the substance bombarded. This was a phenomenon generally known and often spoken of; yet nobody tried to explain it: the question 'why?' was not asked. Rutherford began a study of the problem. Soon he found that the bombardment with alpha rays causes a breakdown of nitrogen atoms, always present in air. Hence the long path. Rutherford designed his experiments in an extremely simple manner. His device is shown in Figure 4. The

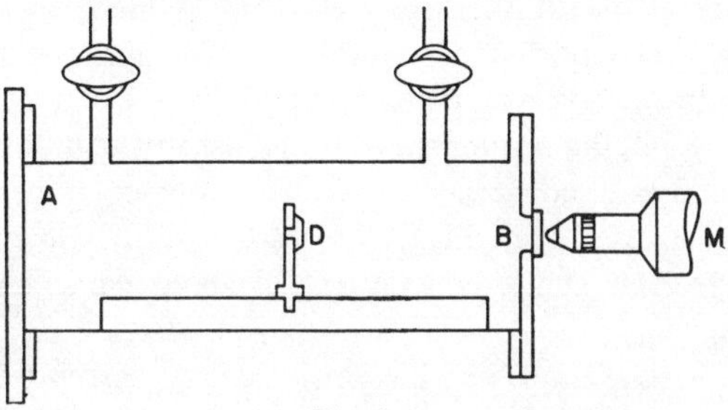

Fig. 4.

hermetically sealed chamber *A* can be filled with a gas via two taps at different pressures (*D* – source of alpha particles; *B* – screen for the observation of scintillation through the microscope *M*). The screen side facing the source of alpha particles is covered with a silver sheet which absorbs a considerable portion of their path energy. After filling the chamber *A* with nitrogen, Rutherford found that most of the scintillations vanish at a certain pressure. This took place when the alpha particles emitted by the radioactive source lose their entire energy for the ionization of air and do not reach the screen. But the remaining scintillations indicated the existence of a minute number of alpha particles with a path several times greater than that of the alpha particles emitted by the source. If the nitrogen is replaced with another gas such as carbon dioxide or oxygen, such residual scintillations do not occur. The only explanation is that they originate in the nitrogen. Since the residual particles are richer in energy than the primary ones, their formation can only be attributed to a decomposition of the nuclei of nitrogen atoms. The breakdown of nitrogen was thereby proven and the basic task of alchemy solved in principle.

Such simplicity in the formulation of the problem, such a straightforward approach, cannot but astonish any observer, physicist or not. Such simplicity is brilliant indeed, especially when it leads to such revolutionary results.

Many admire Rutherford's extraordinary intuition which, figuratively speaking, told him each time how to set up the experiment and what to look for. Intuition is usually defined as an instinctive process in the mind, something inexplicable which subconsciously leads to the correct solution. In my view this may be partly true, but at any rate it is strongly exaggerated. The ordinary reader is simply unaware of the colossal amount of work done by scientists. He only hears of those specific works which have produced results. Anyone who has closely observed Rutherford can testify to the enormous amount of work he did. Rutherford worked incessantly, always in search of something new. He reported or published only those works which had a positive result; these, however, constituted barely a few percent of the whole mass of work he did; the rest remained unpublished and unknown even to his students. Sometimes in conversations with him one could hear a hint to the effect that he had tried something, but unsuccessfully. He avoided speaking of his future projects and would rather tell about works that had yielded positive results.

One of the best examples of his extraordinary perceptiveness is the discovery of the neutron. The neutron is a material particle equal to the hydrogen nucleus in mass but without any charge. The existence of the neutron was demonstrated experimentally in 1932 by Chadwick – Rutherford's closest student, in Cambridge. Chadwick won the Nobel prize for this discovery. He was engaged in the study of a phenomenon wherein the bombardment of beryllium with gamma rays from polonium yielded extremely penetrating rays. Chadwick was able to show that these were not gamma rays. This radiation was first observed by Bothe and later studied by the Joliot-Curies; but its nature was discovered by Chadwick, who found that it consists of neutrons. The discovery of the neutron was a major event in modern nuclear physics since the neutron is one of the main elementary particles which make up the nucleus of all elements. Twelve years before the discovery of the neutron, however, Rutherford predicted the possibility of its existence in considerable detail. In a lecture read in 1920 before the Royal Society Rutherford said:

> If we are correct in this assumption it seems very likely that one electron can also bind two H nuclei and possibly also one H nucleus. In the one case, this entails the possible existence of an atom of mass nearly 2 carrying one charge, which is to be regarded as an isotope of hydrogen. In the other case, it involves the idea of the possible existence of an

atom of mass 1 which has zero nucleus charge. Such an atomic structure seems by no means impossible. On present views the neutral hydrogen atom is regarded as a nucleus of unit charge with an electron attached at a distance, and the spectrum of hydrogen is ascribed to the movements of this distant electron. Under some conditions, however, it may be possible for an electron to combine much more closely with the H nucleus, forming a kind of neutral doublet. Such an atom would have very novel properties. Its external field would be practically zero, except very close to the nucleus, and in consequence it should be able to move freely through matter. Its presence would probably be difficult to detect by the spectroscope, and it may be impossible to contain it in a sealed vessel. On the other hand, it should enter readily the structure of atoms, and may either unite with the nucleus or be disintegrated by its intense field, resulting possibly in the escape of a charged H atom or an electron or both.[2]

Thus Rutherford predicted long ago the main trends of nuclear physics as it developed after the discoveries made by Chadwick and Joliot-Curie.

To me this process was no intuition. This was a process of deep thinking and thorough experimentation. We all knew that Rutherford was engaged in a long, persistent search for the neutron. But he did not find it where he looked for it. Much depended on luck in this state of affairs. Theory could not tell in advance why one should work precisely with beryllium and polonium rather than any other substances. A persistent search was necessary.

Rutherford's death is a heavy blow to scientists all over the world. Science has lost the greatest pioneer of physical research since the days of Faraday. Throughout his life, replete with scientific discoveries, Rutherford worked on the most fundamental problems of modern atomic theory.

In effect, Rutherford was the creator of a whole new science – nuclear physics.

As long ago as 1896, at a very young age, he began his studies of radioactivity, which had only just been discovered at that time, and for the next 40 years he gave mankind a multitude of new discoveries and ideas of fundamental importance in atomic physics all over the world.

His influence on international science was greatly strengthened by the large number of students of all nationalities, including several Soviet scientists, who worked in his laboratory. Rutherford's selflessness and extraordinary individuality gained him not only respect and admiration but also a deep love. Thus he created around himself the most important school of physicists that has ever existed. And we understand all too acutely why so many scientists feel his death as a great personal loss.

---

[2] *The Collected Papers of Lord Rutherford of Nelson*, ed. Sir James Chadwick. 3 vols. London, Allen and Unwin, 1962–65. Vol. 3: *Cambridge*, 1965, p. 34.

29. HISTORY OF A RUTHERFORD PORTRAIT, 1933–1934

10th March, 1933.

Professor Niels Bohr, F.R.S.,
Blegdamsvej, 15,
København, 0,

Dear Professor Bohr,

I am writing to you at the suggestion of Lord Rutherford to find out your views on the following matter:

As you know, we have just built here a new laboratory with a grant made by the Royal Society, the building has just been opened, and is called the Royal Society Mond Laboratory. You will find the details of the new laboratory in the booklet issued at the opening which I am sending under separate cover. The possibility of my being able to bring to life my experiments with strong magnetic fields has been due entirely to the support of Rutherford, and to acknowledge the great interest he has taken in our work I asked an artist to make a carving of him which has been let into the wall of the entrance hall of the laboratory. As you will see from the picture of the laboratory, it is a modern building, and I thought it appropriate to ask an artist of the modern school to do the carving. I chose Eric Gill since he and Epstein are the two leading sculptors of the modern school in this country.

From the enclosed photograph you will see how he chose to represent Rutherford. It is a fine bit of carving, but a number of people do not approve of it. There are two opinions on the carving – firstly the "conservative" people consider that there is no likeness whatever to Rutherford, and that it is an insult to him, and request its removal; the others, with whom I agree, think that in order to get a complete likeness of a man there exist such methods as photography and taking of masks, but in modern art the idea is to produce a creation of the artist inspired by the model. Eric Gill, when he was told that the carving was not considered to be a good likeness, told the story

---

From the Mond Laboratory Archive, 1933–34, with errors as in the original.

of how when Lorenzo de Medici complained that Michael Angelo's portrait of him was not like him, Michael Angelo replied "it will be like you in a hundred years' time". Actually Gill thinks that he has made the protrait too like and would have liked to simplify it still further.

I personally take the following attitude:– that this portrait is meant to honour Lord Rutherford, and only at his personal desire would I be prepared to remove it. Lord Rutherford, on being approached by the 'conservative' people, speaking to me, said that he does not understand anything about Art, and is even unable to judge the likeness, although he finds the nose in the portrait too pronounced and more of an Assyrian type. He fails in any case to see any offence in the portrait, and said to me, "You had better write and ask Bohr's view – he knows me well, and also takes a great interest in modern art – I should like to know what he thinks".

Well, I am sorry to trouble you, but would like to ask you for your Solomon's judgment.

Accept our most kind regards to yourself and Mrs Bohr,

Yours sincerely,
[Signed P. Kapitza]

UNIVERSITETETS INSTITUT
FOR
TEORETISK FYSIK

BLEGDAMSVEJ 15, KØBENHAVN Ø.
DEN March 15th 1933.

Dear Kapitza,

It is indeed a very difficult if not impossible task your kind letter has put before me. Even if I have no sufficient qualifications to give a proper judgment in such matters, I have sufficient experience to know, how impossible it is to judge a piece of art from a photograph and without having seen its surroundings. With this reservation, however, the carving of Rutherford looks to me most excellent, being at the same time thoughtful and powerful. I can therefore in no way support the critics of the portrait, and if Rutherford does not object to it and you are satisfied with it, I think that it fulfills its object. I hope that it will remain in its place in many years to come to witness the good work which we all know will be done in your new laboratory. I look forward to see the booklet about the laboratory which I have not yet received, but I am sending these lines off at once, as I am leaving to-night on a skiing trip to get a much needed recreation.

With my best wishes and kindest regards to Mrs.Kapitza and yourself and all common friends in Cambridge from my wife and

yours sincerely,

N. Bohr.

Kapitza to Bohr 20th March, 1933.

My dear Professor Bohr,

I thank you very much for your letter which gave me great joy and satisfaction, It was really very good of you to give us your view so promptly

I quite realize that it is a difficult task to judge a piece of sculpture from a photograph, but in this case I think the photograph gave a correct impression. We hope that soon we shall have the great pleasure of seeing you here in Cambridge, and when you look at the carving your views will not be changed.

Accept my most sincere thanks and best wishes, . . .

Kapitza to Gill

Eric Gill, Esq., [21 March 1933]
Piggotts,
North Dean,
HIGH WYCOMBE.

Dear Gill,

I am sure you would appreciate it if you knew all the fighting I have had lately over your carving. Apparently it irritated some of the most important people in the University so much that they requested that it should be removed. There were big discussion and I had to fight hard and give lectures on Modern Art and its meanings and explain such elementary things as the difference between a photographer and an artist. I do not think I have succeeded in changing their minds, but I think I have succeeded in saving the piece of art, which I appreciate very much.

Rutherford, who found himself in a awkward position, as without his authority nothing could be done and he candidly admits that he is no judge of Art, suggested that the judgment should be left to one of his most famous pupils – Professor Niels Bohr of Copenhagen – who is a lover of modern art. A photograph of the carving was taken and sent to him, and with the only reservation that he has not seen the actual carving himself, but judging by the photograph, he thinks it to be "a very powerful and thoughtful piece of work".

Everything is now all right I hope, but I am still surprised to see the extremely formal approach to art existing among most of the people in this country. I had not actually experienced this narrow-mindedness before, but you probably meet with it often. I see they are going to make a fuss about your work on the B.B.C., and I very much sympathise with you for being attacked through being an individualist in your work.

I should like to have an opportunity of seeing you some time, so if you come to Cambridge don't forget that we can always put you up and it would be a great pleasure to us.

With my most kind regards and best wishes,

Yours sincerely,

Gill to Kapitza

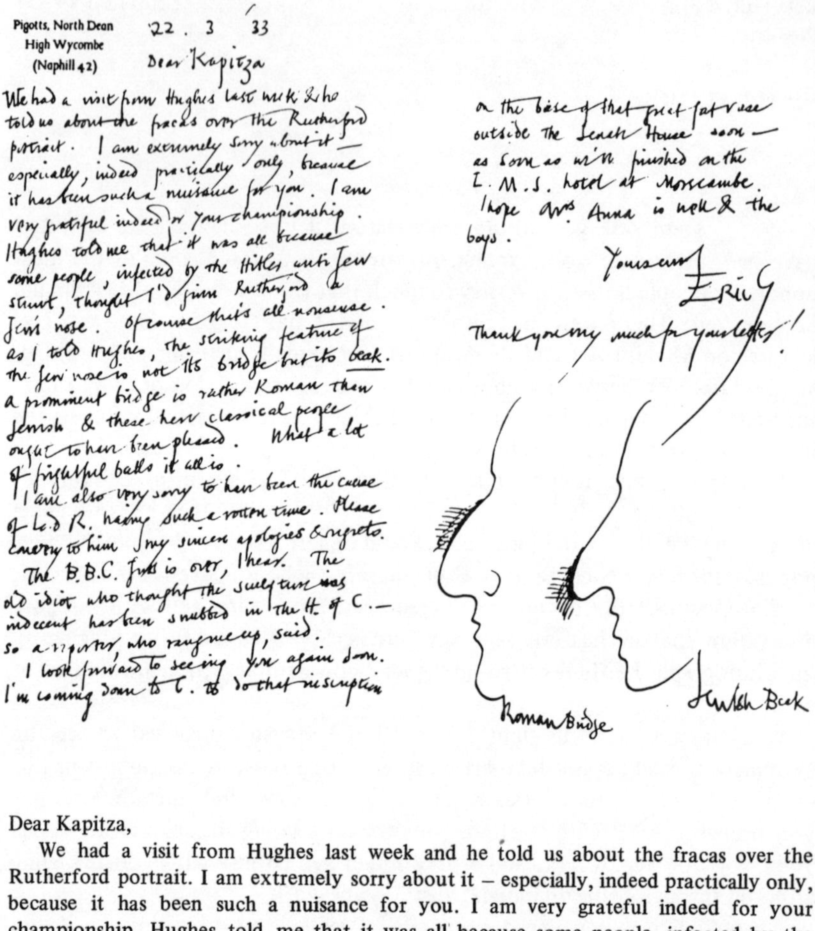

Pigotts, North Dean
High Wycombe
(Naphill 42)

22. 3. 33

Dear Kapitza

We had a visit from Hughes last week & he told us about the fracas over the Rutherford portrait. I am extremely sorry about it — especially, indeed practically only, because it has been such a nuisance for you. I am very grateful indeed for your championship. Hughes told me that it was all because some people, infected by the Hitler anti Jew stunt, thought I'd given Rutherford a Jew's nose. Of course that's all nonsense. As I told Hughes, the striking feature of the Jew nose is not its bridge but its beak. A prominent bridge is rather Roman than Jewish & these here classical people ought to have been pleased. What a lot of frightful balls it all is.

I am also very sorry to have been the cause of Ld R. having such a rotten time. Please convey to him my sincere apologies & regrets.

The B.B.C. fuss is over, I hear. The old idiot who thought the sculpture was indecent has been snubbed in the H. of C. — so a reporter, who rang me up, said.

I look forward to seeing you again soon. I'm coming down to C. to do that inscription on the base of that great fat vase outside the Senate House soon — as soon as we've finished on the L.M.S. hotel at Morecambe. I hope Mrs Anna is well & the boys.

Yours ever
Eric G

Thank you very much for your letter!

Dear Kapitza,

We had a visit from Hughes last week and he told us about the fracas over the Rutherford portrait. I am extremely sorry about it – especially, indeed practically only, because it has been such a nuisance for you. I am very grateful indeed for your championship. Hughes told me that it was all because some people, infected by the Hitler anti-Jew stunt, thought I'd given Rutherford a Jew's nose. Of course that's all nonsense. As I told Hughes, the striking feature of the Jew nose is not its bridge but its *beak*. A prominent bridge is rather Roman than Jewish, and these here classical people ought to have been pleased. What a lot of frightful balls it all is.

I am also very sorry to have been the cause of Lord R. having such a rotten time. Please convey to him my sincere apologies and regrets.

The B.B.C. fuss is over, I hear. The old idiot who thought the sculpture was indecent has been snubbed in the H. of C. – so a reporter, who rang me up, said.

I look forward to seeing you again soon. I'm coming down to C. to do that inscription on the base of that great fat vase outside the Senate House soon – as soon as we've finished on the L.M.S. hotel at Morecambe. I hope Mrs. Anna is well and the boys.

Yours ever,
ERIC G.

Thank you very much for your letter!

Kapitza to Gill

Eric Gill, Esq.,
Piggotts,
North Dean,
HIGH WYCOMBE.

24th June, 1933.

Dear Gill,

The noise about your carving of Rutherford is gradually quietening down. We had a special meeting of the Buildings Syndicate to consider the question. This had to be done as otherwise the conservative people would have felt themselves completely ignored and it is a little too dangerous as the conservative people are always the most important in this world! Still, we had to make them feel that we were not ignoring their opinion – that would offend them more than anything else!

The Buildings Syndicate definitely decided on my suggestion to do nothing with the carving unless you and Hughes consent. So we decided that when you are here in Cambridge to carve the inscription on the Greek vase we will have a small meeting consisting of yourself, Hughes, myself and two or three members of the Buildings Syndicate (the majority is ours!) which I hope will boil down to a general discussion of Modern Art. Let me know when you are coming, and we shall be very glad if you will stay with us while you are doing this job.

We have had rather a troublesome time with our second boy who has been seriously ill which has upset the household. In August and September maybe I am going to Russia.

I hope your work is going well. Looking forward to seeing you when you come,

With most kind regards and best wishes,

Yours most sincerely,

Kapitza to Bohr

3rd February, 1934.

Professor Niels Bohr,
Blegdamsvej, 15,
Kobenhavn,
DENMARK.

Dear Bohr,

All the troubles with the Rutherford carving are now, I hope, over, and it will be left on the wall.

It is impossible to exaggerate the importance of the role you played in saving its life. I should be glad to acknowledge your support by asking you to accept as a present an author's copy of the original. I have spoken to Eric Gill about it to-day, and he is quite willing to make one in July when he returns from Palestine.

I propose that the copy should be of the same size and carved in the same stone as the original, provided you do not suggest any alteration to make it suit the surroundings of the place where you propose to keep it.

I hope you and Mrs Bohr are keeping well.

Yours sincerely,

Bohr to Kapitza

February 9th 1934.

Dear Kapitza,

I was very happy from your kind letter to learn that your troubles with the beautiful and forceful stone carving of Rutherford are now over, and I am very thankful indeed to you and to the artist for your generous offer to present me with a copy of the carving which I and the whole institute shall greet as a most welcome symbol of what physics and I myself personally owe to our great master. We shall find a fitting place for it in the institute, and I quite agree with your suggestion that the copy should be of the same size and carved in the same stone as the original, as we can always when it is put in place remould the surroundings so as to suit it.

With kindest regards to Mrs. Kapitza and yourself from my wife and me with my heartiest thanks.

30. RECOLLECTIONS OF LORD RUTHERFORD

I am greatly honoured by the invitation of the President of the Royal Society to speak to you on my reminiscences of Lord Rutherford. But this is a very difficult task and I accepted it after long deliberation. At first sight I thought that to speak about the scientific achievements of so great a scientist as Rutherford would be easy. The greater the achievements of a scientist the more exactly and briefly can they be described. Rutherford created the modern study of radioactivity; he was the first to understand that it is the spontaneous disintegration of the atoms of radioactive elements. He was the first to produce the artificial disintegration of the nucleus and finally he was the first to discover that the atom has a planetary system. Each of these achievements is sufficient to make a man a great physicist. But nowadays these achievements and their fundamental values are well known not only to research students but even to schoolboys. Equally we all know the very simple and beautiful classical experiments by means of which Rutherford made his great discoveries. To come from the Soviet Union to speak about all this to the Fellows of the Royal Society would scarcely be appropriate.

You are all well aware that from research into radioactivity there grew up an independent science which is now called nuclear physics, and of all the papers published on physical research one fifth relate to the investigation of nuclear phenomena. Both nuclear energy and the use of artificial radioactivity in science and technology are developing quickly and simultaneously. All these fields absorb the main bulk of the monetary resources spent on science and which now reaches the sum of thousands of millions of pounds, dollars and roubles. And all this for the last 30 years grew out of one modest domain of physics which in the old days was called radioactivity and the father of which is justly called Rutherford. To speak of the development of nuclear technology and physics which came from the work of Rutherford and his school is very interesting and very instructive. But I am sure that such Fellows of the Royal Society as our President Professor Blackett, Sir James Chadwick, Sir John Cockcroft, Sir Charles Ellis and Sir Mark Oliphant who in the old days were the most active members of Rutherford's school and who

Proc. Roy. Soc. A **294**, 123–137 (1966).

themselves in this domain have made fundamental discoveries and researches are certainly more qualified than I to speak on these matters.

To only way in which I can satisfy the interest of the Fellows of the Royal Society is to speak of Rutherford the man, of how I remember him during my 13 years' work in the Cavendish Laboratory, of how he worked, how he trained us young scientists and also of his relations with the scientific world. My task is therefore to draw you a portrait of a great scientist and of a great man. Frankly, this is the job of a writer and not of a scientist. If I have now decided to do so, this is mainly for the following reasons. When I look back and see myself as a young man coming to England in 1921 and starting work in the Cavendish Laboratory and, after 13 years, growing into a scientist, I feel that these years of my work were the happiest, and for all that I have been able to achieve I feel immensely grateful for the attention and kindness which Rutherford showed me, not only as a teacher but as a very kind and sympathetic man for whom I have a sincere affection and with whom I eventually became great friends.

However imperfect my recollections of Rutherford may be, this is the only way in which I can express my deep gratitude to this great and remarkable man.

As is well known, Rutherford was not only a great scientist but also a great teacher. I can recall no other scientist contemporary to Rutherford in whose laboratory so many outstanding physicists were trained. The history of science tells us that an outstanding scientist is not necessarily a great man, but a great teacher must be a great man. Therefore my task is even more difficult: I must give you a portrait not only of a scientist but of a man. I will attempt to make my portrait of Rutherford as alive as possible and for this purpose I shall illustrate my talk with episodes which I most vividly remember. From my many recollections I shall select the ones which characterize different sides of Rutherford's nature. I hope this will help you to reconstruct a lifelike picture of Rutherford in your imagination from all these fragments.

I would like to begin my recollections with a small episode which happened in 1930 in the Cavendish Laboratory. At that time a small conference was being held in Cambridge to commemorate the centenary of the birth of Maxwell, the first director of the Cavendish Laboratory. He was succeeded by Rayleigh, J. J. Thomson and Rutherford, four great physicists of the last and present century. After the official part of the meeting in which some of Maxwell's pupils talked of their reminiscences, Rutherford asked me how I liked the speeches. I answered that they were very interesting, but I was surprised that all the speakers spoke only of the positive side of Maxwell's work and personality and made a 'sugary extract' (*sic*) of him and I said that I would like to see Maxwell presented as a living figure with all his human traits and faults which of course every man possesses however great his genius. Rutherford as usual laughed and said that he charged me after his death to tell future generations what he was really like. Rutherford was joking and I was laughing too. And now when I try to fulfil his behest and I imagine Rutherford as I have to present him before you, I see that time has absorbed all his minor human imperfections and I can only see a great man with an astounding brain and great human qualities. How well I now understand Maxwell's pupils who spoke about him in Cambridge.

There are numerous books and articles on Rutherford as a scientist. It is widely recognized that the simplicity and clarity of his thinking, his great intuition and great temperament were very characteristic of his creative ability. Studying the works of Rutherford and observing how he worked I think the basic characteristics of his thinking were great independence and hence great daring.

The basic method by which science is developing consists of experimental investigations into natural phenomena and the continuous verification of the consistency of the results of our investigations with our theoretical conceptions. The progress of our knowledge of nature appears in cases when we find contradictions between theory and observation, and these contradictions, as they compel us to develop our theories, enable us to widen our knowledge of nature. The more acute these contradictions are, the more they lead us to further fundamental changes in understanding the laws of nature, on the basis of which we may use nature for our cultural development. In science, as in history, definite stages of development demand their particular kind of genius. A definite period of development requires men with corresponding mental abilities. In the history of the development of physics, as in any other experimental science, the most interesting periods are those in which we are brought to revise our fundamental scientific conceptions. Then not only deep thinking and intuition are required from the scientist but also a daring imagination. As an illustration I shall remind you of two well-known cases in the history of physics. They made a great impression on me personally. The first case concerns Franklin's creation of the study of electricity. On the basis of this study Franklin stated that electricity has a material origin and can impregnate metal and freely move in it. In his day this concept was in fundamental contradiction to the concept of the continuous nature of matter. But Franklin's view was eventually accepted as it gave a simple and complete explanation of all the electrostatic phenomena observed in his day.

It is only recently, 150 years later, after J. J. Thomson discovered the electron, that Franklin's concepts were completely justified. But the most striking thing in this story is how it could have happened that Franklin, who had never before done any scientific work, could, in the course of a few years in a small remote American town when he was already a middle-aged man, find the right way by which this most important branch of science should be developed. And this happened in the middle of the eighteenth century when science was developing on the level of Newton, Huygens and Euler. How could Franklin achieve such results which were beyond the reach of professional scientists?

The other similar case in which the fundamental concepts of electricity had to be revised in the light of experiments is also well known. This is Faraday's concept of the electrical field. It is difficult to find a more revolutionary and original idea than Faraday's. He advanced the concept that electrodynamical processes must be explained by the phenomena happening in the space surrounding the conductor. I mention this case mainly because Faraday was a scientist who had no traditional scientific education, even though at that time its level was high for an average English scientist. I mention these two well-known cases only to show that at a particular stage of the development of science, when new fundamental concepts have to be found, wide erudition and conventional training are not the most important characteristics of a scientist required to solve this kind of problem. It appears that in this case imagination, very concrete thinking and, most of all, daring are needed. Strict logical thinking which is so necessary in mathematics hinders the imagination of a scientist when new fundamental concepts must be found. The ability of a scientist to solve such scientific problems without showing a local trend of thought is usually called 'intuition'. Possibly there is a way of thinking which takes place in our subconscious but the laws by which it is governed are at present unknown. If I am not mistaken, even Freud, a pioneer in the study of subconscious processes, was not aware of it. But if intuition exists as a powerful creative thinking process then doubtless Franklin and Faraday mastered it thoroughly. I am sure that Rutherford mastered it too and he has rightly been called the Faraday of our time.

When at the beginning of our century Rutherford started studying radioactivity it had already been proved experimentally that these phenomena contradicted the most fundamental law of nature, the law of conservation of energy. The explanation of radioactivity which Rutherford gave, namely the disintegration of matter, at once provided not only the key to the understanding of these phenomena but also led all investigation in the right direction. The same thing happened when Rutherford created the planetary model of the atom. At first sight this model completely contradicted laws of classical electrodynamics since in its circular motion an electron was perpetually bound to lose its kinetic energy by radiation. But the experiments of scattering the α-particles, performed by Rutherford's pupil Marsden in 1910, definitely showed the existence of a heavy nucleus in the centre of the atom and Rutherford imagined the collision of particles so clearly that even these contradictions with the fundamental laws of electrodynamics could not prevent him from establishing the planetary structure of the atom. We know that only three years later Bohr, on the basis of the developing quantum theory of light, evolved his brilliant theory of the structure of the atom which not only justified Rutherford's planetary model but also quantitatively explained the spectra of atomic radiation.

The peculiar character of Rutherford's thinking could easily be followed when talking to him on scientific topics. He liked being told about new experiments but you could easily and immediately see by his expression whether he was listening with interest or whether he was bored. You had to talk only about fundamental facts and ideas without going into the technical details in which Rutherford took no interest. I remember, when I had to bring him for approval my drawings of the impulse generator for strong magnetic fields, for politeness sake he would put them on the table before him, without noticing that they were lying upside down, and he would say to me: 'These blueprints don't interest me. Please state simply the principle on which this machine works.' He grasped the basic idea of an experiment extremely quickly, in half a word. This struck me very much, especially during my first years in Cambridge, when my knowledge of English was poor and I spoke it so badly that I could only vaguely explain my ideas, yet in spite of this Rutherford caught on very quickly and always expressed very interesting opinions.

Rutherford also liked talking about his own experiments. When he was explaining something he usually made drawings. For this purpose he kept small bits of pencil in his waistcoat pocket. He held them in a peculiar way – it always seemed to me a very inconvenient one – with the tips of his fingers and thumb. He drew with a slightly shaky hand, his drawings were always simple and consisted of a few thickly drawn lines, made by pressing hard on the pencil. More often than not the point of the pencil broke and then he would take another bit from his pocket.

A number of physicists, especially theoreticians, like to discuss science and apparently the process of argument is a way of thinking. I never heard

Rutherford argue about science. Usually he gave his views on the subject very briefly, with the maximum of clarity and very directly. If anybody contradicted him he listened to the argument with interest but would not answer it and then the discussion ended.

I greatly enjoyed Rutherford's lectures. I followed the course of general physics which he gave to the undergraduates as Cavendish Professor. I did not learn much physics from this course since by that time I already possessed a fair knowledge of the subject, but from Rutherford's approach to it I learnt a great deal. Rutherford delivered his lectures with great enthusiasm. He used hardly any mathematical formulae, he used diagrams widely and accompanied his lectures with very precise but restrained gestures from which it could be seen how vividly and picturesquely Rutherford thought. I found it interesting that during the lecture he changed the topic as his thoughts, probably following some analogy, turned to a different phenomenon. This was usually connected with some new experiments made in the field of radioactivity which fascinated him and he then proceeded to speak with enthusiasm on the new subject. In this case he usually put his assistant in a difficult position by asking him to give a demonstration which was not part of the original planned version.

About the same time I also attended the lectures of J. J. Thomson in his special course on the conductivity of electricity through gas. It was interesting to notice how differently these two great scientists approached scientific problems. If Rutherford's way of thinking was inductive, then the way of thinking of J. J. Thomson was deductive.

I think it useful when training young scientists to ask them to follow a course of lectures, even an elementary one, but delivered by an eminent scientist. Listening to these lectures they will learn something that they will never find in any textbooks. In this connexion I remember a conversation which I had with Sir Horace Lamb. He was telling me how he had attended Maxwell's lectures. Maxwell, he said was not a brilliant lecturer; he usually came to lectures without any notes. When he was doing mathematics on the blackboard he often made mistakes and sometimes got muddled. From the way in which Maxwell tried to disentangle and correct his mistakes Lamb learned more than from any textbooks he ever read. Lamb told me that for him the most precious parts of Maxwell's lectures were those in which he made mistakes. No doubt the mistakes of a genius are sometimes as instructive as his achievements.

When I came to Cambridge Rutherford did no more experimental work by himself; he worked chiefly either with Chadwick or with Ellis. But in both cases he took an active part in experiments. The setting up of the apparatus was done mainly by his laboratory assistant, Crowe, whom he treated rather severely. But I sometimes saw how Rutherford himself, despite his slightly shaking hands, dealt quite skilfully with the finewalled glass tubes filled with radium emanation. Although Rutherford's experiments are well known, I cannot refrain from saying a few words about them. The most attractive thing about these experiments was the clarity of setting the problem. The simplicity and directness of approach to the solution of the problem were most remarkable. From my long experince as an experimenter I have learned that the best way of correctly evaluating the capacity of a beginner as well as of a mature scientist is by his natural inclination and ability to find a simple way of solving problems. There is an excellent saying by an unknown French author which applies perfectly to Rutherford: 'La simplicité c'est la plus grande sagesse.' I should also like to quote the profound saying of a Ukrainian philosopher, Gregory Skovoroda. He was by origin a peasant and lived in the second half of the eighteenth century. His writings are most interesting but probably quite unknown in England. He said 'We must be grateful to God that He created the world in such a way that everything simple is true and everything complicated is untrue.' Rutherford's finest and simplest experiments concerned the phenomena of scattering by nuclear collisions. The methods of observation of scintillations by counters were worked out by Rutherford in collaboration with Geiger in 1908. Since then more than half a century has passed and this method and the Wilson chamber invented about the same time remain the fundamental methods for studying nuclear phenomena, and only the optical and resonance methods for determining nuclear moments have since been added. And up to now all nuclear physics possesses no experimental possibilities other than those used by Rutherford and his collaborators. The present development of nuclear physics is proceeding not by the invention of new experimental possibilities of investigating nuclear phenomena but thanks to the possibility of investigating nuclear collisions of a *larger* number of elements: and these collisions are studied in the domain of larger energies which are reached mainly by the use of powerful modern accelerators. But even now the way which leads us to the knowledge of the nucleus is still the method discovered by Rutherford, and he was the first to appreciate its fundamental value. I am referring here to the investigation of the collision of nuclei. Rutherford always liked to say 'Smash the atom!'.

Even now, in the process of investigation nuclear collisions, there is one great weakness: the necessity of using statistical methods in the interpretation of experimental results. Great care is required to deduce correct general laws from limited statistical data. Someone once said about statistics: 'There are three kinds of lies: a simple lie, an impudent lie and statistics.' In fact this was said about the application of statistics to social problems. But to some extent it is true of statistics in physics. I do not think that in any other branch of physics so many mistakes and faulty discoveries were made as in the course of the interpretation of statistical data obtained from experiments on nuclear collisions. Nearly every year new particles of resonance levels are still discovered, some of which may not exist. Rutherford was well aware what danger lies concealed in the interpretation of experimental data of statistical origin, especially when the scientist anticipates definite results. Therefore Rutherford was very careful to exclude the personal element and took the following precautions during the course of these experiments: the counting of scintillations was usually done by undergraduates who did not know the purpose of the experiment; the curves were drawn by persons who did not know what results were expected. As far as I remember, Rutherford and his pupils never made a single such mistake, while in the same line of investigation a number of mistakes were made in other laboratories. I remember that in those days the most critical approach in the interpretation of statistical data was that of Chadwick on whose judgement Rutherford usually relied completely.

I did not work with Rutherford because my investigations were not connected with nuclear physics and therefore I did not see him working in his laboratory. But I know that up to the very end of his life the main bulk of his time was taken up by his personal scientific research. I expect he gave the same amount of attention and strength to directing the work of young research pupils working in the Cavendish Laboratory. The detailed guidance of scientific work he left to one of the senior scientific workers, usually Chadwick. But he himself always took an interest in the choice of problem for experiment and of the experimental approach. Until the research student began obtaining results Rutherford showed no marked interest in his work. He never bothered about detailed guidance.

He often came to the laboratory but only for a short time; just to make remarks like: 'Why don't you get a move on – when are you going to get some results?' When I started working in the Cavendish Laboratory such remarks made a great impression on me, especially as they were made in a thundering voice and with a severe expression. But eventually I found out that such utterances were automatic, maybe customary for a New Zealand farmer who when going through the fields found it useful to stimulate the workers with a few 'kind' words. That it was actually so was proved by an episode which happened a few years later in the Cavendish. One day it was necessary to break a hole through a stone wall to put through a cable needed for some experiment. The work was urgent and it happened that at that time there was a building strike and it was exceptionally difficult to find a bricklayer who would consent to work. Finally a man was found and he started work but after a while he came and said that he refused to go on. When asked why, he replied that twice a gentleman had passed by him and both times had asked him when he would finally start work and get the job done. These remarks offended the workman. When asked who this gentleman was, his description showed without a doubt that it was Rutherford. When we reproachfully pointed out to Rutherford that during a strike one should be a little careful we were surprised that Rutherford denied having said anything to the bricklayer. Obviously when he likewise grumbled at us in the laboratory for our slow work, he did it unconsciously; it was a kind of conditional reflex.

The greatest quality of Rutherford as a teacher was his ability to direct research work in the right direction, then to encourage the beginner and to give just appraisal of his achievements. What he valued most in a pupil was independent thought and originality in his work. Rutherford did his utmost to develop in his pupil an individuality. I remember how in the first years of my work in the Cavendish I once said to Rutherford: 'You know that the work of X is pretty hopeless; don't you think he's wasting his time and apparatus?' Rutherford replied that he too knew that the man was working on a hopeless problem, 'but', said Rutherford, 'it is a problem of his own and even if the work cannot be accomplished it will lead him to another original research problem which will be successful'. The future showed that Rutherford was right. As I said, Rutherford would do his utmost to develop in his pupils independence and originality of thought and as soon as a pupil showed these qualities Rutherford would pay close attention to his work. As an example of Rutherford's ability to direct the research of his pupils I remember the story, as Rutherford told it himself, of the discovery made by Moseley. In 1912 Moseley worked with Rutherford in Manchester. He was very young and Rutherford spoke of him as one of his best pupils. When Moseley came to Manchester he at once accomplished some minor research work and then eventually he came to Rutherford and told him of three different topics he would like to investigate. One of these researches was the classical work which had made Moseley's name so well known – the dependence of the wavelengths of Röntgen rays on the position of atoms in the periodical system. Rutherford at once advised Moseley to choose this work for his investigation. The future showed that Rutherford made the right choice, but he always pointed out that the idea of the experiment belonged to Moseley.

Rutherford in the Cavendish Laboratory for high magnetic fields.

A meeting of the governing board of the Magnetic Laboratory: Rutherford (right), Cockcroft (left), Kapitza (centre).

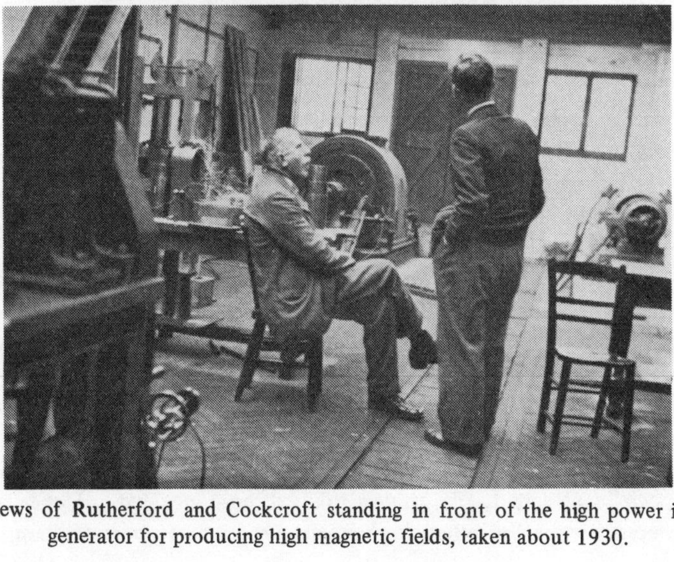

Two views of Rutherford and Cockcroft standing in front of the high power impulse generator for producing high magnetic fields, taken about 1930.

Rutherford was very particular to give credit for the exact authorship of any idea. He always did this in his lectures as well as in his published works. If anybody in the laboratory forgot to mention the author of the idea Rutherford always corrected him. He was also very particular not to give a beginner technically difficult research work. He reckoned that, even if a man was able, he needed some success to begin with. Otherwise he might be disappointed in his abilities, which could be disastrous for his future. Any success of a young research worker must be duly appreciated and must be duly acknowledged.

Once, in one of our outspoken talks, he told me that the most important thing a teacher must learn is not to be jealous of the successes of his pupils – which is not so easily done as the teacher gets older! This profound truth made a great impression on me. No doubt the greatest quality of a good teacher should be generosity. Rutherford was undoubtedly very generous and I think this is one of the main secrets which explains why so many first-class scientists came from his laboratory. There was always an atmosphere of freedom and efficiency there.

Rutherford well understood the importance that his pupils had for him. It was not merely that young research students increased the scientific productivity of the laboratory, but, as he said, 'My pupils keep me young'. This is very true, since pupils do not permit a teacher to lag behind new achievements in science. How often do we notice that when a scientist is ageing he starts opposing new ideas and underestimates the significance of new trends in science. Rutherford, with great ease and generosity, always accepted new ideas in physics like wave quantum mechanics, while a number of distinguished scientists of his generation were sceptical of the same ideas. Such conservatism is characteristic of scientists who work by themselves without having pupils to be directed and encouraged.

Rutherford was very sociable and loved talking to the scientists who came to visit him and the Cavendish Laboratory. Usually there were many such visitors. His attitude to other people's work was kind and considerate. In conversation Rutherford was very lively; he was fond of jokes and often made them himself. He laughed easily, his laughter was sincere, loud and infectious. His face was very expressive: you could see at once what mood he was in, good or bad, or whenever he was worried by anything. You always knew he was in a good temper when he good-naturedly teased the person he was talking to. The more he teased him, the more he liked him. This was particularly noticeable when he talked to Bohr or to Langevin to both of whom he was especially attached. His kindest jokes often concealed a deeper sense. I remember one occasion when he brought Professor Robert Millikan to my room in the laboratory. Rutherford said to me, 'Let me introduce you to Millikan; no doubt you know who he is. Show him your installation to produce strong magnetic fields and tell him about your experiments. But I doubt whether he will let you speak as he himself will tell you about his own experiments!' There followed loud laughter in which Millikan joined with rather less enthusiasm. Rutherford then left us, and I soon found out that his prophecy was correct.

I shall not describe the way in which Rutherford read his papers. I always liked them very much as regards both their content and their exposition. He attached great importance to the way in which his papers were presented and evidently prepared them very carefully. He taught me how to read papers to the Royal Society, and one of his instructions I still remember very clearly: 'Don't show too many slides. When it is dark in the lecture room some of the audience take the opportunity to leave!'

Rutherford's interests were not limited narrowly to physics; they were much wider. He was well read, he liked books on geography and history and liked to discuss what he had read. He absorbed all knowledge enthusiastically and always extracted the essentials.

Later on, when I became a Fellow of Trinity and used to accompany him home after dinner on Sundays, we often discussed politics. On the first day I started work in the Cavendish I was surprised to hear him saying to me that in no circumstances would he tolerate my making Communist propaganda in his laboratory. At this time this remark came quite unexpectedly. It not only surprised me, but also shocked me and to a certain extent even offended me. Undoubtedly it was a consequence of the current atmosphere of acute political struggle and was connected with the propaganda which existed in those days, only four years after the Russian revolution. Before coming to England, I was so absorbed by my research work in Russia that I was completely unaware of what was happening in Western Europe and could not appreciate the scale of the bitter political controversy which then existed. Later on when my first experimental research was published I presented Rutherford with a reprint and I made an inscription on it that this work was proof that I had come to his laboratory to do scientific work and not to make Communist propaganda. He got extremely angry with this inscription, swore and gave me the reprint back. I had foreseen this and I had another reprint in reserve with an extremely appropriate inscription with which I immediately presented him. Obviously Rutherford appreciated my foresight and the incident closed. Rutherford had a characteristically hot temper but cooled down just as quickly.

Eventually we had many conversations on political questions; we were especially concerned about the growth of fascism in Europe. Rutherford was an optimist and thought that all would soon be over. We now know that this was not the case. Rutherford, like most scientists who work in the exact sciences, had progressive political views. I involved Rutherford in some political activity on two occasions.

The first of these was connected with Langevin. In his younger days Rutherford had worked with Langevin in the same room at the Cavendish. A deep friendship developed between them. Indeed it was practically impossible not to be friendly with a man of such brilliant intelligence and exceptional moral qualities. In Paris my friends, pupils of Langevin, were greatly shocked that Langevin, undoubtedly the best French physicist, had not been elected to the French Academy as a result of his left wing political views. Langevin had taken part in a number of progressive organizations, had been the founder of the League of the Rights of Man (*Ligue des droits d'hommes*) and had fought anti-semitism in the Dreyfus case. I told Rutherford of the difficulties Langevin had encountered in France and asked him whether a man who held such leftist views as Langevin could be a Foreign Member of the Royal Society. Rutherford said something I could not quite follow, then started to tell me what a really good man Langevin was, and then recalled that during the war Langevin had been very active in inventing supersonic beams propagated in water by which he had established communication between England and France across the Channel. At this point the conversation ended. I learnt later that at the next election in 1928 Langevin was elected a Foreign Member of the Royal Society and this was much earlier than his election to the French Academy.

The second example occurred much later, when Hitler started to come into power. We were very anxious about the fate of such distinguished physicists as Stern, Franck, Born and a number of others in the conditions of active and increasing anti-semitism in Germany. About this time Szilard came to England and we were faced with the question of how to get these scientists out of Germany without raising suspicion. I spoke to Rutherford and he was very willing to help, writing personal letters to these scientists, and inviting them to come to lecture in Cambridge.

Rutherford took an interest in a great variety of people, but he particularly liked people with strong personalities. When Rutherford was elected President of the Royal Society and often had to attend dinner parties with distinguished politicians, businessmen and statesmen, he was fond of telling stories afterwards about the conversations he had had with them and always gave descriptions of them. I specially remember that Churchill made a great impression on Rutherford. His description of Churchill was, like all his descriptions, short and clear, and in due course I found out that it was quite correct. I well remember that Churchill in those days already regarded Hitler as a real danger to peace and called him 'a man riding a tiger'. Possibly this conversation somewhat altered Rutherford's optimistic view of the future. Rutherford's interest in understanding human psychology and his kindness to others was undoubtedly felt by them. This explains why Rutherford's excessively direct way of speaking which was sometimes not very tactful, was completely compensated for by his kindness and cordiality.

Of course Rutherford's correct evaluation of people and his understanding of them was due to the fact that he was a subtle psychologist. People interested him and he had the faculty of understanding them. His assessments of people were always very outspoken and direct. As in his scientific work, his description of a man was always brief and very accurate. I was always convinced that his descriptions were correct. Possibly his approach to people was also a subconsicious process and could be called intuitive.

I should like to illustrate his interest in psychology with the following two episodes. In Cambridge there was a small but progressive theatre which produced Chekov's play *Uncle Vanya.* Rutherford went to this play and was greatly taken with it. As in all Chekov's works, it deals with a psychological problem complicated by the fact that all the people in the play are highly intellectual and therefore their acceptance of life is very complex. In the play a certain retired professor comes to live on the estate of his wife. Uncle Vanya, who manages the estate, has devoted his whole life to supporting the professor. Soon Uncle Vanya finds that the professor is a fake celebrity, scholastic and pedantic in his work. Against a background of complex psychological situations Uncle Vanya fires a pistol at the professor but misses him. I remember how vividly, clearly and simply Rutherford told me this plot and his sympathy was completely on Uncle Vanya's side. The fact that Rutherford was so attracted by the play shows that he undoubtedly enjoyed disentangling complicated psychological cases of this kind.

A great impression was made on me by the following case which demonstrates Rutherford's skill in handling complicated psychological problems. I think enough time has now passed and I can tell you about this case which involved the then well-known physicist, Paul Ehrenfest.

Ehrenfest was born in Austria. On one of his mountaineering excursions he met a Russian woman scientist and followed her to Russia where he married her. In Russia he published a number of outstanding theoretical works on thermodynamics. Eventually he was invited to Leiden University to take the chair of theoretical physics vacated by the great Lorentz, creator of the electronic theory of metals and one of the founders of the theory of relativity. In Leiden, Ehrenfest and his house became one of the world centres of theoretical physics. Ehrenfest's main quality was his precise critical mind. He was not only a very good teacher of young scientists, who were very fond of him, but his criticisms were regarded as profound and of such high quality that leading theoretical scientists like Einstein and Bohr often came to Ehrenfest to discuss their work. Ehrenfest always noticed even the smallest contradiction or mistake. His critical remarks were made very readily, with great spirit and even sharply, but always very good-naturedly. The quality of his criticism was greatly appreciated. Despite our difference in age we became friends and I often visited his very hospitable and very charming family and more than once was present at his scientific discussions.

Ehrenfest's exceptionally strong critical mind evidently acted as a restraint on his creative imagination and he did not succeed in producing scientific work which he himself would have considered of sufficiently high standard. In those days I did not know that in his acute nervous condition Ehrenfest suffered greatly when he could not in his work attain the level of the friends he criticized. I learnt about his feelings in the following manner. In the beginning of 1933 I received a long letter from him in which he described in detail his state of mental depression and spoke of the futility of his achievements. He had come to the conclusion that it was not worth living any more. The only way to save himself, he thought, was to leave Leiden and settle somewhere away from his friends. He asked me to help him to find a chair at some small university in Canada and to ask Rutherford, who doubtless had connexions in Canada, to assist him. I was, of course, very upset by this letter. We all liked Ehrenfest and all knew that his influence as a teacher and critic in the development of modern physics was colossal. I translated the letter from German into English and came to Rutherford, who had little personal acquaintance with Ehrenfest. I handed Rutherford the letter and told him that we were very worried about Ehrenfest's future as, without any doubt, the letter showed that he was mentally unbalanced; perhaps, I said, this state was only temporary and everything possible should be done to help him out of his state of depression. Rutherford said I must not worry and he would handle the case himself. I do not know what Rutherford wrote to Ehrenfest but shortly afterwards I received a letter telling me that he was once again in a happy frame of mind. He said that Rutherford had explained what a great role he was playing in physics and he added that of course there was now no need for him to go to Canada. This story shows how skilfully Rutherford dealt with a very complicated psychological case, probably better than a professional psychiatrist.

A few months later, while I was on a visit to Russia, the state of depression returned to Ehrenfest and on 25 September 1933 he committed suicide.

I should now like to recall quite a different and rather amusing case characteristic of Rutherford's attitude to the young. Once Rutherford called me into his study and I found him reading a letter and roaring with laughter. It appeared that the letter was from some Ukrainian schoolboys. They had written to say that they had organized a physics club and were proposing to continue Rutherford's fundamental work on the study of the nucleus of the atom and ask him to be an honorary member and to send them reprints of his scientific work. In the part of the letter in which they described Rutherford's achievements in nuclear physics, instead of using the correct term in physics they used a corresponding term which in slang has a different meaning. In this way the description of the structure of the atom acquired a property of the living organism. Its character is such that one does not speak about it in polite society, and it made Rutherford laugh heartily. I explained to Rutherford that the schoolboys were apparently not very well versed in English and the writing of the letter was mostly done with the use of a dictionary and the mistake was bona fide. Rutherford said that he appreciated this. He sent the boys a reply, thanking them for the honour of being elected a member of the club and promising to send them reprints.

Finally, I should like to discuss a question I have come across several times in descriptions of Rutherford's activities. The question is: did Rutherford foresee the great practical consequences which would emerge from his scientific discoveries and investigations into radioactivity?

The immense reserves of energy which are hidden in matter was understood by physicists a long time ago. The development of his view took place side by side with the development of the theory of relativity. The question which was not clear at that time was: would it eventually be possible to find technical means of making practical use of these reserves? We know now that the actual possibility of obtaining energy from nuclear collisions was becoming more and more real as nuclear phenomena were better understood. But up to the last moment it was not certain whether it would be technically possible to produce nuclear reactions with a great yield of energy. I remember only rare occasions on which I discussed this question with Rutherford and in all these conversations he expressed no interest in it. From the beginning of my acquaintance with Rutherford I noticed that he never took any interest in technical problems and I even had the impression that he was prejudiced about applied problems. Possibly this was because such problems were connected with business interests.

I am by training a chartered engineer and naturally I always took an interest in solving technical problems. During my stay in Cambridge I was approached several times to help in solving technical problems in industry. In these cases I used to take advice from Rutherford and he always said to me; 'You cannot serve God and Mammon at the same time.' Of course he was right. Once I remember Rutherford telling me about Pupin who as an able young physicist had come to Cambridge and done successful scientific work in the Cavendish Laboratory. Pupin was somewhat senior to Rutherford so they met only occasionally. Eventually Pupin turned to commercial activity in the U.S.A. and made a lot of money. Rutherford spoke disapprovingly of Pupin's activities. So I think that Rutherford's opinions on the practical applications of nuclear physics had no real value as they lay outside the scope of his interests and tastes.

In connexion with Rutherford's views on industry I remember a conversation I had with him during a high table dinner at Trinity College. I do not remember how the conversation started, maybe it was under the influence of Lombroso's book, *Genius and Madness*. I was telling my neighbour that every great scientist must be to some extent a madman. Rutherford overheard this conversation and asked me, 'In your opinion, Kapitza, am I mad too?'

'Yes, Professor', I replied.

'How will you prove it?' he asked.

'Very simply', I replied. 'Maybe you remember a few days ago you mentioned to me that you had had a letter from the U.S.A., from a big American company. (I do not remember now which one it was, possibly General Electric Co.) In this letter they offered to build you a colossal laboratory in America and to pay you a fabulous salary. You only laughed at the offer and would not consider it seriously. I think you will agree with me that from the point of view of an ordinary man you acted like a madman!' Rutherford laughed and said that in all probability I was right.

The last time I saw Rutherford was in the autumn of 1934 when I went as usual to the Soviet Union to see my mother and my friends and unexpectedly was deprived of the possibility of returning to Cambridge. I did not hear his voice again, nor hear him laugh. For the next three years I had no laboratory to work in and was unable to continue my scientific work and the only scientist with whom I freely corresponded outside Russia was Rutherford. At least once every two months he wrote me long letters which I greatly valued.

In these letters he gave me an account of life in Cambridge, spoke about the scientific achievements of himself and his pupils, wrote about himself, made jokes, gave good advice and invariably cheered me up in my difficult position. He understood that the important thing for me was to start my scientific work which had been interrupted for several years. It is no secret that it was only due to his intervention and help that I was able to obtain the scientific installation and apparatus of the Mond Laboratory and in three years time I was able to renew my work in the domain of low-temperature physics.

I am sure that in the course of time all Rutherford's letters will be published but even so I should like here and now to quote three short extracts which require no comment.

On 21 November 1935 he wrote:

'. . . I am inclined to give you a little advice, even though it may not be necessary. I think it will be important for you to get down to work on the installation of the laboratory as soon as possible, and try to train your assistants to be useful. I think you will find many of your troubles will fall from you when you are hard at work again, and I am confident that your relations with the authorities will improve at once when they see that you are working wholeheartedly to get your show going. I would not worry too much about the attitude or opinions of individuals, provided they do not interfere with your work. I daresay you will think I do not understand the situation, but I am sure that chances of your happiness in the future depend on your keeping your nose down to the grindstone in the laboratory. Too much introspection is bad for anybody! . . .'

On 15 May 1936 he wrote:

'. . . This term I have been busier than I have ever been, but as you know my temper has improved during recent years, and I am not aware that anyone has suffered from it for the last few weeks! . . .

'. . . Get down to some research even though it may not be of an epoch-making kind as soon as you can and you will feel happier. The harder the work the less time you will have for other troubles. As you know, "a reasonable number of fleas is good for a dog" – but I expect you feel you have more than the average number!'

You see what short and clear and fatherly advice he gave me. The last letter is dated 9 October 1937. He wrote in great detail about his proposed journey to India. In the last part of the letter he said: '. . . I am glad to say that I am feeling physically pretty fit, but I wish that life was not quite so strenuous in term time . . .' Ten days before his death he did not feel that it was so near.

For me the death of Rutherford meant not only the loss of a great teacher and friend; for me, as for a number of scientists, it was also the end of a whole epoch in science.

Obviously we should attribute to those years the beginning of the new period in the history of human culture which is now called the scientific-technical revolution. One of the greatest events in this revolution has been the use of atomic energy. We all know that the consequences of this revolution may be very terrible – it may destroy mankind. In 1921 Rutherford warned me not to make any Communist propaganda in his laboratory, but it now appears that just at that time he himself together with his pupils were laying the foundations for a scientific-technical revolution.

We all hope that in the end people will have sufficient wisdom to direct this scientific revolution to the benefit of humanity.

But nevertheless the year that Rutherford died there disappeared forever the happy days of free scientific work which gave us such delight in our youth. Science has lost her freedom. Science has become a productive force. She has become rich but she has become enslaved and part of her is veiled in secrecy.

I do not know whether Rutherford would continue nowadays to joke and laugh as he used to do.

31. THE ROLE OF AN OUTSTANDING SCIENTIST IN THE DEVELOPMENT OF SCIENCE

The privilege of opening this Colloquium gives me a special pleasure for I not only admire Rutherford as a fellow scientist, for his fundamental contribution to the study of radioactivity and atomic structure, but I also have the honor of being one of his students. In the development of my academic work I owe very much to his kind regard for me. During thirteen years spent at the Cavendish Laboratory I learned a great deal from him, not only as my teacher but also as a founder and leader of one of the most outstanding schools of physics of our time.

The scientists who have gathered here today are going to present a number of interesting lectures on Rutherford. Most of the speeches will be delivered by Rutherford's collaborators who started their academic careers as I myself did at the Cavendish; we shall be listening to Allibone, Feather, Lewis, Devons, and Shoenberg. Quite a few of us are still around, and it is a great pity that neither Blackett nor Chadwick, Oliphant or Ellis could make their appearances at this Colloquium. They will take part in anniversary celebrations in England, held by the Royal Society in London and in Cambridge in October.

Opening today's session I do not intend to talk about Rutherford as a scientist or Rutherford as a teacher, but would rather draw upon his example in the light of a more general problem – that of the role of a great creative mind in the development of science.

This problem has been discussed many times, as it is by far one of the most crucial in the field of the organization of scientific research. In a somewhat simplified manner it may be stated as follows: science is the domain of human inquiry into the laws of nature; these laws are unique, hence the course of scientific development is predetermined and it is absolutely beyond individual capability to interfere with it. Therefore, the genius of Rutherford or any other great scientist would not suffice to change the course of scientific development in any significant way. But, if that be so, can we not then replace a genius with a team of less gifted people whose academic promise rests

---

A Lecture Delivered at the International Colloquium Commemorating the Centenary of the Birth of Rutherford, August 20, 1971.

entirely on proper scientific management, in other words, can we not replace quality with quantity? This proposal has always been maintained because of its alleged relative practical ease and safety as opposed to the troubles one usually has with geniuses who do not tend to be particularly easygoing.

I have heard this opinion from different public figures in position of high authority. There is some truth to it, since well-organized institutions do indeed contribute to the advancement of science, but I doubt that research institutions could function purposefully without major executives or leading scientists. For example, as we all know from history, troops without a good commander never win. The question for us to consider: is it at all possible for an army of scientists to succeed in conquering nature without its own great commander?

It is well known that in science we proceed by finding new natural phenomena and disclosing the laws that govern them. Most commonly this happens as a result of new findings in research methodology. Creation of something new, not hitherto existing, we attribute to human creativity, which is considered to be the highest human spiritual activity. Creativity is what defines the genius of a person, whether he is a scientist or writer, an artist, a musician or even a commander or statesman.

The importance of creativity is well illustrated by the following example, which, as far as I know, was also suggested by James Jeans a long time ago. Let x be a number of typewriters and let each typewriter be assigned a typist monkey, who can only strike keys – in other words, a creature with no literary aptitude whatsoever. The question is: how many of those monkey-typewriters would it take to arrive accidentally at the text of *Hamlet*, one of the most remarkable masterpieces of world literature, created by the genius of Shakespeare? The problem is easy. Here is the solution. Suppose, if we use all the keys and shifts of a modern typewriter, coincidence of the first letter would require 100 independent strikes by the monkeys and an equal number of typewriters. Then, in order for n initial letters to coincide with the text of *Hamlet*, the number of monkeys must be $x = 100^n = 10^{2n}$. The result is rather appalling. In order to have 40 initial letters, i.e., less than the first line of *Hamlet*, the number of monkeys x must be approximately 10^{80}, and that is the number of atoms in the entire universe as it appears to contemporary astronomers.

Of course this is not a complete solution yet – what is typed by the monkeys is still to be deciphered and identified as the desired text. Comprehension of a typewritten text would take a human mind, if not creative, then at least capable of critical evaluation of the text's literary merit.

This example is not realistic of course: nobody has yet suggested staffing research institutions with monkeys. However, it is a good example for illustrating the necessity of a careful search for creative research fellows, since any dubious endeavor in science almost immediately, as in this monkey case, diminishes the probability of a successful outcome to zero.

If we could assess human creativity numerically, that would solve an important problem of the organization of scientific research, namely, how to predict the solvability of a particular scientific problem in terms of the creative potential of selected personnel. To our regret we cannot yet solve this problem numerically, but our everyday experience proves that the efficiency of a research institution depends entirely on its creative potential. Mathematically speaking, this relation is represented as in the monkey case, by an exponential function, collective creative potential serving as its index of power. This index is powerful and therefore even a small decrease causes a dramatic impoverishment of the creative activity of the whole institution. But the reverse is also true – with the appearance of even one great scientist, the efficiency of the whole institution rapidly increases.

Indeed the history of science demonstrates how remarkably the development of science is fostered by well-matched schools of scientific research workers usually founded by great scientists. A striking example of such a school is the one founded by Rutherford at the Cavendish Laboratory.

In considering the efficiency of research activities, one should not forget another factor crucial to the successful creative activity of the people involved in science as well as the arts – healthy public opinion on creative achievements. In Jeans' problem it corresponds to the requirement for competent people who could select monkey-written tests according to their literary merit. Hence efficient research both in science and the arts is virtually impossible without broad cultural public involvement (response). In this connection I would like to invoke one particularly striking historical example – unprecedented achievement of human creativity during the Italian Renaissance, though not in science but in art. Art historians have long been confronted with the question: how did Italy, then a small country, produce a series of such outstanding artists as Raphael, Titian, Michelangelo, Leonardo, Correggio, Botticelli, Tintoretto and others during a rather short period of time? The following five centuries have never seen such a galaxy of geniuses. So the question is: was it a mere coincidence or was it a historical law? I think that Taine in his remarkable work, *Philosophy of Art*, gives a meaningful explanation of the sudden emergence of this galaxy of geniuses. He shows that during the Renaissance creative talents could develop so successfully because of the contemporary public attitude towards art. In economically flourishing Italy certain historical circumstances led to (or provided for) broad public acclaim and appreciation for art and support for its most talented representatives.

Accordingly a galaxy of such outstanding physicists as Maxwell, Rayleigh, Thomson and Rutherford, who succeeded one another in the post as director of the Cavendish Laboratory at Cambridge, could not have come into being had it not been for the cultured scientific milieu which properly evaluated and supported scientific activities.

Historical experience proves that the number of people endowed with enough creativity to exercise considerable influence on the development of science or art is extremely limited. This is obvious from considering the ratio of the volume of all published works to the volume of works that actually leave their impact on the development of science. The same is true with respect to pictures that can be called masterpieces. Marx explained the extremely high cost of masterpieces created by great artists by the fact that those costs included expenses related to the abundance of pictures of no artistic value. There is a similar rigorous selection of worthy works in literature and music.

Apparently in order to foster the development of science and art, a country must possess a great variety of scientific and artistic works from which that small portion that alone serves the advancement of science and artistic culture can be chosen. It is for the sake of this selection that healthy public opinion capable of just and sound judgement on the best works has to exist.

Thus, healthy organization of scientific research in a country is guaranteed not only by favorable conditions for scientific works, but also by providing the conditions for its proper assessment. At the present time this purpose is best served in many countries by special public organizations such as academies of sciences, scientific societies and councils, etc. Because of the international importance of science a more objective evaluation has become possible by way of creating international public opinion. This is achieved through the broad collaboration of scientists at symposiums, congresses, through the translation of scientific articles into foreign languages, etc.

With current growth of the role of science in the technological, economical and cultural development of a country, scientific research work consumes a considerable portion of state funds, and the efficient organization of scientific research has become an urgent task of state importance.

Organization of scientific research cannot be left to spontaneous development; we have to study the regularities of collective scientific work and be able to select gifted people. And this must be based on the study of activities of great scientists and scientific managers, one of whom was Rutherford.

The most complicated and challenging tasks in the organization of scientific research are the selection of truly gifted youth and the establishment of conditions in which their talent can be developed to its fullest potential. This requires the ability to assess the creative potential of the young, when they are just starting their academic careers. The major common mistake here is to take their comprehension and erudition for creative capability.

One episode in Rutherford's biography is particularly instructive. When he was still a young scientist in New Zealand, the university got a fellowship for one of its alumni to go to Cambridge to continue academic work. I cannot recall who was selected as the first candidate, but Rutherford was only the second. It is known that only by mere chance the first candidate did not go and Rutherford took his place. We know from the history of science that such mistakes are not unusual in selection processes and are caused most often by the inability to assess the creative capabilities of a young scientist and by overrating his ability to learn factual material.

From this point of view the study of the early works of such a great scientist as Rutherford is of tremendous interest as it reveals the evolution of his creative genius. These works are by now almost forgotten because their methodology is out of date and empirical data is now much more precise. But what an important source it is for tracing the development of Rutherford's creative genius!

We see from these works that from the very beginning of his career Rutherford could not be regarded as one of the scientists of great erudition. But his creative imagination, daring scientific hypothesizing, and his intuition were major factors that determined the success of his academic pursuits.

Of course we all know it very well now because of Rutherford's fundamental discoveries. The major challenge to scientific managers is to discern talent in such scientists as Rutherford while they are still young.

There is presently rather little interest in the original works, in the great classics of science. Usually people learn about their achievements from textbooks, monographs and encyclopedias. This is of course quite justified from an academic point of view, but for a scientist who has to become a leader of the young, the scientific manager of collaborative research, the major factor determining success of his activities will be personnel selection on the merit of creativity. One of the most effective ways of learning how to assess the creative potential of youth is the analysis of original works by great scientists.

This is not to be neglected. I personally learned a lot from the works of such scientists as Maxwell, Rayleigh, Curie and Lebedev; besides, it gives aesthetic pleasure. Manifestations of human creative genius are always beautiful and one cannot help admiring them!

My own experience proves that it is through the assessment of the creative capabilities of young scientists that the major talent of a manager of an academic institution reveals itself. Without it a scientist cannot select a strong scientific collective environment for his school.

Rutherford was undoubtedly one of the most gifted scientific managers and his major talent was his ability to select scientists according to their creative capabilities. Rutherford could also properly assess the character of a scientist's talent, which is also very important for the successful development of creative genius.

Answering our initial question about the role of an individual in the development of science and summarizing our argument, we come to the conclusion that, although the course of scientific development is predetermined, it is sustained only by the work of a very few especially gifted individuals. It is the quality of selection of scientists of tremendous creative ingenuity that is a major factor guaranteeing the high level of scientific development. It is very important for successful scientific development to create favorable conditions for the advancement of scientists' natural talents; for this purpose creative work must be made attractive. This is the responsibility of public organizations, which, while adequately assessing scientific achievements, would also let the scientists feel how important and useful their activity is to mankind. In science, public opinion must be mobilized on the international level, since scientific achievements belong to all mankind.

People like Rutherford cease to be the national pride of the countries where they were born and where they worked – they become the pride of the whole of mankind.

PART SIX

32. LOMONOSOV AND WORLD SCIENCE

FOREWORD

Lomonosov has always been regarded in his country as the founder of Russian science. In the Soviet Union his work and biography are taught in the schools. Outside his country Lomonosov is less known; for the foreign readers of my address 'Lomonosov and World Science' I shall therefore dwell briefly on this great man.

Lomonosov lived betwen 1711 and 1765. His work took place immediately after the reign of Peter the Great (1672–1725), the greatest reformer of Russia, the tsar who decisively steered the cultural development of his country along European channels. That was a difficult task indeed after several centuries of Tatar rule over Russia prior to the days of Peter the Great. The process was given a powerful impetus by the Tsar, and Lomonosov is rightly regarded as the founder of all major branches of Russian science. Lomonosov's scientific and public activity was extensive indeed. To provide but a general idea of its scope I shall mention some of the milestones of his endeavors: Lomonosov was the founder of higher education in Russia, and he created the Moscow University. The Russian Academy of Sciences was founded by a royal project prepared by Peter I; initially, however, it was composed entirely of foreign scientists (including such illustrious names as Bernoulli and Euler). Lomonosov was the first major Russian scientist to join the Academy of Sciences, and he contributed a great deal to its growth and development.

Lomonosov devoted particular attention to metallurgy, geology, geography, and the study of Russian natural resources. He made numerous important original contributions in these fields. Lomonosov also showed interest in Russian history and language, the vocabulary of which at the time

---

[On the occasion of the 200th anniversary of the death of M. V. Lomonosov, we are printing the text of a lecture presented on Nov. 17, 1961 at the general meeting of the Division of Physico-Mathematical Sciences of the U.S.S.R. Academy of Sciences, devoted to the 250th anniversary of the birth of M. V. Lomonosov. – Ed., *Usp. Fiz, Nauk.*]

[From *Sov. Phys. Uspekhi* 8, 720–728 (1966), with a new foreword by the author, and some modification of the English translation.]

was not rich enough for the development of literature and poetry. Apart from his extensive scientific and public activities, Lomonosov was also a poet and painter. His odes and verses are still being read with delight. He also developed a mosaic technique (including a glass foundry, and smalt preparation), and used to make mosaics, mostly portraits which still adorn the walls of the Academy of Sciences.

This vast scientific and public activity is well known. Until recently, however, little has been said about Lomonosov's scientific research, his greatest passion. Lomonosov the researcher was essentially an experimentalist who devoted particular attention to chemistry and physics. But these works remained largely forgotten until Professor B. N. Menshutkin's studies in the early twentieth century. It became evident that Lomonosov was engaged in immense original work in physics and chemistry, work correctly designed and often anticipating contemporary Western science. This is evident, for example, from his comprehensive theory of the kinetic concept of heat and the law of the conservation of matter, which he discovered 17 years before Lavoisier. Lomonosov's works in astronomy and atmospheric electricity are remarkable original contributions. Today there is no doubt that in terms of the scope and achievements of his scientific work Lomonosov can be placed together with the greatest scientists of his time, such as Newton, Galileo, Copernicus, or Leibniz. But the history of world science shows practically no trace or influence of Lomonosov. This discrepancy is the subject of my address.

Apart from his scientific and public activity, Lomonosov had a romantic, fascinating personality.

Lomonosov was born in the family of a well-to-do farmer-fisherman on the White Sea coast, far from Moscow. He worked with his father from childhood and was expected to continue in the family tradition. But at the age of 19, the youth became fired with a passion for science; penniless and against his father's will Lomonosov went on foot from Arkhangel'sk to Moscow (a thousand kilometers!), entered a church school and half-starved his way to graduation, whereupon he managed to join the Academy of Sciences, which sent him to Germany for university studies. After returning from Germany he became a professor of chemistry at the Academy of Sciences, where all his work took place. These few details illustrate quite clearly Lomonosov's ambition, temperament and will. In addition, he was endowed with a great physical strength and courage. As the story goes, once he took care of three bandits who tried to rob him in the woods near St. Petersburg. Two of the bandits fled, but he took the clothes from the third with him as a trophy. His temperament and the heated manner in which he strove to achieve his aims resulted in personal disputes and enmities. Lomonosov fought hard against religious prejudices which he regarded as an impediment to the cultural growth of the country. He entered into major conflicts with the leading clergy. On the other hand, Lomonosov's talent, honesty and devotion to science and knowledge attracted the progressive people of the time.

One hardly needs a vivid imagination to visualize Lomonosov's remarkable portrait on the basis of this all too brief outline, the portrait of the founder of Russian scientific culture, a fascinating man highly respected in our country to the present day.

LOMONOSOV AND WORLD SCIENCE

It is a pleasure to speak of Lomonosov, as it is a pleasure to commune with one of the original geniuses in the history of human culture. It is difficult to speak now of Lomonosov, since we are all well acquainted from grade school with his image and his works. Here it is hard to say anything new, since the life and works of Lomonosov have already been studied and discussed on all sides for 200 years. Our greatest writers, publicists, scientists, and statesmen have spoken and written about Lomonosov: Radishchev, Pushkin, Belinskiĭ, Dobrolyubov, Chernyshevskiĭ, Herzen, Pisarev, Aksakov, Menshutkin, Plekhanov, Walden, Vavilov, Fersman, Komarov, and many, many others. While some aspects of Lomonosov's work have been criticized, still, all without exception have spoken of him with great reverence, and have acknowledged his colossal influence on the development of our national culture: language, literature, education, technology, and science. The great progressive significance of Lomonosov was recognized both in the pre-revolutionary times and now. Even in the last century, without fail they solemnly celebrated the festival dates of his birth and death. In our time, these celebrations are coming to have an ever larger, international scope.

The first monument to Lomonosov was erected at his birthplace in Archangel; it was made by our very great sculptor Martos. The subscription was begun in 1825, and the monument was unveiled in four years.

In 1865 (the centenary of Lomonosov's death), the Academy of Sciences instituted a yearly prize of 1,000 rubles in his name. This prize was awarded alternately in the humanities and in the natural sciences. In our time the Academy of Sciences has also instituted a prize and medal in Lomonosov's name.

The only thing left undone for 200 years was the publication of M. V. Lomonosov's complete collected works. This has been carried out in the last few years.

Few of our scientists or statesmen furnish such rich biographic and historiographic material as Lomonosov. When one becomes acquainted with this material, one cannot but regret that no good portrait of Lomonosov has come down to us. The portraits and engravings that are usually reproduced were done posthumously, and are copies of one particular original painted by an unknown and untalented artist. Only the bust done by Shubin, who personally knew Lomonosov, gives us his living and spirited image.

In studying the material on Lomonosov, one is most deeply disappointed by the fact that none of our great writers have ever drawn a picture of him as a person. Of course, even among major scientists there are many in the world whose interests are bounded by the walls of their laboratories. Usually the picture of such scientists as persons is not very interesting. However, when the work of a major scientist and great original personage such as Lomonosov encompasses the development of the culture of an entire country, and moreover, in one of the most interesting moments of its history, a living picture of him becomes of great general human interest.

The greater the man, the more contradictions he has in himself, and the more contradictions in the problems that life puts in his way. The scope of the contradictions is a measure of the genius of a man. Both the contradictions in Lomonosov's own character and those under which he lived were exceedingly great.

It is hard to find a greater contradiction than in the fate of the 'peasant from Archangel,' who lived and worked among the top ranks of the official and noble classes of the court. Lomonosov was a progressive statesman, who lived and worked among the top ranks of the official and noble classes of the court. Lomonosov was a progressive statesman, who saw the necessity of public education and science, and combatted superstition and prejudice, but in order to carry out his work, he had to rely on the high noblemen of the court. In spite of his peasant origin, he understood the necessity of flattery and praise of the ruling potentates, and handled this problem in his own way. With the brilliance of his personal qualities, he gained the friendship and protection of the most influential noblemen of that time, Shuvalov, Vorontsov, and Orlov.

When Peter 'opened a window' into Europe, the wind blew in more to us from the West than culture and science. Along with the true scientists such as Euler and Bernoulli, who brought us the most advanced Western science, the wind brought us a large number of foreign scientists who were mediocrities or even adventurers. The latter were interested only in material benefits and in keeping their privileged position in Russia, which gave them the opportunity to enrich themselves with ease. Naturally, they hindered the growth of Russian influence in the Academy of Sciences. It is well known how Lomonosov, while relying on the authority of the foreign scientists, had to combat the predominance of the foreigners. With his acute mind, Lomonosov made an excellent estimate of the complexity of the conditions under which he worked. It demanded great endurance and tact on his part, but this was counter to his irrepressible temperament and impassioned character. Here arose the sharp conflicts well known from Lomonosov's biography. In the end, Lomonosov's genius was still able to win this complex battle, but no one has yet drawn a good picture of this battle.

Lomonosov understood the great value of the development of science in Russia and the necessity of improving higher education. He put much work into founding the university in Moscow. He attracted the youth into scientific work, but could not himself devote as much time to scientific work as he would have liked. Apparently he was not a teacher by nature. His extreme individualism did not make a disciplined teacher of him. Consequently it happened that, while he had put so much of his effort into disseminating science in Russia, nevertheless he left no students after him. Menshutkin, the greatest expert on Lomonosov's scientific work, says that "he founded no school, and among his students after his death, the only one going into science was W. Ya. Rumovskiĭ," who was later professor of astronomy of the Academy of Sciences.

We could prolong the list of contradictions in Lomonosov's life, but to draw a living image of Lomonosov, who combined all these contradictions within himself, is a problem awaiting a major writer.

Now I would like to spend some time on one of the contradictions in Lomonosov's life. Although it is well known, it hasn't yet been properly explained. I think that it is important to us even now.

Many times Lomonosov said that his work as a poet and writer, as reformer of the Russian language, historian, statesman, geologist, or administrator was of little satisfaction to him, and that he saw his main calling in scientific work in physics and chemistry. It would seem that scientific work in chemistry and physics must have been his major activity, since he occupied the position of adjunct in physics from the very beginning of his membership in the Academy of Sciences in 1741, and he was appointed professor of chemistry within four years. One might naturally assume that under these conditions Lomonosov's genius must have left a very strong imprint, both in our national and in world science. However, we know that this did not happen, and repeatedly this has puzzled many who have studied the history of science. In a speech given before the Academy of Sciences at the Lomonosov festival in 1911, Academician P. I. Walden took up this problem in detail, and pointed out the "tragedy in the fate of Lomonosov's scientific works, which left no visible traces in chemistry and physics." Walden gave a series of data confirming the ignorance of foreign historians on Lomonosov's scientific work.

One does not find Lomonosov's name at all in Heller's (1889) or in Rosenberger's (1882–1890) detailed history of physics. The French historian of chemistry, F. Hoefer (1860), wrote only a few lines about him, which are not devoid of curiosity. I repeat them verbatim: "Parmi les chimistes russes qui se sont fait connaître comme chimistes, nous citerons Michel Lomonosov, qu'il ne faut pas confondre avec le poète ce nom." ("Among the Russian chemists who have become known as chemists, we mention Mikhail Lomonosov, who must not be confused with the poet of this name.")

However, while in the West, Lomonosov's scientific work as a physicist and chemist was hardly known, even among us it remained either unknown or forgotten until very recently. Among all the extensive material dealing with Lomonosov up to the beginning of this century, there are only two festival articles on Lomonosov as a physicist, both printed in 1865. One is by N. A. Lyubimov, giving an untalented rehash of several of Lomonosov's works. The other, consisting of only five pages, is by N. P. Beketov. In both of the great Russian encyclopedias, both Brockhaus and Granat, just as in the Encyclopedia Britannica and in the French Larousse, they say nothing about Lomonosov's achievements as a physicist and chemist. Up to the appearance of Menshutkin's studies, there was not a single reference to Lomonosov, even in our basic physics text by O. D. Khvol'son with its meticulous citation of the literature.

On the other hand, in discussing Lomonosov's work, A. S. Pushkin said in his notes, *Puteshestvie iz Moskvy v Peterburg* (Journey from Moscow to St. Petersburg) (1834): "Lomonosov himself did not value his poetry, and concerned himself much more with his chemical experiments than odes on a high official Saint's day." Pushkin spoke of Lomonosov as a great man of science; his striking words made history: "He himself, rather, was our first university." Pushkin saw Lomonosov's genius as a scientist. Pushkin's opinion is very important to us as that of one of the persons most educated and deeply understanding of Russian life. Besides, Pushkin could still meet people who had seen and heard Lomonosov alive. Thus, even his contemporaries recognized Lomonosov to be a great scientist. However, it is characteristic that none of them surrounding him could describe exactly what Lomonosov did in science, beyond the fact that one had to consider him a great scientist.

Thus it continued until the beginning of our century, when the professor of physical chemistry Boris Nikolaevich Menshutkin began as a scientist to study Lomonosov's original scientific works in chemistry and physics. Menshutkin translated Lomonosov's works from the Latin and German, and critically studied not only Lomonosov's major works, but also his correspondence and personal notes. Beginning in 1904, Menshutkin systematically published this material. Later this work was continued by S. I. Vavilov, T. P. Kravets, and a number of other scientists. Thus we found out only after 200 years on what and how Lomonosov worked. Now, through knowing the path along which science has developed since Lomonosov, we can correctly evaluate his scientific work in chemistry and physics. Thus, only now has it been ascertained that Lomonosov's scientific work was most advanced for his time, and undoubtedly should have left a deep imprint in the development of world science. The intuitive estimate that Pushkin made more than a hundred years ago that Lomonosov was a great scientist was correct. All this makes us wonder more than ever: how could it happen that all this scientific work of Lomonosov took place without leaving any trace, not only abroad, but in our country? I must speak of this with sorrow, since both our science and world science suffered a considerable loss thereby. Of course, such an isolation of Lomonosov's scientific work from world science could not have happened by chance, but had its historic causes. I think that we have had many such cases in which the discoveries and achievements of Russian scientists did not exert a due influence on the development of world science. Hence, the contradictions between Lomonosov's very great achievements in science and the lack of due influence on the development of world science are of interest even today. I shall take up this problem in further detail.

In order to analyze the relation between Lomonosov's work as a scientist and the science of his period, I must draw a picture, though in very general outline, of the conditions under which the natural sciences were developing in the first half of the 18th century. I might mention that in the history of human culture, only the 16th century can be considered to begin the intensive growth of the natural sciences. Before that time, mankind also knew of great scientists, for instance Pythagoras, Archimedes, and Avicenna, but they were solitary creative geniuses. Science developed slowly then. Only since the 16th century has science begun to grow at an increasing rate, because scientific work became a collective creation of men carried out on an international scale. The first great advances in this collective creation by scientists are well known: they were the rapid growth of astronomy and mechanics. The participants were the Pole Copernicus, the Dane Tycho Brahe, the German Kepler, the Italian Galileo, the Englishman Newton, the Frenchman Descartes, the Dutchman Huygens, and many, many other less-known scientists.

Also today, the collective labor of scientists on an international scale is a fundamental factor ensuring the rapid growth of science. It has been made possible not only by the growth in the material welfare of people and the development of means of communication between countries, but mainly, by the invention of printing in the 15th century. All scientists know well that even today without books one can neither publish nor preserve scientific experience and scientific advances. Without this, of course, science cannot fully develop. At the same time, science became separated from the church, as was necessary for science to develop on a sound materialistic basis.

Since that time, the leading statesmen began to understand the importance of the development of the natural sciences for the growth of human culture. As early as the beginning of the 17th century, Francis Bacon clearly formulated the vast importance of the experimental study of nature and the inductive method of generalizing these experiments, leading to a knowledge of the laws of nature. Bacon, an important statesman who attained the post of Lord Chancellor, was convicted of bribery in 1621. He spent the end of his life in semi-exile, where he wrote the philosophical works that immortalized his name. Thus, dishonor during his life turned to glory after his death. In Bacon's writings, in an unfinished work that he called the *New Atlantis*, he revived the story of Plato's Atlantis. This island lives and is guided by scientists. We can find in the description of the island both scientific institutes and other aspects of the organization of scientific life recalling our own governmental organization of science.

Bacon gave the meaning of science as a powerful force directing the growth of the country's culture along the right path in the following beautiful image, in which science is contrasted with empiricism: "A limping cripple traveling along the right road can outrun a racehorse if the latter takes the wrong path. Furthermore, the faster the racer runs, once he has lost the way, the farther the cripple will leave him behind." Also Bacon proclaimed physics as the "mother of all sciences," which first points out the path of development of human culture. I am giving this description in so much detail because Bacon was widely read in those times, and his *New Atlantis* was put out in many editions. His views were propagated in the ruling circles of the advanced countries, and at this time the development of science began to be considered to be a governmental concern. At the same time, scientific work had expanded so much that the need arose for coordinated work. Therefore, as early as the 17th century, they began to establish academies of science in many countries, or analogous scientific societies. Periodical scientific journals and memoirs began to be printed.

When Peter I visited Europe, he quickly grasped the significance of science in the development of a country, and of course, he could not but understand that Russia also needed science in order to become an advanced, cultured country. Here occurred the famous conversations of Peter with Leibniz on this subject, and the idea arose of founding an Academy of Sciences in Russia. Our Academy was estblished in 1725 in the reign of Catherine I, just after the death of Peter. As is well known, the Academy was made up of foreigners in order that they should train Russian scientists. We know that Lomonosov was fortunate – he arrived in St. Petersburg at the right time to become one of the first Russian students in the Academy of Sciences. Of course, however, the Academy of Sciences was even more fortunate in that their first Russian student was Lomonosov. He got his higher scientific traning in Germany, where he studied for five years, mainly under Professor Christian Wolff. In 1741, Lomonosov returned to St. Petersburg, where he had to begin his scientific work under very unfavourable conditions.

Around that time, the Academy of Sciences already had existed for almost twenty years, Anna was reigning, Biron ruled, and Peter's idea of the development of Russia's own science began to take second place. When the Academy of Sciences was founded, among invited foreigners there were only two real major scientists, both of whom had become famous. They were Leonhard Euler and Daniel Bernoulli. However, attention to them had continued to decline, and in 1741, when Lomonosov returned from Germany to St. Petersburg, both of them had already left the Academy, first Bernoulli and then Euler. Interestingly, Euler had left St. Petersburg three days before Lomonosov returned from Germany, and he returned to St. Petersburg only in the reign of Catherine II, when attention to scientists had begun to increase again. However, this was a year after Lomonosov's death. Thus while Lomonosov corresponded much with Euler, they never met personally, unless we count Lomonosov's possible attendance at Euler's lectures before he left for Germany.

Thus, in the Academy of Sciences in the field of his studies in physics and chemistry, Lomonsov was left in almost complete isolation. He had to follow the development of science in the literature, which was scanty then. He had no personal contact with important scientists, since from the time Lomonosov became a scientist, he never went abroad, while foreign scientists never came to St. Petersburg to confer with him, for the Academy of Sciences of that time was of no interest.

In spite of this isolation from world science, Lomonosov was still able to concentrate his studies on the most pressing problems of chemistry and physics of that time. As a scientist, he combined the thinker and the experimenter within himself. His remarks on the relation between theory and experiment are of interest, and are quite to the point even today: "some theoreticians, who without any preliminary experiments, abuse their leisure to dream up an empty and false theory and burden the literature with it . . .".

Lomonosov put experiment in first place in the study of nature. This was his characteristic trait as a scientist. Therefore he put much effort into establishing a laboratory, and worked there assiduously. However, the milieu of that time valued Lomonosov little as a scientist, but much more as a poet. For one of his eulogistic odes, Lomonosov received 2000 rubles from the Czarina. This was more than his salary for three years in the Academy of Sciences (660 rubles per year).

They also valued Lomonosov as a historian, as the founder of the literary Russian language, for his grammar, for his translations, and they valued him as a statesman concerned with the development of education and technology in Russia.

The meaning of his scientific enterprises in the laboratory was incomprehensible to the officials and the court. In order to justify his laboratory enterprises, Lomonosov wrote in 1753 to Count Shuvalov: "I suppose that I might be permitted to spend several hours a day on physical and chemical experiments instead of billiards . . . ". Thus, Lomonosov had to justify his scientific work in that he spent leisure time on it instead of games and billiards. Of course, the justification of the expense of government funds on the laboratory were the practical results, for instance, the production of mosaic glass and the solution of various technical problems.

It is amazing how much Lomonosov did in the field of experimental basic science, in spite of these unfavorable conditions. First, he very widely covered various fields of physics in his studies. He studied the liquid, solid, and gaseous states of substances. He meticulously developed thermometry, and precisely calibrated his own mercury thermometers. Using them, for example, he determined the expansion coefficient of gases upon heating with a remarkable accuracy for his time. If we compare his data with the modern values, we find that he made an error of less than 3%, which was 10 times more precise than the value accepted then. This shows Lomonosov's extremely good technique as an experimenter. It would take too much time to enumerate the rest of Lomonosov's achievements in the field of experimental physics and chemistry, which were performed on the same high level, and this is not our topic. Those interested in this problem can read B. N. Menshutkin's excellent monograph on Lomonosov's work in physics and chemistry, which was published in 1947.

Undoubtedly, these studies by Lomonosov must even in themselves establish him in the ranks of the most important experimentalists of that time. Interestingly, Lomonosov's electrical experiments, in which he further developed Franklin's work, are better known, not for their scientific results, but for the fact that they resulted in the death of Richman, who was killed by a lightning discharge. These studies led Lomonosov to propose an interesting hypothesis on the nature of the electric discharge in clouds.

He also carried out a series of optical studies, which amounted to constructing more refined optical instruments, e.g., a reflector telescope, with which Lomonosov observed a rare phenomenon in 1761, the transit of Venus across the Sun's disk. These observations were also a major contribution to science. He observed the deformation and blurring of the edges of the disk of Venus. Thus he was the first to show that Venus must have an atmosphere. It is interesting to note that modern astronomical textbooks say that such a demonstration was made only in 1882, i.e., 121 years later,when Venus again transited the Sun's disk.

Lomonosov's most important achievement was the experimental proof of the 'law of conservation of matter.' Lomonosov's discovery of the law of conservation of matter is now well studied, and it has been thoroughly established beyond a doubt that Lomonosov was the first to discover it. In 1756 he perfomed a classical experiment, in which he showed that lead plates are oxidized upon heating in a sealed vessel, but without change in the total weight of the vessel. Lomonosov's experiment was analogous to Lavoisier's famous experiment, but Lavoisier's came 17 years later. I shall not repeat in detail this entire story, for most of you know it all. Undoubtedly, this discovery of one of the most fundamental laws of nature should have established the name of Lomonosov in the history of science among the greatest world scientists.

However, all of these studies of Lomonosov were not only poorly known abroad, but most of them were not known to us until the cited studies of

Menshutkin. Evidently, under these conditions, Lomonosov's work in physics and chemistry could not exert its due influence on the growth of either world science, or our own.

Why did this happen?

It would seem that the first reason why Lomonosov's work was poorly known abroad might be that he put no stock in the priority of his discoveries, and did not publish enough. People put no less stock in priority in scientific work then than now. It suffices to recall the dispute on the priority of discovery of the differential calculus between Newton and Leibniz, which turned into a major diplomatic incident. Here Leibniz's career suffered greatly.

The material that has come down to us shows that Lomonosov too put stock in priority, and hence published his studies either in Latin or in German. He had an excellent command of both languages. An indication that Lomonosov was concerned that his scientific work should be known abroad is the following. In 1753, when Richman was killed by lightning, the general meeting of the Academy of Sciences was adjourned, but Lomonosov asked for an opportunity to give his lecture on electricity, "before it lost its novelty." Therefore, the president of the Academy of Sciences, Count Razumovskiĭ, ordered a bill passed on the coronation festival: "that Mr. Lomonosov's new works should not be late in reaching the scientists of Europe, and thereby that his work should not be missing among the electrical experiments performed up to now." Lomonosov's lecture was sent out thereafter to many foreign scientists. It is also known that Lomonosov wrote about his experiments to Euler and a number of other scientists. We should note that in those days they considered personal correspondence between scientists to be one of the most effective methods of spreading scientific information, and used it widely. Thus, there are no grounds to think that scientists could not have known of Lomonosov's work, either here or abroad. They knew of them, but did not pay due attention.

Some of Lomonosov's biographers have advanced the hypothesis that the lack of attention to Lomonosov's studies arose from the fact that his ideas were extremely advanced. I think that his ideas were extremely advanced. I think that this hypothesis is also ungrounded. Actually, Lomonosov's lively and bold mind encompassed almost all fields of natural science found in the circle of interests of the 'natural philosophers' of that time. In breadth of grasp, it is hard to name another scientist contemporary to Lomonosov with such varied interests and knowledge. Lomonosov's theoretical ideas in the fields of science immediately where he conducted his experimental work (the study of heat, the states of matter, and chemistry) impress one with their coincidence, even in detail, with the pathway along which these fields developed after Lomonosov, and are developing even today. The fact that the kinetic nature of heat was quite evident to Lomonosov is very striking to the modern reader. He graphically associated the heating of a body with an increase in the translational and circular (rotational) motion of atoms and molecules, which of course he called by other names. The false notion of the existence of 'caloric' prevailed in physics then. While these views of Lomonosov were advanced, he was not their lone proponent, since, e.g., Bernoulli also shared them. Lomonosov developed these views extremely systematically and logically. For example, he came very close to the concept of absolute zero. In the discussion 'on heat and cold' in Sec. 26, he speaks of the "highest possible degree of cold, involving complete rest of the particles and cessation of any motion in them."

The following curious fact may serve to illustrate Lomonosov's conviction of the correctness of his conception of the physical nature of heat. In 1761, Lomonosov wrote the note, 'On the Multiplication and Preservation of the Russian People.' In this note he discussed the various causes for the high mortality in Russia, and proposed a series of measures for combating it. Thus, in Sec. 7 he writes that children should always be baptized in warm water: "the priests carry out the prescriptions of the Prayer Book that the water should be natural without adulteration, and take heat to be an added material, without thinking that in summer they themselves perform baptism in warm water, which in their opinion is adulterated, and are thus contradicting themselves; and especially, in their thoughtlessness, they do not know that even the coldest water still contains very much heat. However, there is no need to explain physics to the ignorant priests."

Interestingly, this note was never published in Tsarist times, since the thoughts expressed in it were extremely revolutionary.

Lomonosov's ideas guiding his studies in the field of chemistry were also perfectly correct and advanced. They were always based on atomic concepts, and approached the idea of the molecular structure of chemical compounds. He considered it necessary to use the quantitative method in scientific studies in chemistry. He developed precise methods of weighing. He considered it important to use as pure reagents as possible. This is the quantitative approach to studying chemical reactions that led him to the necessity of proving experimentally the law of conservation of matter. All this gives us complete reason to consider Lomonosov to be the founder of the introduction of physical research methods into chemistry in the sense that it existed in the 18th century.

Lomonosov did less in the field of electricity. He knew of the experiments of his contemporary Franklin, and he repeated them, but Lomonosov directed his major interest to problems concerning atmospheric electricity. He related its origin to the rising and falling air currents that always accompany thunderclouds. This opinion is considered correct even today, but the very mechanism of the origin of the charge of a cloud proves to be so poorly amenable to research that it has not been settled finally even now.

In the field of wave optics, Lomonosov, along with Euler, correctly supported the wave theory of light proposed by Huygens, whereas the authority of Newton, who stubbornly insisted on his erroneous corpuscular theory of light, stood in the way of its acceptance. However, Lomonosov fell into error in the further development of the theory of light. The same happened to Euler.

Lomonosov's most important error in one of the fundamental problems of physics is of great interest.

As we know, Galileo discovered one of the most remarkable laws of nature. He established that the mass of a body, independently of its nature, is proportional to the force of gravity, or simply to its weight at a given point in space. Newton showed that this law is obeyed very exactly. Newton's experiment is very simple, precise, and convincing. In a doorway in his room at the college he hung two pendulums of the same length, but made of different materials. It turned out that the pendulums always oscillated in strict synchronism, regardless of the materials suspended. This could happen only if the mass of a body is strictly proportional to its weight. Lomonosov thought that this was wrong. He began to make remarks on this topic in 1748, and continued until 1757. All of these statements were made at a time considerably later than Newton's pendulum experiments. However, all this time Lomonosov, amazingly, stubbornly opposed this law. Thus, in 1755 Lomonosov proposed that the Academy of Sciences offer as a prize problem the experimental testing of the "hypothesis that the matter in a body is proportional to its weight." The posing of this problem met objections in the Academy of Sciences as contradicting the views of the great Newton, and Euler was brought in as the arbiter. Euler, who was usually on Lomonosov's side, did not support him in this case, and opposed the posing of this problem. We should note that Lomonosov's only student, S. Y. Rumovskiĭ, also did not share Lomonosov's views, as we see from his letters to Euler in 1757. Rumovskiĭ, who later became an academician, studied mathematics for two years with Euler in Berlin, and of course, knew Newton's mechanics well.

Perhaps, Rumovskiĭ was then able to show Lomonosov his error, since after 1757 I have found no sign of Lomonosov's having raised his problem anew.

Nothing is so instructive as the error of a genius. It seems to me that in this case the error did not happen by chance, but had a deeper reason. It would take far more time than I have to discuss this topic with confidence.

I assume that the reason for Lomonosov's error involves a philosophical concept to which he erroneously decided to grant universal meaning. This concept of Lomonosov consists in the idea that motion is always conserved in nature, is never created nor destroyed, but is only transmitted from one body to another, and only by direct contact. We know that this idea is correct in the case of elastic collision of spheres. Now we also know well that if we consider the collisions between atoms and molecules as collisions between elastic spheres, we can construct a complete and correct picture of the kinetic nature of heat. Thus we can understand why Lomonosov, who, on the one hand accepted the atomic structure of matter, on the other hand treated the interaction between atoms as subject to the laws of collision of elastic bodies. Thus he was the first to be able to construct correctly an almost complete picture of thermal phenomena based on kinetic concepts. As I have said, he not only came close to defining absolute zero, but also came very close to formulating the law of conservation of energy, of course not in a general form, but only for the transformation of kinetic into thermal energy.

Lomonosov's error lay in the fact that he ascribed a universal character to his idea, and began to think that only one means of interaction between bodies exists in nature, and this was through contact. Lomonosov rejected the possibility of action at a distance by gravitation or electric interaction. In developing these ideas, he thought that when a body has gained velocity under the action of gravity, then it is necessary that the medium surrounding the body should lose velocity. The medium having such a property of generating motion was of course hypothetical, and Lomonosov postulated its existence in nature. He tried also to describe electric interaction between bodies analogously. It is not hard to understand that on the basis of these ideas Lomonosov not only could not draw a clear picture of phenomena involving interaction of bodies at a distance, but they led him to deny the existence of a universal relation between the weight and mass of bodies.

It is hard to understand how Lomonosov, in developing these views, could fail to treat the described pendulum experiments of Newton. Perhaps he either did not know of them, or did not understand them, for I have not been able to find anywhere any mention by Lomonosov of these experiments. When one becomes acquainted with Wolff's textbook of physics, which

Lomonosov studied and translated into Russian, it strikes one that due attention is not paid to Newton's work in mechanics. There is also no mention of the described pendulum experiment. Interestingly, the only problem of mechanics that Wolff devotes attention to is precisely the collision of spheres. I have compared the writings of Christian Wolff with those of other physicists of that time; he impresses me as a scientist with limited powers in physics. He owed his fame, as we know, to works on abstract philosophical topics. Apparently, Wolff did not instill in Lomonosov the elements of concrete mathematical thinking, without which it is hard to grasp Newton's mechanics. As I have pointed out, Lomonosov had no opportunity to meet such scientists as Bernoulli and Euler, who not only had an excellent knowledge of Newton's mechanics, but also were themselves famous for developing it for a continuous medium. I can say with assurance that if such a communication had existed, Lomonosov would not have made this error.

The saddest thing in Lomonosov's fate was that he could spend only a small fraction of his energy and time on his experimental studies. With all his great erudition and exceptional imagination, he had no opportunity to subject all his hypotheses to experimental test. Thus it happened that in the fields in which Lomonosov worked experimentally, his theoretical and philosophical concepts were on the right track. But wherever he was divorced from practice and tried to gain the truth deductively, he often lost the right track. If he had been established in circumstances where he could have developed his experimental work more extensively, e.g., if he had had many students, then certainly he would have made far fewer false hypotheses. With his exceptional imagination, Lomosov could have been the director of a large scientific school. However, the conditions for founding such a school did not exist in Russia at that time.

Thus, the explanation that Lomonosov was not recognized as a scientist because he was divorced far from reality is ungrounded.

It is *à propos* here to recall that generally in the history of Russian science, the Russian scientists were often isolated from world science. I think that we should seek a general cause that is more profound than those mentioned. However, before going on to discuss it, I consider it useful to mention briefly another unrecognized Russian discovery highly reminiscent of Lomonosov's case.

A very major discovery in physics was made in our country at the very beginning of the 19th century, which also did not have its due influence on world science. This happened in 1802, when Vasiliĭ Vladimirovich Petrov discovered the phenomenon of the electric arc discharge in a gas, which he called the 'voltaic arc.' Now we all know well the entire subsequent vast role of the arc discharge, both in science and in technology. However, now in our time it is hard to judge by their merits all the difficulties of the discovery of this fundamental phenomenon, as was done first by Petrov. It was done eleven years after the discovery of the galvanic current, and only three years after Volta had invented the galvanic pile. Of course, little was known about the galvanic current within these three years. Petrov himself not only had to make batteries consisting of 4,200 copper and zinc disks stacked in a pile more than three meters long, but he also himself had to make the wire, insulating it with sealing wax.

Petrov observed the arc discharge not only at normal pressure, but at reduce pressures, by passing the current through an evacuated bell jar. We can now consider his discovery of this type of discharge to be the discovery of a plasma. Although Petrov's studies were published in individual pamphlets and sent out to many of the scientific establishments of that time, still the discovery of the arc discharge is usually credited to Davy, although he did it only in 1810. Petrov performed a number of other interesting studies in luminescence and chemistry. Apparently he was the first to decompose water by electrolysis, but none of these studies exerted their due influence on world science.

Petrov's biography is very instructive. The son of a parish priest, he began his work as a humble teacher in Bernaul in a provincial school, and later gained the position of professor of physics in the Medico-surgical Academy in St. Peterburg. Like Lomonosov, Petrov was a solitary scientist, and he also left no school after him. His studies and he himself remained unknown in the history of science, not only abroad but also here. No portrait of Petrov has been preserved, and it was only recently found where he is buried. I have no doubt that for his scientific discoveries, Vasiliĭ Vladimirovich Petrov should have occupied one of the most prominent places as a major experimental physicist, not only in our science, but in world science.

One often has occasion to hear that the inattention to the achievements of Russian scientists is to be explained by the fact that in the West they usually considered the culture of the Slavs to be second-rate, and not worth considering in the history of world culture. Undoubtedly, in the 18th and 19th centuries they quite often had this attitude to the Slavs in general, and the Russians in particular. However, I do not think that it can explain the question at hand, since the history of science shows that the evaluation of the scientific attainments of major scientists has always extended beyond national boundaries. They recognized Copernicus, even though he was a Slav.

It suffices also to recall that Euler repeatedly spoke highly of Lomonosov's work. Besides, this does not explain why we ourselves had so underestimated the scientific work of Lomonosov, Petrov, and a number of other Russian scientists.

I think that we must seek the explanation in the conditions under which science was developing in the country. It is not enough for a scientist to make a scientific discovery for it to influence the development of world culture. It is necessary that certain conditions should exist in the country, along with a necessary connection with the scientific community abroad. If these conditions do not hold, then even such remarkable scientific studies as those of Lomonosov and Petrov cannot affect the development of world culture.

It is these conditions, which were necessary in Lomonosov's time just as they are important today, that I want to take up now.

As I have said, from the 17th century the natural sciences began to grow considerably faster than before, owing to the collaboration of scientists on an international scale. This could happen only because these sciences, when developed on an experimental basis, are the same for all minkind. This characteristic unity of materialistic science has made possible its development in a broad international community of scientists. The pattern that the international collaboration of scientists follows is well known, and remains the same today as in the time of Lomonosov. The different countries have their own groups of scientific workers, who are found in universities, academies, and other scientific institutes. Since each scientific field or problem can grow only in one direction lest it lose this true direction, it is necessary to move slowly and spend much effort in exploratory work. Collaboration in scientific work consists in the fact that these laborious exploratory studies are distributed among collectives of scientists working on a given problem. The work of a scientist done outside a collective usually remains unnoted.

Life continually shows that such a collective effort of scientists can occur only through personal contact, either within a country or on an international scale. In order that a scientist's work become known, he must not only publish it, but in addition he has to convince people of its correctness and prove its importance. All this can be done successfully only with personal contact. Both in Lomonosov's time and today, in order that a scientist may influence the collective work with his studies, he needs personal communication, he needs a lively exchange of opinions, and discussion. None of this can be replaced by printed work or correspondence. It is not so easy to explain why this is. I think that most of us know from our own experience how necessary personal contact between people is in harmonizing creative work.

Only when you see a person, see his laboratory, hear the intonation of his voice, see the expression on his face, do you trust his work and want to collaborate with him. For this very reason, no textbook can ever replace the teacher.

Today, the necessity of personal contacts between scientists is accepted as something self-evident, both by our own and by foreign scientists. They are continually increasing, and now occur on a broad scale through congresses and conferences.

In Lomonosov's time, personal encounters of scientists were already widely developed. Usually it happened as follows. A leading center of scientific work in a given field of knowledge would be founded in some country. Naturally, such a center would attract to itself other scientists, often those working alone. In the 18th century science was strongest in England. This is explained by the wealth of the country, which was exceptional for that time. The country's patrons supported science, and it could develop most broadly. For example, Franklin went there, a man who had been a solitary scientist in America, similar to Lomonosov. He gained recognition for his remarkable studies on electricity when he reported them at the Royal Society in London. Also Leeuwenhoek gained full recognition for his microscope studies after a trip to London, whereas they were first treated with doubt.

The tragedy of isolation from world science of the work of Lomonosov, Petrov, and others of our solitary scientists consisted only in the fact that they could not be included in the collective work of scientists abroad, since they had no opportunity to travel abroad. This is the answer to the question that we have asked, that of the reason for the lack of influence of their work on world science.

Now we still have to take up the question of why Lomonosov's scientific work was not recognized for so long in our own country. It is quite clear that for a scientist to be recognized, the society surrounding him must be on a level to understand and evaluate his work in its essence. Of course, neither the administrative official apparatus nor the high nobles surrounding Lomonosov could understand the meaning of his scientific work, and hence, it only became possible to recognize his work in physics and chemistry when a scientific community had arisen in our country.

We should consider this lesson of history in order to evaluate the great role played by the community in the organization of science. Today this is very important to us, since we have the problem before us of building the most advanced science.

As we know well, any creative labor needs a relation to society to develop successfully. A writer, actor, musician, or artist creates at his maximum and develops his talent only when he is related to a community. The creative work of a scientist also cannot develop successfully outside a collective. Furthermore, just as the level of art in a country is determined by the tastes and culture of society, the level of science is determined by the degree of development of the scientific community. The tragedy of Lomonosov was redoubled by the fact, as I have said, that our country did not have a scientific community at that time. The lack of a sound critical collective hindered Lomonosov from being able to see where he was going on the right track in his investigations, and where he was in error.

Hence, Lomonosov could not manifest the full power of his genius. He painfully experienced the lack of understanding and recognition of his work in his own country as well as abroad. He did not gain the full satisfaction in his creative work that he deserved by virtue of his genius.

It is not hard to see that an advanced scientific community must exist for the development of an advanced science. If we had not created our own advanced scientific community, then we could not have built an advanced science in our country, no matter how many Lomonosovs were born among us. The creation of a sound, advanced scientific community is a major problem that we still pay insufficient attention to. This is more difficult than the training of selected talented youth for scientific work or the building of large institutes. The creation of a sound community includes the education of board strata of people involved in scientific work. They must be trained to be interested broadly in science, to esteem and love the sciences, to be able to evaluate objectively the advance of our science, and to support all that is actually important and best in science. It is only when a scientific community is able to evaluate a scientific advance correctly that it can help the scientist follow the right track.

Only an advanced scientific community can evaluate the cognitive power of a scientific advance independently of its immediate practical significance.

All the natural sciences can develop in the right direction only by relying on a sound scientific community. As I have noted, we discovered and recognized Lomonosov at the beginning of this century, not by chance, but only because a sound scientific community had begun to grow among us at about that time. The community in physics developed in our country when the material conditions for scientific work had improved, and it became possible for our major scientists of that time, Lebedev, Rozhdestvenskiĭ, Lazarev, and Ioffe, to found their schools.

Now under socialism, when science has been made the basis of development of society, the influence and significance of our scientific community is growing rapidly. We must remember that in order that our science may be the most advanced, our scientific community must also be the most advanced. It must be leading and authoritative, so that its judgements and evaluations will be recognized on a world scale.

33. THE SCIENTIFIC ACTIVITY OF BENJAMIN FRANKLIN

He snatched the lightning from the heavens, the scepter from tyrants.
(Turgot on Franklin)*

Franklin was born in America in the city of Boston, in 1706, and died when he was 84 years old. His activity covered nearly the whole of the eighteenth century and was closely associated with the rapid development of natural and social sciences which was underway at that time. This was an epoch of enlightenment, the epoch that preceded the period of radical social upheavals in Europe.

The name of Franklin went down in the history of the world culture not only as the name of a great scientist, one of the founders of the study of electricity, but also as the name of a great progressive statesman and public figure of America who took an active part in the struggle for its liberation from the colonial conditions.

His contemporaries were unanimous in describing Franklin as an exceptionally fascinating man, widely educated, of broad humanitarian outlook, an interesting and witty speaker. Franklin often travelled and spent many years abroad, mainly in England and France. There he had extensive contact with the progressive-minded people of his time, and toward the end of his life he became the most popular figure in Europe. In his native country, America, Franklin is still one of the most esteemed persons of the entire history of the U.S.A. The versatile work and life of Franklin are well known, and many studies are devoted to these.

His fundamental scientific discoveries in the field of electricity were made by Franklin at the middle of the eighteenth century, before the work of

---

Discourse given on January 17, 1956, at a meeting of the Academy of Sciences of the U.S.S.R., the Soviet Committee for the Defence of Peace, the M. V. Lomonosov State University of Moscow, the Union of Writers of the U.S.S.R. and the Soviet Society for Cultural Contact with Foreign Countries, held in celebration of the 250th anniversary of Benjamin Franklin.

* [Turgot, quoting Manilius, said of Franklin: "*Eripuit caelo fulmen sceptrumque tyrannis*". For source, and probable variant due to Turgot, see Carl van Doren, *Benjamin Franklin*, New York, Viking Press, 1938, p. 606.]

Galvani and Volta, i.e., before the epoch of the galvanic current, and therefore they refer to the initial period of mastering this powerful force of nature by science.

In the course of 200 years since the time of Franklin's research the study of electricity advanced so much that nowadays Franklin's work is studied in secondary schools only in classes where one starts to learn physics. All of us have been acquainted since our young days with the fundamentals of static electricity, although some of us might have forgotten what, in fact, Franklin did perform. For example, do all of us remember that the terms 'positive' and 'negative' poles were first introduced in science by Franklin?

A detailed description of the scientific work of Franklin would hardly arouse interest now, but it appears to me that the very history of the development of Franklin's research in the field of electricity is not only interesting but also useful for contemporary scientists. This follows from the fact that the path of the development of science, i.e., of the cognition of nature, followed by humanity is a single one. In our searching for scientific facts we are frequently led astray, which results in a waste of time. Therefore the less our deviation from the right way, the more rapid and economic the growth of our knowledge and of our mastering the forces of nature. In studying the history of science we find factors that favour a rapid development of science. From this standpoint the story of the development of Franklin's scientific work is of great interest.

Franklin carried through his studies on electricity within a short time interval: only for seven years, from 1747 to 1753. He began to do scientific research when he was 41 years old. By that time Franklin already had become a man of substance. A printing press, a journal, a famous almanac, and other printed matter produced by him in Philadelphia, which was at that time still a provincial town, were a great success. Franklin began to take interest in scientific work completely accidentally, after he had to attend a popular lecture with demonstrations on electricity. Such lectures enjoyed then a wide popularity, because a number of electric phenomena such as the repulsion and attraction of charged bodies, the electric spark, or the disagreeable feeling caused by a discharge through the human body were at that time new and completely out of the ordinary, and served as very good material for popular scientific lectures.

Not long before Franklin attended the lecture on electricity, the Leiden jar was discovered and for the first time yielded a method of condensing an appreciable amount of electricity. The possibility to carry out experiments with a considerable amount of electricity in store at once made the demonstrations of electric phenomena more striking.

Franklin came to be very much interested in experiments on electricity, and in the course of seven years he devoted the major part of his time to scientific research. This work became then of major significance in the development of the study on electricity, and was recognized widely. In this short time interval Franklin was acknowledged as the leading scientist of this time. Most great scientific societies or academies recognized the scientific merits of Franklin by electing him to membership, while a number of universities conferred on him the honorary doctorate.

Naturally, the question arises how it could happen that Franklin, who was never earlier engaged in physics, was able to lead the development of an entire branch of science in only a few years of work, in a small American town far from the centers of the international science, and being an already mature man, standing by himself?

Moreover, this took place by the middle of the eighteenth century when science was led by highly competent scientists such as Newton, Huygens and Euler. Dilettantism here is out of the question. But how could Franklin achieve such results which could not be reached even by professional scientists?

It appears to me that the explanation is to be sought for in the fact that Franklin was the first to understand the essence of electric phenomena and, hence, to discover the right trend of further investigations in this field. Many of us were witnesses to an analogous turning-point in the development of one of the most important fields of physics – radioactivity.

After Becquerel's discovery of the phenomenon of radioactivity in 1896, very abundant experimental material on the physical phenomena associated with the radioactive property of matter was accumulated during several subsequent years. The various experimental data were lacking in coherence, since the essence of the very phenomenon of radioactivity was not understood. Rutherford was the first to find out that physical phenomena associated with radioactivity were at once accounted for if radioactivity was assumed to be a process of disintegration of matter. In order to see this, it was not necessary for Rutherford to be of great erudition; the most important factors were his imaginativeness, perspicacity and audacity. At such initial stages in the development of a science, the precision and punctuality normal for professional scientists may rather hamper the advancement of such daring assumptions.

At the initial stage in studying electricity, it was necessary that such a challenging step should be made. And Franklin made it.

Before Franklin started his research, abundant experimental material was already accumulated, but the facts were incoherent; the hypothesis that he suggested not only united these facts into a consistent pattern but also pointed out the right path for further investigations.

Franklin expounded his basic hypothesis in a letter sent to Peter Collinson in 1749. It gives a clear picture of the processes taking place when a body is being electrified. This picture has remained till now basically a correct one.

A passage from this letter reads:

> The Electrical Matter consists of Particles extreamly subtile, since it can permeate common Matter, even the densest Mettals, with such Ease and Freedom, as not to receive any perceptible Resistance.*

Nowadays we call these 'extremely small particles' the electrons. Further, Franklin considered every physical body as a sponge saturated with these particles of electricity. Electrification consists in that a body having an excess of electrical particles is charged positively, whereas if a body is deficient in these particles then it is charged negatively. Franklin proved this by a very graphic experiment.

Imagine two men standing on wax cushions, i.e. on insulators. One of these is electrifying a glass rod by rubbing it. Then, if he touches the other man with it, they both become electrified with respect to the earth, which is simply proved by the fact that either of these men produces a spark when touching an object connected with the earth. If immediately after getting electrified the two men standing on insulators touch each other, then this will give rise to a spark between them; thereupon their charge with respect to the earth will vanish. This is proved by the fact that now no spark will arise when they touch an object connected with the earth.

Franklin's hypothesis started from the material nature of electricity and simply explained these experiments. When the insulated man touched the other one with the glass rod, one of these men lost electric matter whereas the other one gained it in the same amount. One of these was charged positively and the other one negatively. When they touched each other, then a discharge took place and, since the constant amount of electric matter was preserved, the previous equilibrium was established. Of course, Franklin had then no possibility to study the material character of electricity experimentally and

* [From *Opinions and Conjectures concerning the Properties and Effects of the Electrical Matter, arising from Experiments and Observations made in Philadelphia, 1749*, Section One; as reprinted in *The Papers of Benjamin Franklin*, ed. L. W. Laberee (New Haven, Yale University Press), 1961) Vol. 4, pp. 9–10.]

hence he could not determine who in fact gained the electrical matter and, consequently, was charged positively, and who lost it, i.e. was charged negatively. Therefore he assumed arbitrarily that the electrified glass was charged positively, perhaps supposing that the woollen stuff was rubbing the electricity into the glass. Only at the close of the last century, after the discovery of the particles of electricity – electrons – did it become clear that it was not the positive electrode that accumulated the electric particles, as Franklin assumed, but the negative one. To retain the adopted symbols for the positive and negative polarity, a negative charge was assigned to the electron.

I will cite one more experiment of Franklin, which is also of great scientific interest.

The property of mutual repulsion of bodies whose charge is of the same sign was generalized by Franklin to charges on metal conductors. He assumed that the charges, being repelled from one another, would tend to come to the surface of the electrified metal body. He proved his assumption by the following experiment.

A metal tea-pot was put on an insulator and electrified. An experiment was to be devised which would prove that the charge was distributed on the external surface of the tea-pot. To this end, inside the tea-pot a chain was put which, by means of an insulated holder, could gradually be drawn out of the tea-pot. The degree of electrification of the tea-pot was determined from the repulsion of two balls suspended on filaments. The experiment consisted in raising the chain from the tea-pot by means of the insulated holder and in observing how the degree of electrification of the tea-pot decreased as the chain was drawn out.

Franklin reasoned as follows. So long as the chain was inside the tea-pot its surface increased the inside surface of the tea-pot, whereas as the chain was drawn out, its surface increased the external surface of the tea-pot. Franklin concluded that if charge is distributed only at the external surface of an electrified conductor, then the degree of electrification decreases with an increasing external surface. In fact, this was observed when the experiment was carried out.

I recounted these two experiments not only because they are striking in their simplicity but also because they are most fundamental in their results. Franklin described all his experiments in his letters to his friend Collinson in England.

In these letters a large number of various experiments are described which have become classical: the production of an electric wind, the properties of the flow of charges out of a spike, and so on. In the same letters, Franklin gave a correct explanation of a number of already known electric phenomena in terms of his hypothesis, for example a picture of the accumulation of electric charges in the Leiden jar; and based on this he constructed a plane condenser. Collinson reported Franklin's studies to the Royal Society. Afterwards he published them in the form of a separate book which represents the basic scientific work of Franklin. The book ran into a number of editions and has been translated into many languages.

I shall not describe other of Franklin's experiments, and shall mention only those proving the electric nature of lightning. These experiments became famous in Franklin's life-time and made him illustrious. Although the hypothesis that lightning and the discharge obtained from electricity produced by rubbing are the same phenomenon though on a different scale was already put forward before Franklin, no experimental evidence of the correctness of this hypothesis was available.

The clarity and accuracy of Franklin's concepts of electric phenomena allowed him to devise an experiment which for the first time proved conclusively the electrical nature of thunderstorm discharges. The idea of Franklin's experiment was the following.

Assume that a long vertical metal rod insulated from the ground is placed between a storm cloud and earth. If the storm cloud has an electric charge, then a charge of the opposite sign is supplied to the upper part of the rod. If a spike is put at this upper end of the rod, then the supplied charge will flow out and the rod will get charged by electricity of the same sign as the storm cloud.

Franklin supposed that the presence of this charge could be detected from a spark arising when the conductor is touched by the free end of a filament connected to earth. Franklin's assumption that the rod should be placed on a hill nearer to the cloud in order that the experiment may succeed turned out to be wrong. Since no such hill was present near his house, he thought that he would not succeed in carrying out this experiment. He described in detail how it should be carried out and suggested that somebody else might perform it. As to himself, he decided to carry out an analogous experiment but in a somewhat different way needing no hill.

For this experiment, instead of a metal rod he decided to use a string, lifting it by a kite. Since a thunderstorm is always accompanied by wind one can fly the kite and since a thunderstorm is also accompanied by rain the string getting wet becomes a conductor and can replace a metal rod. In order that the string might be charged easily its upper end was provided with a possibility for the induced charge to flow out. For this purpose Franklin put spikes at the corners of the kite frame. In order to insulate the string from earth a silk band protected from rain was tied to the lower part of the string. A metal key, from which Franklin obtained a spark during a storm, was suspended on the end of the string near ground. In such a way he proved in the presence of his friends and acquaintances the electric nature of the thunder discharge. The experiment with the kite was performed by Franklin on April 12, 1753; he was then the first to reveal that storm-clouds are as a rule charged negatively.

The French scientist D'Alibard constructed, exactly according to Franklin's description, an insulated rod at Marley, and on May 10, 1752 in the course of a storm he obtained the electric sparks experimentally for the first time, and thus the electric nature of storms was proved successfully somewhat before Franklin but by his method.

The technical details of these experiments of Franklin as well as of other ones are very interesting, because they show his great experimental inventiveness. In studying the history of the development of Franklin's studies one is astonished by the quickness with which Franklin's concepts entered science. Despite the opposition of a number of eminent scientists, for example, Abbé Nollet and Benjamin Wilson, Franklin's ideas in a very short time made their way into science. Of course, scientific truth is a single one and will always make its way through, but it depends on men and not on the truth whether this way will be a rapid and straight one. In this respect Franklin's activity can also nowadays be quoted, speaking in contemporary language, as an example of how one has to put one's scientific achievements into practice.

Franklin attempted to make each of his researches at once a property of the largest possible circle of a people. At his home in Philadelphia he organized a Philosophical Society from among the local citizens, where he carried out demonstrations and read lectures. He often stayed abroad, where he made wide contacts with the scientific community. He kept an intense scientific correspondence with a number of leading French, Italian and English scientists even when America was at war with England. He taught himself the French, Italian and Spanish languages, and also knew Latin.

His ability to struggle for new ideas manifested itself particularly clearly when he had to get the lightning conductor into use. But we shall speak about it later on. We return now to the further development of Franklin's research on electricity. In connection with Franklin's ideas, scientists from many countries were engaged in experimental studies of the nature of electricity. In our country, at Petersburg, Lomonosov and Richman constructed rods for the study of atmospheric electricity, and called them 'the thunder machine'.

Unfortunately, although the studies of Lomonosov not only in the field of electricity but mainly in the field of chemistry where he first discovered the law of conservation of matter were of fundamental importance, they could not influence the development of international science to the extent that they undoubtedly deserved.

I think that the main reason for this lies in the fact that social conditions in which Lomonosov lived and did his creative work did not allow him to make contact with scientists from other countries and to visit abroad. The isolated circumstances of Lomonosov and Richman undoubtedly also hindered any effect of Russian science upon world science.

Richman's fate was particularly sad. In his studies Richman rightly pointed out that a further development of Franklin's experimental research should proceed to find a quantitative description of the electric phenomena. Searching for a method of quantitative measurement of the charge of the electrified rod of a 'thunder machine' in the course of a storm, Richman imprudently bent down and approached too closely to the conductor, wishing to make a quantitative record. He was killed on the spot by an electric discharge through his head. This happened in 1753.

After Franklin's work, the greatest step in the development of the science of electricity was the move to a quantitative description of electric phenomena. This was done by Coulomb and not earlier than in 1785. It is well known how with his torsion balance he discovered his fundamental law of interaction of electric charges. Coulomb found that the force of interaction depends on the square of the distance between the charges.

The subsequent theoretical studies of Gauss, Laplace and Poisson developed this basic law of nature into the consistent theory of the electrostatic field which is nowadays so widely applied. But in the history of the development of the study of the electric field there is a relatively little known page related to fundamental studies of Franklin, which is interesting to note.

Nearly 100 years after the work of Coulomb, in 1877, Maxwell published an article on some unpublished studies of Henry Cavendish in the field of electricity. Maxwell as the first director of the Cavendish Laboratory at Cambridge was given the archives of Henry Cavendish. In these archives he discovered a manuscript of Cavendish, completely ready for publication, providing experimentally the inverse square law discovered by Coulomb. The experimental evidence in the work of Cavendish differed essentially from that of Coulomb; the method was more simple and the evidence more accurate than those by Coulomb. There was no date on the Cavendish manuscript, but Maxwell dated it at any rate not later than from 1775, consequently at least ten years before the discovery of the law by Coulomb.

In his work Cavendish started from the fact that it can be shown theoretically that on a hollow metal conductor the total electric charge can be distributed at the external surface only if these charges are repelled by one another according to the law of the square of the distance. But the experimental evidence of the charge distribution at the external surface of the conductor had already been given by Franklin in the experiment with the electrified tea-pot and chain, about which I have already spoken; it was necessary only to find a way of making this more accurate.

Therefore Cavendish carried out the experiment in a more exact way. Instead of a tea-pot he used a hollow metal sphere, and instead of a chain he put another concentric sphere inside the first one. Both spheres could be either insulated or grounded, according to what was needed. Cavendish chose concentric spheres because this form made it possible to elaborate the obtained experimental result quantitatively. The Cavendish experiment consisted in providing evidence that the charge imparted to the external sphere was distributed only over it and did not pass on to the internal sphere.

At Cambridge Maxwell arranged a reproduction of the Cavendish experiment but with a more perfect measuring device, and showed the Coulomb law to be accurate to within nearly one part in a million, whereas by the method of Coulomb's torsion balance this law could be checked with an accuracy to within not much more than one percent.

Here the question arises why in the course of 100 years, such first-class scientists as Gauss, Poisson, Laplace and so on, the founders of the theory of electric field, had not noted that the simple Franklin experiment with the tea-pot could already serve as an experimental evidence of the validity of the Coulomb law – one of the fundamental laws of the electrostatic field?

How could it happen that for 100 years Cavendish's research remained unknown? In his article Maxwell also pointed out that in this same Cavendish work, ready for publication, besides Coulomb's law, Ohm's law was also formulated and roughly checked! And this was done 70 years before Ohm himself discovered this law!

It is natural to put the question how it could happen that such a great scientist as Cavendish, who was called by many people the 'Newton of contemporary chemistry', could leave unpublished this work on electricity, which he surely could not but consider as fundamental?

History will hardly ever find an answer to this question, but it is most likely that Cavendish simply forgot to send it to the press.

This explanation seems at first to be unlikely because, one would think, his friends were bound to know these studies and remind him of them. But here lies the peculiarity in the character of Cavendish – he had no friends and no colleagues, and in general avoided people. He was a very rich man, brother of the Duke of Devonshire, but his way of life was an exceptionally unsocial one, and he took interest only in science. Even his household servants were not allowed to appear in his chambers in his presence. His meal was served on the table before he came into the dining-room. Due to this isolation from people, the scientific work of Cavendish, the results of his greatest scientific achievements, made in England, had no effect on the development of world science.

Much later French [and German] scientists discovered these laws of nature independently. They communicated their knowledge to mankind and these fundamental laws of nature are justifiably called by the names of Coulomb and Ohm.

Besides purely scientific studies, we owe to Franklin a universally recognized achievement – his invention of the lightning rod. The history of bringing this invention to life is also very instructive. This is a long story and many studies are devoted to it. Therefore I will only briefly recount how Franklin invented and introduced the lightning rod.

I have already mentioned that Franklin proved lightning to be none other than an electric spark arising between clouds and the earth when these are of opposite charge. After the essence of thunder discharge was elucidated, naturally the question arose how one could rationally combat the destruction and fire caused by lightning. It became clear that when a building, ship or any other object is struck by lightning then the damage is due to the fact that an intense electric current passing over a poorly conducting surface produces destruction and fire. Hence, if the electric discharge arising when a building is struck by lightning is allowed to pass through a well-conducting medium like metal, then no destruction will be produced. It became understandable why buildings with metal roofs and drain pipes were less endangered by thunder discharges. For instance, Solomon's temple in Jerusalem was not once in the course of a thousand years endangered by lightning, since it was covered by polished metal plates.

It is natural that after Franklin's studies revealing the nature of thunder discharges, a number of persons arrived at the idea of the possibility of protection from lightning by removing electric discharge through well-conducting metal rods.

It is not unlikely that a modest priest called Prokop Divish, starting from his understanding of the processes of electric discharges and using a chain connected to ground as the conductor for removing the current, established independently on the roof of his house in a small provincial Czech town in 1754 an arrangement resembling closely Franklin's lightning conductor. This undertaking ended sadly, because the inhabitants of the little town, seized by a superstitious fear, tore down and destroyed the priest's arrangement.

Undoubtedly, Franklin with his sharp practical intelligence, saw before anybody else the possibility for protection against lightning by deflecting the current. But it was much more difficult for him to find the most rational form of lightning conductor and to make public opinion recognize it as an efficient means of combating destruction caused by thunderstorms. He managed this task excellently, and his activity in this sense can also nowadays serve as an example how new technological ideas are to be brought into practice.

Franklin not only took out no patent for his lightning conductor but freely allowed everybody who wanted it to take advantage of it. Moreover, he carried on a large and skilful campaign to introduce it into practice. For lack of time, I cannot recount completely the story of the practical introduction of the lightning conductor; hence I shall dwell on the most striking moment.

It may well be that no other invention gave rise to such a storm of various objections as was raised 200 years ago by that small metal rod which nowadays crowns almost every building and is a standard element of its construction.

Two hundred years ago the objections raised against lightning conductors were of extremely diverse sorts; there were such arguments as: "lightning is a tool of punishment in the hands of Providence, hence it is a sin to oppose it". Another not less 'convincing' argument was the following: "thunderstorms occur when evil demons disobey God". Therefore, the single right way of combating it was assumed to be the ringing of church bells which fought back the evil demons. That is why for a long time it was assumed to be necessary to ring bells during a storm. Since it is obvious that church bell towers are the most vulnerable to lightning strokes, it was rather dangerous to ring the bells during a storm. Even after the invention of the lightning conductor, they were for a long time not set in churches, and church bells were rung as before. One hundred and twenty bell-ringers were killed and 400 bell towers were destroyed in Germany within 33 years at the close of the eighteenth century.

But Franklin concentrated the main struggle for the lightning rod not against superstitious religious objections arising from less-educated sections of the population. The struggle was carried on against the very top of the society of that time. Both scientific and political objections were raised against lightning rods.

When Franklin gave the description of the lightning conductor he pointed not only to its obvious function in providing a free path to the electric current over the metal rod into earth, but also to a possibility of the existence of another process.

Franklin assumed that if a thundercloud is above a building, and if the lightning conductor is provided with a spike, then the electric charge may flow slowly out of it. We now call this phenomenon the silent discharge. It is this phenomenon that will neutralize the charge of the cloud and discharge it. Therefore Franklin argued that a lightning conductor not only protects buildings but may prevent thunder discharges. Scientific opponents of Franklin supposed on their part that the flow of the charge out of the spike would not only not neutralize the charge of the cloud but would even produce more favourable conditions for the occurrence of lightning. Hence in their opinion a lightning conductor was rather harmful, since it made possible the occurrence of thunder discharges which would not at all take place if the lightning rod were absent.

Scientists who held this view assumed it to be particularly harmful and dangerous for a building to be in the neighbourhood of another one provided with a lightning conductor.

Public opinion was very much interested in these problems, and this is well illustrated by the well-known case when at Saint-Omer, France, Monsieur de Vissery installed a lightning rod on his house, and his neighbours were so much frightened and indignant at this that they brought a court action against him. The trial raised a clamour and lasted for several years between 1780 and 1784. It is interesting that the then young lawyer Maximilien Robespierre undertook the defence of the lightning rod, and this famous affair marked the beginning of his reputation. It is also curious that Marat, who considered the lightning rod to be dangerous and was against its establishment, was one of the experts on the part of the plaintiff. After a long struggle and many appeals against the court's decision, de Vissery finally won his case.

Franklin's tactics in this struggle for the lightning rod is also interesting.

As a rule he did not appear in public, but by means of his discussions and his extensive correspondence he continuously influenced leading scientists and public figures. Through such propaganda, he created a powerful group of partisans from among progressive-minded persons of that time, which struggled to bring his invention – the lightning rod – into practice.

In England the struggle for the lightning rod assumed a political character. The English scientist Wilson attempted to prove that it is possible to prevent the harmful effect of a lightning rod if one makes its end blunt and thus hinders the flow of charge. Since the time of this debate coincided with the epoch of America's liberation from colonial rule so that Franklin became a great political figure of the new America and one of the most agile fighters for freedom, each English citizen who provided his lightning conductor with a spike and not with a blunt end was considered as politically suspect.

The King of England, George III, asked the Royal Society to retract its decision in favour of the spike on Franklin's lightning rod. Sir John Pringle, who was then the President of the Royal Society, King's Physician and personal friend of Franklin, gave the following famous reply to the king's demand: "I will always do my best to fulfill the wishes of His Majesty, but I am able to change neither the laws of nature nor the effect of its forces".

For these words he was dismissed from the post of King's Physician and from the presidency of the Royal Society.

In the course of the struggle over the problem of the lightning rod, all methods were used: calumnies, insinuations, against both Franklin and his friends. Franklin remained calm, did not pay attention to the personal attacks and constantly stated that in problems of science the truth is revealed only by experiment.

Indeed, it was experiment that resolved this controversy, but only after several dozen years, when the knowledge of gas discharges and of the electrostatic field reached our contemporary level. At present we know that this controversy was groundless, since for an ordinary lightning conductor it is of no importance whether it is provided with a spike or a blunt end. At a small distance from earth, the geometric form of the end of the lightning rod cannot have an appreciable effect on the electric-field distribution above ground.

However, Dr. Schoenland, one of the leading experts on thunderstorm discharges, has pointed out that the process of neutralization of the cloud charge by means of a silent discharge, predicted by Franklin, can nevertheless take place, but only if the spike of the lightning conductor is at a large distance from earth comparable with the altitude of the cloud. This is the case with lightning conductors placed on the loftiest American skyscrapers; then one can indeed observe from the spike of the rod a silent discharge which does not go over into lightning. Schoenland added that this would undoubtedly produce a feeling of justified satisfaction in Franklin if he could know of this.

Nowadays the lightning rod is an integral part of all buildings, and there are surely innumerable edifices, installations and ships which were preserved by it from fire and destruction. The merit is rightfully given to Franklin's initiative.

I shall only briefly present Franklin's activity in other branches of science, because besides the famous achievements described, Franklin has achievements in other fields, too.

Franklin took interest also in geophysics, made a map of the Gulf Stream, invented a musical instrument based on rubbing glass balls, an economic stove which is still widely used in America and France, street lamps, bifocal spectacles for farsightedness, and many other things. Moreover, owing to his sociable character and lively mind Franklin gave much advice and furthered the development of science. Of course, data on most of these consultations are lost, but some reached us.

Thus, for example, Louis XVI asked Franklin to become a member of a committee of inquiry on the value of a medical treatment proposed by Dr. Mesmer who made use of so-called 'animal magnetism'. It is interesting that the well-known doctor Guillotin, the inventor of the guillotine, took part in this committee. Franklin denied the existence of animal magnetism, but considered this treatment to be harmless, since it distracted wealthy people without prejudice to their health, which cannot always be said of other unsubstantiated medical treatments.

Franklin very much approved of the flights of the brothers Montgolfier.

We should also note a field of Franklin's activity associated with possibilities of developing world science, which Franklin worked on as a great statesman of his time. Franklin considered scientific achievements to be the property of the whole of humanity, and the concern for the development of world science to be above political and military conflicts between different peoples. Thus, at the time of the war against England, when the famous explorer Captain Cook was coming back from the circumnavigation of the globe, Franklin gave instructions to all American ships and corsairs to treat Cook respectfully wherever they came across him during his voyage. For our time it is also of interest that Franklin, as a member of the Congress, persuaded it that the embargo put on all goods of English provenance should not be extended to scientific equipment.

In studying the life of Franklin, we understand more and more fully why this great man, whom the people of America gave to humanity, is universally esteemed and worshipped.

In our epoch of rapid progress of natural sciences, every country gave its own great pioneer of science – in our country this was Lomonosov, in England – Newton, in Italy – Galileo, in Holland – Huygens, in France – Descartes, in Germany – Leibniz, in America – Franklin. What these great scientists have achieved is the pride of all mankind.

And we, the Soviet people, are grateful to the American people who reared the great Franklin and gave him to humanity.

34. THE PHYSICIST AND PUBLIC FIGURE PAUL LANGEVIN

Paul Langevin was not only an outstanding physicist, but also a great progressive public figure and friend of the Soviet Union. In 1924 he was elected Corresponding Member of the Academy of Sciences of the U.S.S.R., and in 1929 he was made Honorary Member. He was also an honorary member of different academies and scientific societies in a number of countries; in 1928 he was elected a Foreign Member of the Royal Society of London, and in 1934 he was elected to the French Academy of Sciences. This honour he merited as a physicist for exceptionally important scientific research.

His main works relate to theoretical physics and the greatest of them to magnetism, where his work still has value and is considered classic, and to acoustics. Langevin found the method and then produced ultrasonic waves, and he first proposed, using the phenomenon of piezocrystals, to excite short acoustic oscillations. A whole new field of science and engineering has grown up on this basis.

It must be said that Paul Langevin's influence on the development of world physics has been very great and is not confined to these two fields. Langevin was also a great teacher; under him were many pupils, of whom two have received world recognition, one of them de Broglie and the other Joliot-Curie.

Although Langevin used to publicize his works comparatively little, yet he was a very generous teacher, gave ideas, inspired and backed up his pupils. In this respect his influence on French physics – if an account could be drawn up, but which unfortunately is not possible – is certainly no less, but even greater than that of the works which he did publish.

I saw rather much of Langevin, I was lucky to gain his friendship and now I think of him with exceptionally warm feeling.

The image of Langevin can, I think, be described in a word: he was a man serving progress in everything; he was progressive in science, progressive in his political views, progressive in his philosophical views, and progressive in his

---

Text of a talk given to a joint session of the Academy of Sciences of the U.S.S.R. and the Society 'Druzhba' (Friendship) on 23 January, 1957, in Moscow.

social activity. This progressiveness ran through all his life. The culture of mankind grows, science moves forward, the social system develops and our philosophical representations of the interrelation between man and the material world become deeper. Whether we like it or not, all goes forward. People are divided into three categories: some stand in front and spend their powers to further science, culture and humanity – these are the progressive people. Others, and these are the majority, go along with progress, at the side, they do not interfere and do not help; and then finally there are people who stand behind and hold back, conservators, cowardly and without imagination.

For those who go ahead it is hardest of all; they make new paths for progress; all possible tests of fate befall them. One such was Paul Langevin, and fate made him face many severe tests. One asks why there are people who choose this road, what compels them to go forward when it is more pleasant and quieter to go at the side, even if not to trail behind? It seems to me personally that there are two reasons. Intelligent men cannot be other than progressive. Only an intelligent man, endowed with boldness and imagination, can be progressive and understand the new and to what it is leading. But this is insufficient. It is also necessary to have the temperament of a fighter. When intelligence is combined with temperament, man becomes truly progressive. One such man was Paul Langevin. Most often in life we observe that only in youth does that temperament light up most and make a person progressive; with age man wants a tranquil life, and therefore youth, especially in student years, is the most progressive part of humanity. With Langevin this did not happen. He was a fighter for progress to the end of his life, and the older he became the more zealously he fought for progress. Now this extraordinary feature in him always struck me and called forth deep sympathy and respect.

Now I want briefly to consider his scientific work from the viewpoint of progressiveness.

His first works related to magnetism and were done in 1907. Here he was the first to apply statistical mechanics, developed and generalized at this time by Ludwig Boltzmann, to the laws of para- and dia-magnetism, which had been discovered shortly before by Pierre Curie, teacher and friend of Langevin.

Now we regard this as nothing out of the ordinary, but if we reconstruct the situation in physics at the time, we see that these works in their essentials were exceptionally progressive. In those years Boltzmann's ideas were coming into scientific life with great difficulty. In 1906 Boltzmann put an end to his life precisely because the fundamental and bold idea on which he based his works in the kinetic theory of matter – the relation between entropy and the probability of existence of molecular states – was not understood or recognized. The leading scientists of the time, for instance, Ostwald, did not want to recognize the atomic theory at all, and a storm raged round Boltzmann's works. Langevin's approach in his first works on magnetism was based on precisely these new Boltzmann ideas.

Slightly later, or at almost the same time, Einstein's work on the theory of relativity appeared. It was published in 1905. Already 50 years have elapsed since then, and now only the most hardened conservatives object to the main ideas of the theory of relativity. But when it appeared then, of course, there was a stream of objections and the biggest objections, of course, were against the law which for the first time had been formulated perfectly clearly and quantitatively – the law of the equivalence of mass and energy. This equivalence is defined by the law according to which the mass of a substance, multiplied by the speed of light squared, can be converted into an equivalent quantity of energy. Some scientists saw in this a violation of the law of the conservation of energy and of the law of the conservation of matter – the basis of physics of those days – and this caused a storm of objections.

Langevin was one of the first to publicize, with great energy, Einstein's ideas in France. Almost at the same time as the discovery of Einstein's law, he had a paper printed in which he indicated the possibility of a difference between the deviation of integral mass numbers in a periodic system from the integral hydrogen multiples associated with the appearance in complex atoms of a surplus energy which increases their atomic weight.We know now that this was the right prediction which only a great scientist could have made, and it has since been verified experimentally more than once. Now these views have a precise theoretical justification. This again shows how Langevin picked up new ideas in science and how he put them into practice. Of course, now we have the atomic bomb which has demonstrated to all mankind the explosive force when matter changes into energy. Simple calculations show that in an atomic bomb only one gram of matter is converted into energy, whilst in the hydrogen bomb no more than one kilogram. Of the possibility of such a colossal effect in the change of matter into energy, Einstein, Langevin and other leading physicists spoke more than once, but there were more than a few others who did not believe it. A clearer demonstration of Einstein's law than the bomb explosions in Hiroshima and Nagasaki is hard to imagine. Yet, despite this, we in the editorial office of *Journal of Experimental and Theoretical Physics* still to this day are receiving articles which attempt to refute the validity of the theory of relativity. In our days such articles are not even considered, clearly being anti-science. This is a second example of how Langevin fifty years ago travelled the right and leading road in physics.

The third instance of his progressiveness in modern physics I observed personally. In 1924 I went to Langevin in Paris. Then he was professor at the Collège de France. When I arrived he immediately said, "My pupil de Broglie has done remarkable work, I want him to tell you about it". He called de Broglie and asked him in my presence to tell me of his new work on the wave nature of electrons; as is generally known, this work has since become classic. I saw then how Langevin was carried away by this work. Quite possibly, without the backing of Langevin, de Broglie would not have acted towards his new idea with the boldness which was necessary to develop it and put it into practice.

That at the time his idea caused great scepticism, can be illustrated by the following example. When I came back from Paris to Cambridge I told the theoreticians about de Broglie's work. Paul Dirac was then still a student; he was taking a short course on magnetism which I was then giving; he sat in the front row and I did not suppose then that he would become the great scientist who found the most general mathematical expression for de Broglie's idea. The principal theoretician in Cambridge then was Fowler. Neither he, nor his colleagues, wanted to recognize de Broglie's views and take them seriously. And when I suggested supplying a seminar paper on this theme, they said to me: "We will not waste time on this". A year or two later when Schrödinger did the work in which he mathematically generalized de Broglie's idea, and his significant, now classic, equation appeared in which he showed that what de Broglie had done was nothing more than the eigenvalue of functions in specific equations, the fundamental importance of de Broglie's work became clear to all.

The story of how Schrödinger produced his equations is very instructive. Schrödinger at the time was working with Debye and he told me the details of how Schrödinger produced his equations. Reading through de Broglie's paper in the *Comptes rendus*, Debye suggested to Schrödinger that he should discuss it at the seminar. Schrödinger replied approximately so: "I do not want to talk about such nonsense". But Debye, as his senior on the staff, said that it was necessary for him to do it all the same. Schrödinger had to agree and he decided to try to present de Broglie's ideas at the seminar in a more comprehensible mathematical form. When he had succeeded in doing this, he had arrived at the equations which made him famous throughout the world and which now bear his name. Debye told me that when Schrödinger gave his lecture at the seminar he himself did not understand what a great discovery he had made. Debye there and then at the seminar said to him "You have done a remarkable work". But Schrödinger himself thought that he had only found a good method of telling a group of physicists what de Broglie had done. And this took place two years after de Broglie's work appeared. Langevin immediately, from the very first, and earlier than all, understood that a new field of physics lay in the ideas of de Broglie. This example once again shows Langevin's amazing feeling for everything progressive. It always struck me in conversation with Langevin how he was able to see the scope of what was happening in science; one really should how to his perspicacity.

In the social field Langevin was just as progressive as in physics. He often said with pride, "I was born in Montmartre". As is generally known, this is the most proletarian part of Paris. His grandfather was a simple metal worker, his father a land surveyor. Langevin himself was born in 1872 in fairly poor conditions, went through the municipal school and then obtained a higher education by grants. He was a widely gifted person and, of course, he did brilliantly. Then he became a pupil of Pierre Curie. Pierre Curie sent him to the Cavendish Laboratory, where he worked in the same room with Rutherford. In those years Cambridge had long been the center of physics. The Cavendish professor then was Sir Joseph John Thomson who became famous for discovering the electron; the research which he started on the passage of electricity through gases was in those years the leading field, like, for instance, nuclear physics in recent times.

At Cambridge, Langevin did his first experimental work and there his scientific career began. After this, for many years, Langevin maintained a great friendship with British scientists, especially with Rutherford. Langevin was a charming person, always gaining the favour of people of any class, and everywhere he had friends. On his return to Paris, Langevin started working in the Collège de France, where after his teacher Pierre Curie he occupied the position of professor. From his very first works on magnetism onwards, he became one of the leading physicists in France.

He also began his political activity very early: he began it approximately in his schooldays. The beginning of his political activity is linked with the famous Dreyfus affair – a shameful court case started by a group of antisemites whom one may describe as the predecessors of fascism. Then Emile Zola interceded for Dreyfus, writing the famous book *J'accuse*.

When Zola was subjected to victimization, Langevin stood out in his defence. This was his first action in the social field. He often, recalling this action, said "Yes, they were good times then, imagine, the fate of one man could interest all the earthly globe". After this followed a whole series of political actions: the list speaks for itself.

In 1920 at a meeting in the Salle Wagram in Paris Langevin in a brilliant speech stood out in defence of the sailors of the Black Sea squadron who refused to fight against the young Soviet Republic. Langevin with just the same keenness as in science foresaw the progressive value of our social revolution and at once openly took to backing it.

In the same year, when he was professor at the Collège, he stood out against the use of students as strike breakers during the transport strike in Paris.

Together with Romain Rolland and Henri Barbusse he unswervingly stood out against fascism.

He stood out in defence of Dimitrov during the Leipzig case.

He was one of the most active defenders of Ernst Thälmann. He was president of the '*Ligue des droites de l'homme*'. He not only headed it, but was one of its organizers.

He more than once stood out in defence of the Spanish Republic. In this list the same words are repeated: he stood out, he stood out, but behind these words are concealed great social activity and great organizational work.

Langevin stood out with a blazing censure of the Munich pact and stood out against the arrest of 27 communist deputies at the beginning of the war.

When the war started the opportunity was presented to me to write to Langevin and invite him to the Soviet Union for the duration of the war. Knowing the hatred which the fascists nourished for him, his fate in France was terrifying and, of course, it was necessary to offer him the opportunity to go to a country where he could be in safety and could continue the struggle for France. In a letter he replied to me that he would come to the U.S.S.R. with pleasure, but at the time he had to finish one more matter; in the University of Paris an anti-semitic movement was then starting and Langevin headed the struggle against it, and until this movement was liquidated, he did not feel it right for him to leave Paris.

When Langevin decided that he was able to leave Paris, it was already too late; Hitler's government refused to allow him through Germany. Paris was occupied by German forces and Langevin was immediately arrested. Two months he was in prison; then they sent him to a small town where he occupied the position of physics teacher in a secondary girls' school and held this post for the first half of the war.

Langevin's family was progressive and they were all fighters against fascism. Langevin's daughter was arrested and sent to Auschwitz, where she was interned for the war. The daughter's husband, Solomon, a well-known communist, was arrested and shot by the Germans. Langevin had to get out of France. This was no easy task – he was already close upon 70. He escaped across the mountains into Switzerland. An automobile accident was staged, he was bandaged up and as a casualty was carried in someone's arms across the mountains. He stayed all the second half of the war in Switzerland where to the extent of his powers he continued to take part in the freedom movement. When he learned of the death, i.e. of the shooting, of his son-in-law Solomon, he wrote Duclos a letter in which he joined the Communist Party to which Solomon had belonged. Thus from 1942 until his death on 19 December 1947 he was an active member of the Communist Party.

This short list of facts, it seems to me, gives quite a vivid picture of Langevin's social and political activity. And from this list it is seen that there was not one great progressive event in Europe and in France in which Langevin would not take an active part. But there were also other fields of social life where Langevin stood out, for example, his most lively participation in questions of peoples' education. I remember, one day in Paris, I said to Langevin I had to go to Strasbourg to give a lecture at Strasbourg University. Langevin replied "Very good, let us go together. I too am going to Strasbourg: I have to give a lecture there on the teaching of the French language in Alsace". I attended his lecture, just as he came to mine, and I heard how interestingly he analysed the question of the teaching of the French language in Alsace. This was no easy task since a complex political question was bound up with it, and, because the sympathies of the population were divided between France and Germany, Langevin had to speak with great tact. Listening to him, I saw how exceptionally skilfully he drew up his report.

Such is the picture of his activity. A man who was engaged in such progressive activity, both in the field of social life and also in science, could not help but be attractive, more particularly for young people.

Langevin was 20 years older than I, but in spite of this age difference, it was exceptionally easy and simple to associate with him. He was an exceptionally charming person and was liked by the masses of France. All, in my opinion, liked him. I do not know a person who would not treat him kindly. Even people of opposite political views treated him kindly. Gentleness, exceptional kindness and sympathy overcame and subdued all. With any person, whether it was the Prime Minister or a student, he spoke just in the same way and both felt at ease.

As an example of different persons' respect for him, I shall quote the telegram which Einstein sent to the Paris Academy after Langevin's death. This telegram is very short; I chose it not because it was written by Einstein, but because it exceptionally well, in my opinion, expresses and concisely expresses in reality who Langevin was:

> The news of Paul Langevin's death dealt me a greater blow than most of the events of these fateful years, so fraught with disappointment . . . There are so very few in any one generation, in whom clear insight into the nature of things is joined with an intense feeling for the challenge of true humanity and the capacity for militant action. When such a man departs, he leaves a gap that seems unbearable to his survivors.*

In conclusion I would like to talk about a small trait in his character which imparts still more charm and humanity to his individuality. Langevin had one foible, he loved wine. Not in the vulgar sense did he love wine, but he loved the bouquet of wine, he loved wine as a connoisseur. He said "One doesn't drink wine, one talks of it". He has taken a bottle of wine, held it in his hands, inhaled its perfume and said that it was Burgundy of such-and-such and it was distinguished by such-and-such properties. He could recite a whole poem about a bottle of wine. He was proud of his knowledge of wine. This was his, as the English say, 'hobby'.

Once in Zürich during a conference I would sit with him in a restaurant at the same table. Every time he very carefully chose a most rare wine and on the spot gave me a lecture on it. His knowledge of wine was not amateurish. The French wine-makers after the grape-gathering invited him to come and assess what the wine would be like in a few years' time. He went and he was very proud of the fact that the Burgundy wine-makers agreed with his opinion. But most of all he was proud of the fact that once in the Var valley in the south of France, when he was tasting wine, he 'discovered' an excellent new wine. So from a plebeian red wine, on his valuation, a vintage wine was made.

And he genuinely rejoiced in discovering a new sort of vintage wine and he was very proud of this. The theory of magnetism, which was his greatest triumph, he did not mention, but he spoke with great passion of the new vintage wine discovered by him in the Var valley.

Well, here is my short sketch of the work and personality of this remarkable, progressive man. I consider it a great fortune for myself that I knew and loved this outstanding man and had many conversations with him.

---

* [As translated in Albert Einstein, *Out of My Later Years*. New York, Philosophical Library, 1950, p. 231.]

35. IN MEMORY OF IVAN PETROVICH PAVLOV

Great scientists are not always great men at the same time. Testimonies of contemporaries show that many geniuses who have made breakthroughs in science actually had narrow minds. Brilliant scientists are few indeed, but those of them who are also great men are even rarer. Ivan Petrovich Pavlov was one of these rare exceptions. This made him into a charismatic personality not only among physiologists.

Pavlov won his fame by founding a whole new branch of physiology. We physicists find it difficult to assess the full depth and finesse of his works. However, these have one aspect in common with our own field of science. We develop precise quantitative measuring procedures as crucial tools for understanding the nature of the phenomena examined. Precise measurements are still unknown in many branches of physiology, and it seems that their introduction would be particularly difficult in studies of higher nervous activity. But Pavlov can be credited with the development of objective, quantitative methods for the measurement and evaluation of psychological phenomena, and this is one of his greatest scientific merits.

The fascinating reports and works of I. P. Pavlov attract not only those who are specialists in physiology; many of us have tested his theories and conclusions about reflexology by experimenting on our own bodies. This scientific liveliness of Pavlov's works was appreciated by large circles of scientists. Pavlov was famous far beyond the boundaries of his own country, and especially in the intellectual circles of Britain. This was due also to the great power of the school of British physiologists, who had a strong and longstanding appreciation for him. While working in Britain I could actually feel the atmosphere of respect for Pavlov's name.

Pavlov made frequent visits to Britain. He liked the British people, who received him enthusiastically and even forgave him his ignorance of the English language. He was awarded all existing academic degrees and titles. Decorated with all possible medals, he was a member of nearly all the scientific societies and *doctor honoris causa* of all major universities. In Cambridge they still reminisce about the solemn ceremony in which he was awarded an

---

Article in the newspaper *Pravda*, 1936.

honorary degree of the university. Tradition there did not allow the presence of students in the Cambridge Senate Hall; so they filled the upper galleries, and one of them dropped on a string the stuffed body of a small experimental dog – a modest, symbolic tribute to the great scientist. This small gift gladdened Pavlov as a sign that his work was known among the students. The idea of this present came from the grandson of Charles Darwin, who later became a well-known physicist and a professor at the University of Edinburgh.

Throughout the many years of his scientific work Pavlov learned no foreign languages with the exception of a little German. Nevertheless his vibrant speech, his highly expressive intonations and gestures somehow reached the minds of listeners of various nationalities. His entire appearance and manner of speaking reflected a lively, colorful, strong and independent personality. This individuality actually guided Pavlov throughout his scientific career. Perhaps it can explain the fact that this great scientist produced his major works at the age of about 40 years, which is later than the majority of scientists; I know of only one man of comparable stature, Faraday, who started the series of his monumental discoveries at the same age as Pavlov. Strong personalities perhaps prefer to build new roads rather than follow a beaten track. Pavlov probably spent many years in trying to break a new way in science and make it into a broad avenue for his followers.

Pavlov's independence and straightforwardness were powerful weapons in the arsenal of his brilliant mind, which enabled him to make such breakthroughs in science. He managed to go on with his work by carefully preserving his independence. The power, daring and uprightness with which Pavlov defended his views and convictions can serve as an example, and not only to scientists.

His strong personality was always felt from the first words of his lectures. Pavlov had no time for small talk or shallow conversation; he was organically opposed to any phlegmatic, careless or apathetic attitude to the surrounding world. In all his occupations, in research or leisure, his personality invariably revealed its passion and deep involvement – the traits of a man who did nothing halfway.

Pavlov was deeply moved by many problems of life, apart from his scientific work. Lately he spoke to me of his hope to live at least another ten years in order to see the future of his science of conditioned reflexes, his country, his granddaughter. Anyone talking with him could see the power and clarity of his mind – qualities seemingly remote from death . . .

36. ALEXANDRE ALEXANDROVICH FRIEDMANN

Alexandre Friedmann was one of our best scientists. Were it not for his untimely death of typhoid fever at the age of 37 he would still be with us. Of course, he would have had many further achievements in physics and mathematics, apart from a prominent academic career. At his young age he was already a professor of world fame among specialists in the theory of relativity and meteorology. While in Leningrad in the 1920's, I often heard such professors as Krutkov, Fredericks and Bursian speak highly of Friedmann as a prominent scientist.

Friedmann made one of the greatest theoretical discoveries in astronomy – he predicted the expansion of the universe.

Friedmann's solution of Einstein's cosmological equations suggested that the radius of curvature of the universe changes with time. Several years after the publication of Friedmann's works, the American astronomer Hubble found that galaxies move away from one another – a consequence of the expansion of the universe. Thus Friedmann with his pen discovered a striking phenomenon on a cosmic scale.

Some people say in this connection that Friedmann did not think much of his own theory and regarded it merely as a mathematical curiosity. They quote him as saying that his job is to solve equations; as for the physical interpretations, let the physicists take care of them.

This brilliant man's joking attitude to his own work cannot possibly change our great appreciation of his discoveries. Even if Friedmann was not convinced that the expansion of the universe, predicted by his mathematical work, actually exists, this does not in any way detract from his contribution to science. We remember how Dirac predicted the existence of the positron on a theoretical basis. Dirac did not believe in the actual existence of the positron; he regarded his calculations as a strictly mathematical feat suitable for the description of certain processes. But the positron was discovered, and suddenly Dirac found himself a prophet. Nobody has ever tried to deny

---

Address at the session of the Department of Physical-Mathematical Sciences of the U.S.S.R. Academy of Sciences in 1963 on the 75th anniversary of the birth of A. A. Friedmann.

his contribution to science just because he did not believe in his own prophecy.

Later Dirac predicted the existence of individual magnetic poles. These were not found, though Fermi thought for some time that they could really exist. If such poles are ever found, due credit will be given to Dirac.

Friedmann did not live to see the confirmation of his calculation by direct observation. But we know today that he was right. And we must appreciate his remarkable achievements.

Friedmann's name has so far been in undeserved oblivion. This injustice must be corrected. Friedmann's name must be immortalized, for he was a pioneer of Soviet physics, a research worker who has made a great contribution to Soviet and world science.

37. LEV DAVYDOVITCH LANDAU

Lev Davydovitch Landau was born on 22 January, 1908, in Baku, a large university town, then the chief petrol-producing centre in Russia. Landau's father was an engineer, working in the petrol industry; his mother was a medical doctor who also worked on physiological research. Lev's older sister, Sofia, subsequently became a chemical engineer.

The family was comfortably off and Landau started his education in a Gymnasium. His exceptional talents were recognized at a very early stage and he finished school at thirteen. He was too young to go to university so his parents sent him to technical school where he studied economics. His mathematical ability became apparent equally quickly – at thirteen he was already familiar with analysis. In 1922 he entered the University of Baku where he simultaneously studied in two faculties (Chemistry and Physics) but he stayed only two years and in 1924 moved the Department of Physics at the University of Leningrad. With the University of Moscow, this was the leading higher education establishment in Russia. He completed the course by 1927, having been appointed a Supernumerary Fellow in the previous year.

In 1926 Landau published his first scientific work, 'On the Theory of the Spectra of Diatomic Molecules'. This thesis, produced when he was 18 years old, already gave proof of a mature scientist; research carried out during the next year (1927) referred to the problem of damping in wave mechanics. This was the first time that a description was given of the quantum states of systems with the aid of density matrices.

During these years Russia was still suffering from the tragic aftermath of war, which had culminated in the Revolution of 1917, and the consequences of yet more difficult years of intervention and civil war. Life had only just returned to normal. In spite of the difficult economic situation and general disorder, scientific life began to develop at a fast rate not only in the universities but also in newly-established scientific research institutions.

This development was chiefly due to the young scientists who had

From *Biographical Memoirs of Fellows of the Royal Society* **15**, 141–146 (1969). [A 'Survey of Scientific Publications' of L. D. Landau, written by E. M. Lifshitz, follows in the same volume, pp. 147–155.]

been educated after the Revolution. At that time there were only two distinguished senior physicists in Leningrad who had considerable influence on the progress of science: D. S. Rozhdestvensky and A. F. Ioffe. Both were experimentalists, Rozhdestvensky is known for his work in the field of spectroscopy and optics; Ioffe was a student of Röntgen and achieved fame for his work on solid state physics. There were no senior scientists of note able to set up a theoretical school. However, P. S. Ehrenfest had lived and worked in St. Petersburg for five years before the war. He was Austrian by birth. After living in St. Petersburg he was invited to the University of Leiden to take the Chair which had become vacant on Lorentz's retirement. Ehrenfest left behind in Russia several talented students who were even then well known for their theoretical work; among them V. R. Bursian, G. G. Weichard, Yu. A. Krutkov and V. K. Fredericks. While Landau was at the University of Leningrad, these young men led the teaching of theoretical physics.

In contrast, mathematics was taught by such important scientists as V. A. Steklov and A. A. Markov who continued Tschebyshev's traditions.

Thus, while Landau was at university, research in the field of theoretical physics was only just beginning and theoretical physicists developed without supervision from distinguished senior scientists. They worked on their own, exchanged experiences and held seminars. All young scientists at that time felt a great need for travel abroad so as to participate in the work of leading scientific centres and to meet distinguished scientists working in the West. Theoretical physicists turned to such schools as Bohr's in Copenhagen, Debye's in Zürich, Sommerfeld's in Munich and Ehrenfest's in Leiden.

In 1929 Landau went abroad for the first time and was away for one and a half years. The tour was very beneficial but his stay in Denmark with N. Bohr was particularly fruitful. Landau was enchanted by Bohr's talent as a teacher, his personal charm and his personality as a scientist. In Landau, Bohr soon recognized not only a gifted scientist but, in spite of a certain abruptness and extravagance of manner, a man of great spiritual worth. Landau regarded Bohr as his only teacher in theoretical physics and I think he also learned his teaching and training methods from Bohr. The success of the school of theoretical physics which Landau subsequently founded in the Soviet Union undoubtedly owed much to Bohr's example. Landau maintained a sincere and warm friendship with Bohr and his family throughout his life. He went to Copenhagen in 1933 and 1934 and also spent much time with Bohr during the latter's visits to the Soviet Union in 1934, 1937 and 1961. During his stay abroad, Landau did his work on diamagnetism of the electron gas and (together with R. Peierls) on relativistic quantum mechanics, which achieved wide fame.

In 1931 he returned to the Physical-Technical Institute in Leningrad founded by Ioffe immediately after the Revolution and by then the chief physical institute in the U.S.S.R. In the thirteen years of its existence the Institute had grown considerably and branched out into other scientific centres all over the Soviet Union. There were three major centres in Tomsk, Sverdlovsk and Kharkov, the capital of the Ukraine. The Kharkov Physical-Technical Institute was organized and administered by I. V. Obreimov. Its main sphere of interest lay in solid state physics and low temperatures. Landau went to Kharkov in 1932 and stayed there for five years. It was the time of development of his scientific and educational career. Under his supervision the theoretical department grew very rapidly. He occupied the Chair of Theoretical Physics in the Physico-Mechanical Faculty of the Mechanical-Machine Construction Institute and, from 1935, the Chair of General Physics at the University.

Some of Landau's main works belong to his time in Kharkov: on the theory of second-order phase transitions, kinetic equation for particles with a Coulomb interaction, the theory of the intermediate state in superconductivity, and others under the wide heading of theoretical physics. This work by Landau and his students made Kharkov the centre of theoretical physics in the U.S.S.R. and the site of several conferences which were also attended by foreign scientists.

In 1935 the Institute of Physical Problems was created in Moscow to allow me to continue research which I had begun in Cambridge. Due to the goodwill of Lord Rutherford I was able to transfer my laboratory equipment from the Mond Research Laboratory and, after an interval of three years, to renew my work on intense magnetic fields. In 1937 Landau left Kharkov to go to the Institute of Physical Problems and a year later was joined by E. M. Lifshitz, his closest student and friend, his co-author of the theoretical physics *Course*. Landau set up a Department of Theoretical Physics at the Institute of Physical Problems and organized seminars on theoretical physics in Moscow.

In 1937 Landau married K. T. Drobanzeva, a Kharkov girl who had trained as a food technologist. The only interruption in his work at the Institute occurred between 1938 and 1939. During the war (1941–43) the Institute, together with its basic staff including Landau, was evacuated to Kazan. After his return to Moscow Landau continued his activities at the Institute of Physical Problems until the end of his life. In 1943 he also went back to teaching at the Physico-Technical and Physical Faculties of Moscow University.

In 1946 he had a son, Igor, who now also works at the Institute of Physical Problems; he is starting on his scientific career but in contrast to his father he is an experimenter.

On 7 January 1962 Landau was involved in a car accident; he was travelling along an icy road, his car skidded and crashed into a lorry. Landau was the only passenger to be seriously injured. The base of his skull, ribs and pelvic bones were broken. He was unconscious for six weeks and his doctors feared his death several times. Distinguished doctors from several countries helped to save his life, among them Dr W. G. Penfield, a Fellow of the Royal Society and a most eminent neurologist, who travelled to Moscow from Canada on this occasion. When Landau had regained consciousness his faculties returned to him very slowly, and he could no longer perform creative work. During the last six years he was only formally listed Head of the Department of Theoretical Physics at the Institute of Physical Problems. His physical condition did not return to normal and he died on 1 April 1968 after a major operation connected with a severe intestinal blockage.

From his earliest years Landau devoted his intellect to science. He worked out an extremely original process of research whose basic characteristics lay in the fact that it was difficult to differentiate Landau's own work from that of his students. It is hard to imagine how he could have been so successful in so many fields of physics without his students. This work was achieved in the course of non-stop discussions and regular seminars at which Landau was the most active participant and often lectured. Unlike the majority of theoretical physicists, his lectures were short, precise and expressed extremely concentrated ideas. Landau's comments on lectures given during seminars and conferences were equally precise and clear; he did not miss an opportunity to point out sharply the lecturer's fault. In his youth he behaved thus to distinguished professors and made enemies in high academic circles. If it had not been for his great talent and devotion to science, his career could have suffered considerably. With his students Landau entered into most friendly relationships and there was no outward form of respect. One could make a joke and laugh at his expense as he did with others.

It was not easy to join Landau's school. One had to pass a number of special examinations for which he devised the syllabus himself, including not only mechanics and theoretical physics, but also mathematics (in a form required for theoretical physics). Examinations were held in several stages and sometimes lasted many months at a time. Landau called the examination programme 'the theoretical minimum', constituting the minimum knowledge required by a scientist to embark on successful research in theoretical physics.

Only a few managed to pass this examination – about forty altogether. To these students Landau generously gave of his time and granted them great freedom in their choice of subjects. Their work was published under their own names.

For his part, Landau gained much from his students. He read no scientific literature himself but this was read by his students who told him about it. He was usually interested only in the basic idea of each new project. If it interested him he drew his own mathematical conclusions, often using a method different from the one employed by the author. As a result Landau penetrated into contemporary theoretical physics to an exceptional depth.

His wish to impart knowledge to others, especially his students, gave Landau the idea while still in Kharkov of creating a course in theoretical physics, which is now widely known as Landau's and Lifshitz's many-volumed treatise. Landau would be unable to write such a course on his own; in spite of his being an excellent lecturer, he was not very good at expressing his scientific work in writing. Among the young physicists working at Kharkov were two brothers: Evgeny and Ilya Mikhailovich Lifshitz. Both were talented scientists with a wide grasp of theoretical physics. The elder, Evgeny Mikhailovich, has, in addition, a talent for written expression of scientific thought. Lifshitz and Landau complemented each other exceptionally well in the work involved in creating a course in theoretical physics. They were also united by the great friendship which continued unfalteringly throughout Landau's creative life. The course was begun in Kharkov in 1935 and served as an aid to examinations in theoretical physics which were at first taken on the synopses of lectures given by Landau to research workers at the Kharkov Physical-Technical Institute. Landau's and Lifshitz's course appeared regularly from 1938. The first volume was entitled *Statistical Physics,* the next one *Mechanics* appeared in 1940; in 1941: *Theory of Fields;* in 1944: *Hydrodynamics* and *Theory of Elasticity;* in 1948: *Quantum Mechanics;* in 1956: *Electrodynamics of Continuous Media.* Each volume was re-edited several times, amended and brought up to date. In order fully to cover contemporary theoretical physics, the authors proposed to write another two volumes: *Relativistic Quantum Theory* and *Physical Kinetics*. The untimely death of Landau prevented this. These volumes will now be written by Lifshitz together with Landau's student, Pitaevskii. All nine volumes of the course in theoretical physics will then be complete.

The significance of this unique course can be gauged by the fact that it has already been translated into nine languages and has been published in England, U.S.A., Germany, Spain, Poland, Yugoslavia, Japan and China.

The distinguishing feature of this course is that it is as useful to the research worker as it is to the student. Its chief merit lies in its close connexion with the demands of present-day physics. An experimental physicist finds in it theory expressed in such a manner as to aid interpretation of experimental data. This feature is not accidental; it is the result of Landau's constant interest in experiment. He enjoyed studying the results of experiments, evaluated them and discussed their theoretical significance. He felt that in scientific research the connexion between theory and experiment had to be clearly exposed. Experimentalists in their turn greatly enjoyed discussing with Landau the results which they obtained and this undoubtedly helped the development of healthy scientific thought in the Institute of Physical Problems. To what extent Landau valued this connexion with experiment is revealed by the following. His theoretical department at the Institute was small (there were no more than ten research workers and aspirants). Although I suggested that the Academy might set up a special Institute of Theoretical Physics on as large a scale as he wished, Landau not only declined, but even refused to discuss the matter. He said that size was not important and he was extremely happy to be classed as a staff member of the experimental institute. He took a vital part in the life of the Institute, regularly attending all scientific meetings and displaying great interest in all the various events which are part of the life of a scientific establishment.

As a scientist, Landau worked very hard, with great passion. His basic strength lay in his clear and completely logical thinking founded on very wide erudition. But this strict attitude did not prevent him from recognizing the aesthetic side of science, which led him to use his feelings not only in his evaluation of scientific achievement but in the evaluation of the scientists themselves. Landau was always ready to give his appraisal which was apt and usually witty, especially in negative criticisms. These witticisms were quickly passed around and finally reached the subject under criticism, which naturally complicated Landau's relationships with people, especially when the subject of his criticism occupied a responsible position in academic circles.

Landau was a very widely educated man. He had a good knowledge of English and German, and read French easily. He read a great deal of fiction, and he was interested in various forms of art, but music left him quite cold. His opinions on these subjects were, as usual, clearly stated and original. He expressed criticism by giving marks, with lively judgement and sharp paradoxes. Landau was interested in politics and his views were liberal and progressive; but here, too, his opinions were sharply defined and he painted events black or white; there were no half tones.

His uncompromising attitudes, typical of all distinguished scientists in their work, extended to Landau's personal relationships. However, those who knew him well were aware that behind the abruptness was a very kind and responsive man, always ready to help people who had been unjustly wronged. In his youth Landau was very shy and even shunned society; any contact with people required great effort. It would appear that with age this shyness waned but Landau was never able to adjust to society. It was only the many-sided talents of Landau's personality which drew people to him, and as he got to know them they would begin to like him and find great pleasure in his company. Everybody at the Institute of Physical Problems loved Landau and his loss is sorely felt by the whole staff and in particular by his students, whose feelings towards him were especially profound.

Acknowledgement of Landau's scientific merit was shown by a number of academic awards in the U.S.S.R. and abroad. He was elected Member of the Academy of Sciences of the U.S.S.R. in 1946; awarded State Prizes on three occasions (1946, 1949 and 1953); the Lenin Prize in 1962; given the title of Hero of Socialist Labor in 1953 for his scientific activity and for fulfilling government projects; twice awarded the Order of Lenin, and several other medals.

He was a Foreign Member of the Royal Society, the Academies of Denmark, the Netherlands, the U.S.A. and several other scientific establishments. In 1962 he was awarded the Nobel Prize for his pioneer research in the theory of condensed state, especially liquid helium. He also received the Max Planck Medal in 1961 from the German Federal Republic, and the F. London Prize from the U.S.A. in 1961.

PART SEVEN

38. HOW IS ATOMIC WAR TO BE PREVENTED?

1. AVOID THE USE OF NUCLEAR WEAPONS! (BERTRAND RUSSELL)

The destructive force of atomic weapons has reached the stage when it has become clear one has to avoid their use at any price. On this count general agreement exists; the disagreements touch upon how to achieve this desired aim. There is, however, one step about which there should be no difficulty in reaching agreement. This is the halting of test explosions similar to those which are being carried out by America, Great Britain and the Soviet Union. Test explosions must not be kept secret, and this makes agreement easier. Here there will be no difficulty in struggling with the difficult problems of inspection which soon arise when further steps are proposed.

Many would prefer the Great Powers to renounce the use of nuclear weapons and to do away with the stocks of existing weapons. But as yet mutual suspicions make it doubtful whether such an agreement would confer any benefit. Each side would suspect the other of unloyal fulfilment of the agreement. Even though unjustified, such suspicions would maintain tension in relations between East and West, a reduction of which is wanted by every rational person. Moreover, neither side would consider itself bound by the agreement in the event of war occurring. Consequently, nuclear arms even if non-existent at the start of the war, all the same will be produced and used by both sides in the course of hostilities. From this consideration I suppose that the only means of preventing *nuclear* war is to prevent *war*.

Recently, enormous relaxation of tension has taken place in the relations between East and West, and one wants to hope that this will continue until the danger of war is made quite insignificant. Then it will become possible to clear away the controversial questions by negotiations, since both sides recognize that it is impossible to find the solution by military means. In the amicable atmosphere which will arise as a result, the renunciation of the nuclear arm will be sincere and no doubts will be left on either side.

I would like to see an international tribunal set up to which peoples would submit their disagreements. This tribunal should consist of an equal number of representatives of East and West and – for balance – representatives of unaligned countries. If such a tribunal existed, and its authority were recognized by all the Great Powers, peace in the world would be ensured.

But I do not think – and I want to repeat this – that the banning of one or another weapon will confer any great benefit so long as the danger of war exists. Banning could only be effective in conjunction with strict measures of inspection, to which neither side wants to agree. Moreover, even if banning were effective in peacetime, it would cease to

---

Article in *New Times* **14** (Moscow) 1956 [in English and Russian editions].
[The following article by P. Kapitza was written as a response to this article by Bertrand Russell which is reprinted here for completeness.]

be so immediately after the start of a war. I will go still further and say that the confidence engendered by banning nuclear arms would reduce the fear of war, and this would make war more probable.

In my view the moral to be drawn from the fact of the destructive force of modern weapons is that both sides must understand that their aims can no longer be achieved by means of war. This implies that in settling controversial questions preference has to be given to political and diplomatic methods, and not to threats of using force. This a long time ago became the rule in the internal affairs of states (excluding, of course, revolutions). It is required to extend to international affairs those methods which have proved their effectiveness in the internal affairs of peoples. Nothing, moreover, can prevent war once and for always. Whatever bans existed on paper, a big war will inevitably call forth the use of all the most destructive kinds of weapons which are only known to science, though their use will bring mankind to catastrophe.

From the moment of the emergence of organized states all diplomacy has been founded on open or concealed threats of war. It is not easy for one who has gone through this old school to adapt to the new requirements of the more dangerous world in which we have to live. But however difficult it might be for state officials to pick up the new way of thinking and acting, they have to learn this – and fast! – if mankind intends to continue its existence.

THE TASK OF ALL PROGRESSIVE HUMANITY (POLEMIC WITH BERTRAND RUSSELL)

Military experts believe that atomic war assumes a qualitatively new character because the destruction caused by the explosion of a thermonuclear bomb is far greater than that obtained with all previous weapons.

The devastation and misery caused by a future atomic war, in neutral as well as combatant countries, would be so great that they can hardly be compared with any of the most terrible natural disasters that have befallen mankind. This is why the progressive portion of humanity must spare no effort to prevent any atomic war.

Bertrand Russell has been a strong proponent of this view in recent years. This activity is very close to my heart. The great authority of Russell as a scientist and thinker cannot fail to exert a major influence in the struggle to prevent an atomic war.

As a scientist who is not engaged in nuclear physics or its applications (although such activities are often ascribed to me, quite wrongly, in the

---

Article in the magazine *Novoe Vremya* **14**, 39, 8, (1956).

foreign press), I would like to make a contribution of my own toward the prevention of an atomic war. Therefore I readily accepted the invitation of the editorial board of *Novoe Vremya* [New Times] to express my views on the article by Bertrand Russell.

Finding the social and governmental measures that can completely safeguard humanity from atomic war is a difficult task indeed. Its successful solution requires a free and sincere discussion on a broad international scale. To make matters worse, the task is further complicated by the fact that it involves political and social factors apart from scientific factors. Therefore, this work requires close cooperation on the part of scientists and public figures. Regrettably enough, such a cooperation has not yet found its way into traditional decision-making by governments. In my view, scientists should not restrict themselves to the scientific aspect of the problem; they must also intervene in the social and political implications. One should not be deterred from any amateurishness in such matters, since only a few people can grasp all aspects of this intricate problem in depth.

In their endeavors to solve the problem, scientists should first consider the devastating consequences of the use of atomic weapons. The destructive effect of any given type of bomb on military and civil installations can be accurately determined. Experience gained from past wars shows that within one generation even a devastated country can rebuild the material foundations for a normal life. But the main horror of nuclear weapons lies in the contamination of the atmosphere with slowly decaying radioactive substances, which has fatal consequences for mankind.

Various forms of severe radioactive poisoning came to be known long ago. But nevertheless it is difficult and perhaps even impossible to predict the biological consequences of an atomic war on mankind. Modern biology bases its conclusions on empirical data obtained by a statistical analysis of clinical experience. Since we are dealing with people, the amount of this experimental test data is of course limited and hardly allows reliable statistical treatment. Therefore a long time will be required for the correct solution of such crucial problems as the capacity of organisms to adapt themselves to radioactive radiation, or the prospects for using drugs to protect people from radiation.

It would be even more difficult to assess the pernicious effect of radiation on heredity – namely, to what extent can radiation lead to a defective progeny. Judging from the data published in the scientific literature, most scientists assume that even the atmospheric poisoning in the wake of a large-scale atomic war will not wipe out all life on earth. One should not forget that even the most lethal epidemics have always spared some naturally immune people, who were thus able to survive with no loss of viability.

On these grounds, it would be wrong to assume that fear of the fatal consequences of an atomic war is sufficient to prevent the appearance of a person with the thought patterns and moral level of Hitler, someone capable of convincing his loyal subjects to ignore the consequences of atomic warfare while fighting to achieve his aggressive ends.

In the struggle to the prevent atomic war one must also assume the possibility that reliable protection from atomic weapons will be found. If this takes place in a country with aggressive intentions, the acquired protection from the immediate effects of nuclear weapons will dispel any hesitation toward an atomic war.

For this reason scientists must find out whether an effective protection from nuclear weapons is possible at all.

History shows that a suitable means of defensive protection has always been found against any offensive weapon. It is difficult to assume that matters would be different with atomic weapons.

I have no doubt that scientists and engineers in various countries are hard at work on this problem. The results obtained in this field are, of course, classified, but some speculations on the available prospects can be made.

It is now believed that the only means of defense is by preventing the entry of the projectiles which carry nuclear bombs. Can the projectiles carrying nuclear bombs be totally prevented from reaching a given territory? In my opinion, judging from the present status of science, such a protection can be achieved, although this represents an extremely complex scientific-technological task that as yet has no obvious or generally accepted solution. If a solution is ever found, it would probably be based on natural phenomena that are still partly or totally unknown. There is no doubt, moreover, that such a defense will require a vast expenditure of energy, which means a great capital outlay. Such activities can hardly be concealed from the watchful eyes of neighbors.

If an effective defense from nuclear weapons is first developed in a country inclined to conduct its foreign policy from a position of strength, its leaders can afford to ignore their obligations. They can easily start an atomic war or at any rate exploit their advantage to extract concessions from others.

The possibility of devising a reliable means of defense must therefore be considered in treaties for the prevention of atomic war. Such treaties must envisage an obligation for exchange of information on the course of research in matters of defense, even if such a provision would complicate the mutual control on compliance with the treaty.

Bertrand Russell is of course quite right in asserting that the best way to prevent a nuclear war is to prevent any war at all.

The struggle for peace is based today on the natural feeling of disgust with war and on the striving for peace, which is so deeply rooted in the working masses. For this reason we greet the defenders of peace throughout the world for their stimulating efforts to prevent war. But their endeavors alone cannot provide a reliable and definitive barrier to war. The emotional element plays an important but not a decisive role in the history of humanity.

Wars represent a frequently recurring phenomenon closely associated with laws of social processes. The causes of war are amenable to a scientific analysis, which means that they can be identified. The main task of scientists is to examine these factors in order to assist in devising measures for the prevention of war. It is the duty of public leaders vigorously to implement these measures in life.

An essential step toward the prevention of atomic war is the identification and eradication of the various social factors which lead to war. This eradication process is already underway. But being a social process involving the coordinated efforts of a large number of people, it has a slow course.

A sudden war may stop this healing process of developing peaceful relationships between countries. Despite the imperfect means of control, any measures and treaties for disarmament, and for banning the production and testing of nuclear bombs, must be welcomed and supported in all ways. Today they undoubtedly represent the best obstacle to war.

In view of these considerations, I cannot agree with Russell's criticism of the proposals for the banning of nuclear weapons; I would like to remind him of the saying that 'the better is enemy of the good'.

39. PHILOSOPHY AND IDEOLOGICAL STRUGGLE

I heard with interest the report by I. T. Frolov on the prospects of the journal *Voprosy Filosofii* [Problems of Philosophy], because it deals with the future development of the ideology of our socialist society. I think we at the Academy of Sciences tend to underestimate these philosophical questions in our time.

Future historians will undoubtedly regard our century as the arena of a struggle between two social regimes: capitalist and socialist. This struggle affects the economic, political, and ideological spheres. The development of our state during the half-century which followed the October Revolution shows the viability of the socialist system. Looking at the most advanced capitalist country – the U.S.A. – we can quite objectively say that our two countries are more or less equal in the main areas of material and cultural development, in education, in the development of science, in defense capacity. The only field where we are lagging behind is industrial-technological development – a lag due mainly to the fact that our productivity of labor has not yet reached the U.S. level.

Thus the matter boils down to the conflict between the underlying ideologies of these two systems. Philosophy determines the ideological principles which govern the relationship between man and society. Capitalist ideology, as we all know, rests on the striving of the individual toward material wealth. By contrast, the socialist ideology stresses the development of society as a whole – a goal which necessitates a balanced development of the individual. Hence the great value which the socialist society attaches to the creative and ethical qualities of the individual, and the efforts of society to promote these qualities. The ideological struggle between the two systems results from these contradictions in the relation toward the individual in the two systems. Today the most progressive social order is the one which allows a most balanced development of the spiritual qualities of the individual as a necessary foundation for a happy life.

Mass revolutionary movements have appeared in recent years in capitalist

---

From an address to a session of the Presidium of the U.S.S.R. Academy of Sciences, 1969.

countries, especially among the youth. These growing movements, which affect the most developed capitalist countries, are led by students. The forces behind them are only partly understood. At any rate, these movements have nothing to do with the material aspects of life; they strive to change the ideological conditions of life in capitalist society. Thus the progressive public opinion of capitalist countries quite spontaneously demands a revision of the ideology of capitalist society. How can this revision be carried out? Who will prepare the program of reconstruction to be followed by progressive humanity for the advancement of mankind? The answers will come out in the ideological struggle now underway at a fast pace between two opposed world outlooks.

Must we openly join this struggle and what would be our role in it? Of course, the ideas and principles provided by Marxism for the construction of the communist society are the only guide for steering this struggle toward the desired goal. This is now accepted by all of progressive humanity. Now efforts are being made to find specific ways for a more effective development of this revolutionary movement. Such ways are being sought in the polemic between ideologists of the new school, Marcuse for example. Trotskyists and others have also joined in the struggle.

Let us admit that our ideologists have remained isolated from this revolutionary process and practically do not exert any influence on it. This anomaly cannot be tolerated, because the fact of the successful existence of our socialist society constitutes an example for this revolutionary movement.

How can we effectively join in these revolutionary processes which take place in capitalist society? If we want to keep pace with the development of progressive thinking and take advantage of the scientific-technological revolution of recent years, we must improve the level of our social sciences. For this reason we at the Academy must pay tribute to the effort of the editorial board of *Voprosy Filosofii* to promote the development of philosophy so as to be able, among other ends, to influence the progress of social thinking in the revolutionary movement now sweeping the capitalist countries.

To exert such an influence, however, we must actively join in the ideological struggle which is taking place there. Our philosophers must participate in this struggle on an equal footing, just as our sportsmen do. Let us remember that in this struggle our ideologists will lose the privilege they enjoy in our country, where they have no opposite views to contend with. In this struggle they will have to show their own worth.

For this reason I propose that the Presidium of the U.S.S.R. Academy of Sciences support the program formulated by the new editorial board of the journal *Voprosy Filosofii.* Under this program the journal will have to increase its international influence, and the Academy Presidium will be expected to devote more time to a discussion of philosophical problems concerned with the ideological basis of our socialist society. The scientific reports read at the Presidium practically ignore such themes. This state of affairs must be corrected.

40. THE FUTURE OF SCIENCE

They have asked me to speak on the future development of science. This is a difficult task, but one which is interesting and useful: only by having a clear perspective of the future can we rightly direct our work in the present.

I would like to remark that I am one of the oldest present at the conference. Throughout my forty years of scientific work I have had the opportunity to observe the numerous changes which have taken place in the development of science and in its tasks. Glancing in one's mind over this phase of science, it is difficult not to notice the radical changes now in the relations to it. In the days of my youth they often talked about 'pure science', 'science for the sake of science'. Of this there is no more talk now. Science has now come to be regarded as a necessary component in the social system, and not just as useful, but as an integral part. The State is paying more and more attention to science as a most important element of State life; now scientific institutions are being placed on a par with other branches of the social system, for instance, education, transport and the army. This was not the case fifty years ago; then chance factors ruled in the organization of science, and its development was based on private initiative.

Today with the expanding of scientific research in all countries on a larger scale, State financial appropriations for the development of science continue to increase in academic and also industrial scientific institutions. Very complex plants of immense size are being constructed, like accelerators for many millions of electron volts, large atomic reactors, artificial satellites launched for investigating outer space and so forth. The solution to such problems cannot be the affair of individuals, but is the result of collective creativity, requiring great organizational effort and resources which are only within the capabilities of large States.

The growing scale of scientific work involves the dividing of science into basic (cognitive) and applied science. I think that this division in the main must be considered artificial for it is difficult to mark the point where basic

---

Verbatim report of a speech in Prague, 1959, at the International Symposium in Prague on the planning of science, adapted by the author for the journal *Science and Life* **3**, 18 (1962).

science ends and applied science begins. The division is bound up with the particular aims being pursued by the scientist, whether cognitive or applied. Therefore basic science is increasingly being concentrated in academic institutes and in universities and applied science in industry. This separation of science is mostly connected with the necessity of financing, planning and supervision of scientific work.

In the work of many great scientists it is difficult to trace when they were pursuing the applied aim, and when the aim of cognition. For example, the great scientist Langmuir worked all his life in industrial undertakings and he solved some of the biggest technical problems in the electric lamp industry, but it is well known that in the course of this research he carried out many fundamental investigations in electronics and vacuum physics.

COLLECTIVE WORK IN SCIENCE AND THE DIRECTOR'S FUNCTION

Now we often discuss how scientific work should be organized on the scale which requires a large, coordinated creative collective. Who ought to direct the work of such a collective – a scientist or an administrator?

Professor Bernal in his address said that administrators play the decisive role and that they are necessary for organized work on big scientific problems. I do not agree with Professor Bernal – not in that such organizers are not necessary for collective work in science, this is right, but, in my opinion, because they should be scientists directly themselves and not administrators. I am best able to express my thoughts by drawing a comparison with other creative fields, namely, with the theatre and cinema.

Once the theatre consisted only of troops of actors, and the producer was an inconspicuous figure. But especially since the development of the cinema, in which thousands and hundreds of thousands of actors take part, the one who determines the success of the film is now the producer. With large-scale collective research, the producer has now become necessary in science as well. What requirements do we expect of him? The main requirement is that he be creative, and not just play an administrative role. He must understand the meaning and purpose of solutions of scientific work and he must rightly assess the creative possibilities of his executives, allocate functions according to talent, and deploy his forces so expeditiously that all aspects of the problem to be solved develop harmoniously.

Since to organize the solving of any new scientific problem one has to find the organizational forms oneself, the director of a big scientific project, even if he himself personally does not work in science, must be a person with great creative talent. I do not know why the director of such a magnificent achievement in science as the launching of the first artificial satellite is not worthy of the Nobel prize, though, maybe, he personally did not do the scientific work associated with the preparations for this unique experiment? Did he not organize it? Such cinema producers as Sergei Eisenstein, or René Clair, of whom one may say that they are great creative directors, and who, as is well known to us, created the most remarkable feature films, were they themselves not actors in this respect? We know cases when a great actor was also a great producer, for instance, Charlie Chaplin. So also in science cases are known when a great scientist is also a great organizer of collective scientific work. Such versatile scientists were, for instance, Rutherford and Fermi. But this, naturally, is a felicitous exception, and not the rule. Doubtless a period of scientific development is now setting in during which the functions of the organizer are becoming greater and greater.

It seems to me that we must now begin specially to educate and train persons to be organizers of large scientific projects, and in order to make this duty attractive, to treat such persons with great respect, and not to regard them simply as some variety of bureaucrat-administrator. In a short paper it is difficult to formulate the attributes necessary to select these people and how to teach them and educate them. They are met with very rarely and, apparently, this is a unique kind of human talent, and so they need very great attention and great care.

Hence I suppose that one of the problems of the future is this educating and developing of a new type of science-organizer, with the duties and status I have just described. This type of science-manager is at present in the initial phase of development, but he will play a decisive part in the large-scale science of the future.

SCALE OF SCIENTIFIC WORK

The next problem on which I want to dwell is the scale of scientific work in the future. Shall we try to look into the question of the volume of resources (personnel, material and finance) which the State will eventually earmark for scientific work? At the present time even in the most advanced states this volume amounts to a small fraction of one per cent of the total budget. But each year this fraction is increasing in absolute magnitude, both in socialist countries and in capitalist countries.

Many economists are now recording great social phenomena associated with the rise in technical culture. And in agriculture, as well as in industry, the part played by physical effort is continually decreasing. Already, on more than one occasion, it has been pointed out in the press that with the growth of electric power supplies and with the introduction of mechanization, and automation, production will use up little of a person's strength; owing to power station energy this work will be done by cybernetic machines, whilst the released creative forces and spiritual energy of people will mainly be directed to science and the arts. It is asked, but what section of mankind will eventually be engaged in science and the arts? Here we can resort to an analogy in the Herbert Spencer style. If one compares the State organism with an animal and compare the weight of that part of the animal's body which does mental work, namely its head, with the weight of all its other parts doing physical work, we get an interesting result. Let us begin with antediluvian animals, for instance, the dinosaurs. These were reptiles with a small head and a gigantic body. We know that in the evolution of life on the Earth the future did not belong to such an animal. The future in the struggle for existence belonged to man, whose head is roughly 5–10 per cent of the weight of his body.

So also in the evolution of human society, culture will continually grow and more and more resources will be expended on it. Thus one half of the population of the State will perform the social functions, whilst the other half will work in institutes, design bureaus and at experimental plants where mechanization and automation does not apply, but an individual approach is necessary for each new problem which arises. Professor Bernal paints a different picture of the creative scientific work of people in the future. He supposes that everyone will allocate one part of their time to mental creative work, and the other part to productive work. To me, personally, Semenov's suggestion seems more probable, since people who are inclined to creative work will give themselves to it whole-heartedly. This gives the persons more satisfaction, and this makes their creative work more productive.

SCIENTIFIC DISCOVERIES OF THE FUTURE

Now I want to dwell on the fields of science which, one may suppose, will arise in the future. In this respect predictions can be made from different premises. I propose to do it on the principle of extrapolation, and so I shall begin with an estimate of the number of new natural phenomena which have already been discovered. I want to qualify the expression 'new phenomenon'; I apply it to such physical phenomena as cannot possibly be predicted or explained on the basis of theoretical concepts already in existence, and therefore the discovery opens up new fields of investigation. So as to make the proposed extrapolation clearer, I shall call fundamental *new phenomena* in physics *basic phenomena* if they have been discovered in the last 150 years.

First of all I want to cite Galvani's discovery in 1789 of the electric current which, of course, in no way followed from the hitherto existing theoretical concepts of the nature of electricity, essentially due to Franklin.

The next discovery coming within our definition is Oersted's discovery in 1820 of the influence of an electric current on a magnetic needle. From our viewpoint Faraday's later discovery of magnetic induction was not new since magnetic induction is essentially the obverse of Oersted's discovery, and thus at the time it could have been predicted. The work of Oersted and Faraday led to Lenz's law, to the formulation of Maxwell's equations and to many other fundamental conclusions, although they were all an elaboration of the basic discovery of Oersted, which possibly could have been predicted theoretically.

The next instance of a new phenomenon was photoelectric emission, discovered in 1887 by Hertz (we all, of course, honour Hertz rather more for discovering electromagnetic waves). This phenomenon also could not have possibly been predicted theoretically. On the basis of studies of photoemission Einstein thirty years later derived his famous equations, having determined the quantum nature of these phenomena. The principle of uncertainty and quantum theory were pre-determined by the discovery of photoemission, and all the remarkable scientific elaborations of this phenomenon only constituted further methodological work.

Next, one may cite Becquerel's discovery of radioactivity in 1896, which also could not have possibly been predicted on the theories then existing, and which laid the foundations of nuclear physics. Then Thomson's discovery of the electron can be regarded as a new phenomenon, laying the basis of modern electronics. Also the experiment of Michelson and Morley, since it produced a result which it was impossible to have predicted theoretically, and one may say that the discovery of the phenomena which established the basic principles of the theory of relativity was also new. It was impossible to have foreseen Hess' discovery of cosmic rays in 1919. I daresay one also needs to mention uranium fission discovered by Meitner and Hahn as a new phenomenon.

What is typical of all these discoveries? Above all their value was only completely recognized after 20–30 years, when it had become clear that they could not be explained by the scientific views of the times, and therefore under their influence new directions arose and developed in the fundamental theoretical conceptions.

Are such discoveries possible in the future? Have all the physical discoveries in Nature already been exhausted? Are there other fundamental phenomena in Nature which are awaiting to be discovered?

If we plot a curve marking off time on the horizontal axis and the number of discoveries along the vertical axis, and if we honestly consider this discovery curve, it is seen to have a tendency to fall to zero. Therefore, by extrapolating this curve, we see that in the near future we shall be witnesses of no more important and 'new' discoveries other than those just cited. These will enable us to understand Nature more deeply and place at Man's disposal new possibilities for cultural growth. Usually people are inclined to consider that everything about Nature that can be known is already known. So it always has been. It is sufficient to read something of the works of Newton's contemporaries to see that even then many considered that with the discovery of the classical laws of mechanics there was nothing more to be learned about inanimate nature.

Although this is very often contrary to our sense, we must not make the same error – to believe that no new discoveries will be made in the future.

You may possibly ask what these discoveries will be? If I could predict them, they would no longer be unexpected and new. Yet, all the same, I want to mention an instance when it seemed that I might have made a discovery, but I did not try to do so. In this example there is an instructional element.

When in 1930's I produced very strong magnetic fields, ten times more intense than any produced before me, scientists counselled me to carry out tests on the influence of strong magnetic fields on the speed of light. Einstein talked with me most persistently of all. He said, "I do not believe that God created a universe such that the speed of light is independent of everything within it". In such cases Einstein loved to refer to God when there was no more sensible argument. From the tests which had already been made in this direction, it was known that if I were to carry out testing with my stronger fields, the effect would still be very small, only of second order. Here, of course, it was impossible to predict the true magnitude of the effect since the phenomenon was new. At the same time the test promised to be exceptionally complicated since similar experiments had already been carried out with fields up to 20,000 oersted, and they showed that even with a very sensitive measuring method the magnetic field does not influence the speed of light.

Another person who was pressing for this experiment and who had even offered financial backing was Oliver Lodge. He also turned to me with advice to carry out this exceptionally difficult and delicate experiment. Yet all the same I refused. Why? I will explain this by the following instructive example, which, perhaps, is not known to many.

As you will remember, the law of the conservation of matter was experimentally discovered by Lomonosov in 1756 and slightly later by Lavoisier. At the beginning of the present century Landolt verified it with great accuracy. He too placed the substance in sealed vessels, weighed it exactly before and after the reaction, and showed that the weight remained unchanged with an accuracy of up to the tenth decimal point. If one takes the energy which is released in the chemical reaction and calculates the weight variation by Einstein's equation in the theory of relativity, it appears that if Landolt had carried out his experiment with accuracy of two-to-three orders greater, he might have been able to observe a change in the weight of the reagent.

Thus we know that Landolt came very close to discovering one of the most fundamental laws of Nature. But let us suppose that Landolt had expended still greater effort on this experiment, that he had worked at it for five years more, had raised the accuracy by two-three orders, and he had observed this change in weight, the majority of scientists would still not have believed him. It is common knowledge that any experiment with limited accuracy is always unconvincing and that to check it, another experimenter has to be found ready to spend another ten years of intensive work on it. Experience suggests that so long as a problem's solution is at the limit of accuracy by known experimental methods, it can be verified only when Nature suggests a new method of solution. On other occasions it has been so: Einstein's law was verified fairly simply by Aston when he invented and developed a new precise method of determining the mass of radioactive isotopes from the deviation of an ionic beam. So we should wait, and in the case described by me, when Nature offers us new procedural possibilities for studying the influence of magnetic fields on the speed of light, probably, a simple and convincing experiment will become available for studying this phenomenon. That is why I refused to conduct these complex experiments.

Concluding this section, I think that one may say with complete confidence that in the not too distant future physicists are in for discovering very much more that is new and interesting and it is appropriate to quote the words of Hamlet: "There are more things in heaven and earth, Horatio, than are dreamt of in your philosophy". As it was three hundred years ago in Shakespeare's time, so it is now, and so will it always be. In essence, it is a question of nothing other than the law of continuous dialectical development of man's cognition of nature.

MOST IMPORTANT SCIENTIFIC PROBLEMS OF THE NEAR FUTURE

Besides discovering new natural phenomena which we are unable to predict, scientists will also direct principal effects to a deeper study of the natural phenomena already discovered, and to solving methodological and applied problems. To find the lines of scientific research which will be foremost in the near future, one needs to define the fields of natural science which are most closely associated with pressing demands of practice.

The most important and interesting problems confronting science right now are well known to all, and I shall not dwell on them in detail. The first of these problems is the conquest of outer space. To achieve this aim the main thing has been done – man has been shot up from the gravitational field of the Earth; the main problem in the future is the use of nuclear energy as the motive force of space vehicles. What practical results will come from solving this problem? The possibility of populating other planets is increased. This, of course, is a task of the very distant future; we are not as yet even crowded on the Earth. But the beauty and fascination of penetrating hitherto unexplored domains is that man cannot foresee what he will find there. The whole historical store of experience unfailingly shows that new opportunities are always opened up for a rise in human civilization by exploration. Undoubtedly this will also be so in this case.

One may remark that right now there are already practical possibilities for utilizing space rockets – for disposing of the radioactive wastes and slag left from atomic reactors. More than once it has been pointed out that in the near future large quantities of radioactive waste from atomic electrical plants will start to accumulate and disposal will be difficult and dangerous; many consider that this will be the main obstacle to large-scale development of atomic engineering. If these radioactive slags are despatched into outer space on rockets, this will be quite safe for mankind and, apparently, it involves no great expense. It may be the solution to get rid of radioactive waste in this way.

One of the permanent, most important problems of the present and future is the production of cheap electric energy.

A most important possible solution of this problem is the controlled thermonuclear reaction. This is the most important problem of modern physics; it will give people an inexhaustible source of energy; the solution depends on creating a sufficiently high-temperature plasma. Scientists are still searching for ways of solving this problem.

A more immediate problem is that of using the heat from fuel efficiently.

As you know, coal is converted into electricity in successive stages: firstly the chemical energy is converted into heat in a furnace and then the heat is converted into steam in boilers, then into mechanical energy in prime movers, finally, into electric energy in the generators. But in this way only 30–35 per cent of the coal's chemical energy can be used, and even this involves heavy capital investment in machines. To make this process more efficient in the future, a new line of development is envisaged. I have in mind the so-called 'hydroelectromagnetic' generators. The idea of these generators was proposed at the beginning of the century. If a fast-flowing jet of a good conducting plasma is passed through a magnetic field, a transverse electromotive force arises. By using this force one can produce current and so convert the kinetic energy of the jet into electric energy. Recently, thanks to the development of jet aircraft and rocket engineering, the process of producing great jets at high temperatures has been well mastered; therefore realization of the old idea of the hydroelectromagnetic generator has become realistic, and large institutes both with us and in the U.S.A. are seriously working on this right now. One may forecast that this generator will be operating effectively as soon as large outputs of the order of several hundred megawatts are attained; the whole installation will be relatively small in size.

Yet all the same the direct conversion of chemical energy into electric energy must be more alluring. In conventional galvanic elements and accumulators this has already been carried out and, as is well known here, an almost complete transition of chemical energy into electric energy is possible, and theoretically the efficiency may be close to unity. The basic problem facing scientists is to create a galvanic element where electric energy would be obtained directly from the oxidation of coal. The realizing of gaseous elements of this kind which will operate at high temperatures and pressures has appeared possible, and here during the last decades there have been notable successes, although the problem has not as yet been brought to a practical solution. Unfortunately, it is doubtful whether in general one will succeed in realizing similar plants for large outputs, since this is bound up with difficulties of principle. The point is that the chemical oxidation of gaseous elements has to be carried out on the surface of electrodes, and not in the volume of the electrolyte, whilst under these conditions a very extended surface is required for large outputs, which is practicable only on a large scale. Therefore this principle of generating electric energy will only have significance for small-scale power supplies.

One must never forget one problem in power engineering: the direct conversion of chemical energy into mechanical energy. Here people lag behind

Nature. The muscle motor is still the most extensive. The quantity of mechanical energy producible by the muscles of all animals is still several times greater than the energy from all the heat engines created by people. The muscle motor, as shown by experiments, also operates very effectively with higher efficiency than engines, turbines or other heat engines. But most surprisingly, one has to confess, until recently scientists have not understood the essentials of the muscular process. There are many hypotheses, but until scientists have succeeded in artificially reproducing the contracting of muscle fibre, it is impossible to consider this process understood. Doubtless, research into the mechanism of muscular contraction will be one of the key problems in future scientific investigations. Physicists, chemists and biologists will take part in this work. What are the chances of the work being completed in the years immediately ahead? The problem can be considered solved if we succeed in reproducing the reversible contraction of synthetic fibre under the influence of a change in the properties of its surrounding medium, i.e. if we can simulate muscular processes. Today the number of types of links of chains which form the fibre of synthetic polymers is continually increasing. The nature and the properties of molecular bonds in polymers are beginning to be understood more and more deeply, and all this gives hope that the secret of muscular contraction will be laid bare on artificial fibre. In any event this is one of the most important problems of the near future and, by solving it, an effective mechanical motor will be placed in people's hands. It is natural to expect that such a motor will be portable and only suitable for small outputs.

Incidentally, many other secrets are hidden in the fibres of polymers which would be very useful for people to know. For instance, it is well known that an electric impulse (signal) can be freely propagated along nerve fibres which we consider to be dielectric. We know that this undoubtedly takes place in nerve fibres, but the mechanism of this interesting phenomenon is as yet not understood at all, and we do not know how to reproduce it. When this problem also has been solved, we shall be able to make signaling circuits, counters and other elements of cybernetic machines without metal conductors. To understand this phenomenon is also one of the great problems of the future.

SCIENTIFIC METHODS IN THE FUTURE

There are many very important scientific and scientific-technical problems which we cannot solve right now owing to the limited methodological possibilities available to us. Procedures may be limited by the complication of the phenomenon itself, which leads to mathematical problems of insoluble complexity, or of the available method of measurement; the measurements may be insufficiently sensitive or too imprecise, or in general the phenomenon may not lend itself to measuring at all.

Success in solving many problems therefore depends on expanding our methodological possibilities. In the main, success here depends on inventing new methods of observation, inventing measuring apparatus operating on new principles and, finally, by inventing methods for theoretical and mathematical generalizations of scientific experiments.

All these inventions can be regarded as scientific discoveries of their own kind; the greatest of them are made as unexpectedly and as unpredictably as scientific discoveries, and they also are manifestations of man's genius. Great methodological inventions, like scientific discoveries, can also lead to the creating of a whole scientific field and lead to the solving of basic problems which have long been confronting science. As an example of the latter one may point to Newton's invention of differential calculus, or Huygens' balance-spring for clocks.

One of the greatest of such modern methodological inventions is the creating of electronic cybernetic machines which has moved many fields greatly forward. The rapid development of cybernetic machines which is taking place right now is making it possible to solve many problems of great importance which until very recently lay beyond the bounds of accessibility (rapid and exact calculation of space vehicle trajectories, atomic, molecular and crystal lattice structures and many other problems). Doubtless, electronic cybernetic machines will intensively develop in the future, and with their aid many other important problems will be solved which have hitherto been beyond our reach.

Right now I want to turn your attention to some most important problems which it is necessary to solve and of which we need to have the theoretical solution; but at present they are so complex that they are having to be solved in a rough empirical or semi-empirical way. One needs to remember that empiricism has far from spent itself as a scientific research-method.

The using of empiricism in these investigations usually involves laborious collecting of great quantities of test data and great complications in systematizing and utilizing the data. Let us take as an example empirical investigations of a kind which are often now carried out, the problem of creating substances with well-defined mechanical properties – strength, heat-resistance, ductility and so on. It is well known that in the domain of the achievements in aviation, cosmonautics, turbine engineering, the strength and heat-resistance of the materials are usually the chief limiting factors. It would be enough to raise the heat-resistance of alloys by several hundred °C and their ultimate strength by 20–30 per cent – this would enable many new technical problems to be solved. However, despite the fact that all the mechanical properties of metals are being measured well and rapidly, there is as yet no quantitative theory which links these properties of matter with its chemical composition and physical structure, although the nature of the forces between the atoms is well known. The mathematical problem is so complex that it cannot even be formulated. Therefore the main way of searching here is empiricism.

Yet it can be shown without difficulty that even empiricism cannot completely solve this problem. About 100 elements are known which form alloys. Let us assume that it takes one page to describe the necessary properties of one metal or alloy – its strength, heat resistance, elasticity, electroconductivity and so on. To describe the properties of the elements themselves 100 pages are required; to describe the binary alloys 10,000 pages are required already. The alloys of ternary systems already fill a million pages. It is easy to see that to investigate and systematically describe the ternary alloys is the limit of possibilities. Thus the empirical method of study has inherent limits. The invention of cybernetic machines of the tabulator type naturally extends our possibilities here also, but all the same one needs to recognize that the problem of scientifically creating new alloys with specified properties has not been solved for more than three constituents, though it is common knowledge that alloys of three or even more constituents are already used in practice and important problems have already been solved by such alloys.

Will this always be so? I do not think so. These multi-constituent alloys may, perhaps, be found by chance, but more probably by the intuitive 'flair' of the talented scientist who, like a master chef, knows how to cook more tastily than the others. If there is intuition, this means there is also justification. It is the job of science to elucidate these justifications, though the method of solving such complex problems has not as yet been found, and this, undoubtedly, is one of the problems of the future.

There exists one other, less known problem which will be of great interest in the near future – it also is being solved empirically at present. It is the creation of superconducting alloys at normal temperature and with a sufficiently high critical magnetic field, i.e. the field which breaks this superconductivity. It is well known that in superconductors electric current flows without loss and, therefore superconductivity is already beginning to be widely used for designing high-quality radio-oscillatory systems, for coils with strong magnetic fields, for miniaturization of storage devices in electronic computers. But the main difficulty with the practical use of superconductivity is that all these devices operate at the very low temperature of liquid helium (4·2 K). Therefore most practical value would attach to the discovery of materials possessing superconductivity at room temperature. This would cause a revolution in modern electrical engineering since it would enable power to be transmitted without losses. But meanwhile the theory indicates that superconductivity cannot exist in pure metals at temperatures above the Debye temperature and, consequently, at present one can only await the discovery of such materials in the alloys, the theory of the superconducting properties of which is not as yet understood at all. Here also arises the problem of empirically studying multi-constituent compounds, about which I have already spoken.

One of the biggest problems to be solved in solid-state physics is the creation of polymers with specified properties. Polymers in living nature are always the main 'building material' which performs functions of all possible kinds. Our age will not only be the age of the utilization of atomic energy, but also the age when man learns to produce polymers and to employ them as widely in practice as the fundamental 'building material'.

The variety of polymers is unlimited, there may be more of them than the alloys. The mechanical, electrical and magnetic properties of polymers are also diverse. Science is confronted with the problem of creating polymers with specified properties. Here also empiricism is inadequately powerful, as in the example of the alloys. Possibly, owing to the great regularity of polymer structure, the theoretical generalization will be found sooner for them than for alloys, which will save all empirical research into finding polymers with specified properties.

Successes in inventing new methods of experimentally investigating natural phenomena have been exceptionally big during recent years. Apparently, one has to attribute this to the rapid development of theoretical and industrial electronics and to those new possibilities which the development of nuclear engineering and physics has opened up.

It is interesting to mention that today frequency variations are measured accurately to the 16th decimal point (using the Mössbauer effect), and time to the 11th decimal point (by molecular generators). Using electron beams and different methods of magnification, one can see molecules and so on. Such achievements in inventing new methods of observation and measurement are now going on in an unending stream. There are no signs that the development of our experimental possibilities will cease in the future. The next achievements in inventing methods of observation or measurement are as difficult to foresee as the new discoveries. But all the same I would like to point out a field of measurements where man is lagging behind Nature and where one ought to expect new inventions in the near future.

Today a physicist has instruments at his disposal which are many times more sensitive than our sense organs. A microphone hears better than the human ear, a photo cell sees better than the eye over a wider spectral region. Seismographs are more sensitve than our sense of touch and, of course, man determines temperature quite badly compared with a thermometer.

Only one feeling – this sense of smell, i.e. the determination and detection of small quantities of impurities of organic matter, is more perfect in the animal than by existing instruments.

Now let us take connoisseurs of wine. The great physicist Langevin was one. I well remember – it made a great impression on me – how once before dinner at a congress in Zürich in 1925 he tried the wine and once again correctly determined not only the brand of the wine from its taste, but also its vintage. He was an acknowledged connoisseur and he was very proud of this, perhaps, even more than of his successes in physics. But there are no such physical instruments that could do what he did even approximately.

The most sensitive method for determining inorganic impurities at present is considered to be radioactivation analysis. In this way one can detect impurities in amounts of one in 10^8 to 10^9. Comparing this with the smelling of a dog, the dog appears to detect much smaller quantities of impurities and in this case the dog's senses can identify them. One asks why man has not devised other instruments such as could catch the negligible quantity of atoms accessible to the smelling of the dog?

It is well known that the olfactory organs are the most complex of all the sense organs and the nature of the phenomena on the basis of which they function, has not as yet been discovered. Thus, to 'surpass a dog's smell' is one of the problems of physicists in the future.

Let us remember that there is another field where Nature has invented a better mechanism than man. This is the mechanism of memory of the brain. This mechanism is many thousand times more compact and effective than the storage devices in modern computing machines. The nature of the mechanism of memory of the brain is also unknown to us.

All these problems are problems of the future, and here physicists must take away Nature's interesting and fascinating secrets.

THE FUTURE OF BIOLOGICAL SCIENCES

Many big problems in the biological field associated with queries in agricultural engineering, zootechnics and medicine are well known, and I shall not dwell on them.

In the foregoing sections I have indicated some of the biological problems which it is important to solve in the future – the nature of musuclar contraction, propagation of electric signals in nerve fibres, the mechanism of memory of the brain and the mechanism of smell. The successes which have been achieved in solving these problems are undoubtedly due to the penetration of physics and chemistry into biology which is going on right now. The investigations give full grounds to suppose that not only will these problems be described by known laws of inanimate nature, but that these processes can be reproduced artificially and used in practice. By finding out the mechanism of these processes not only is the biological significance revealed, but physics and chemistry is also enriched. Therefore in the future we can expect a fuller merging of physics and chemistry research with biology.

The fuller understanding of biology's own very complex processes will undoubtedly deepen our knowledge of inanimate nature. Therefore the time has now come to speak of the favourable influence of biology on the development of physics and chemistry. The development of the chemistry of polymers and the study of their physical properties which is going on so intensively right now are instances of this favourable influence. For example, from the structure of the polymers used in Nature it is seen that Nature is a better 'design engineer' than man, and at present there is much for us to learn. One ought to mention that in some respects scientists have surpassed Nature by creating processes such as do not go on in Nature. Thus, for example the uranium chain reaction, used for producing nuclear energy, does not go on in natural conditions. We can with great confidence assume that in the near future the solution of the great biological problems will be determined by the development of 'composite problems', i.e. problems involving not only biologists, but also chemists, physicists and even mathematicians. Thus the expansion of big collective research projects about which I spoke at first, will also apply in solving the problems of the biological sciences.

One of the most important and interesting composite problems in the field of biology, where physicists, chemists and mathematicians have to participate, is genetics. The immense successes which have been achieved in genetics in recent years are already widely known. Today scientists have not only learned to produce artificial mutations, but have begun to acquire a detailed understanding of their physical essence. This has become possible thanks to determining the structure of chromosomes, the deciphering of the code by which in the genes the information is stored necessary for the development of a particular organism and, finally, to understanding the mechanism itself of the process of reproduction.

Of course, the main applied problem facing geneticists is to change the form (species) of organisms according to the demands of practice. As yet there is still a long way to go before these tasks are implemented, but ways of solving them are envisaged. The present means of influencing chromosomes – by irradiation or chemical compounds – can only produce random mutations. As yet one can produce desirable mutations only in the most simple of organisms – viruses, microbes, fungi. All these organisms rapidly multiply and the number of possible mutations in them is not so large; therefore, by working out an effective method of artificial selection one can create a species with certain properties for applied purposes. Precisely in this way they are now producing the most active antibiotic drugs. By irradiating fungus cultures they cause mutations in them and select those exhibiting the most activity in relation to a specified bacillus. But, of course, by such methods it is impossible to produce desirable mutations in complex organisms, just as it is impossible to improve a complicated mechanism by hitting it at random with a hammer.

But if scientists succeed in finding a method of producing mutations in desired directions, then, of course, man will have in his own hands a method of changing species on a scale which will be incomparably more potent than is now possible by selection and hybridization. Here due research is imminent. It is not hard to predict that the know-how to alter species will first be acquired on the simplest organisms, and then be extended to the more and more complex.

Let us suppose that scientists ultimately succeed in finding a method of producing artificial directional mutations which will alter the human species. An interesting and very controversial issue will then arise – the issue of actually altering the human species. This opens up the prospect likewise of changing the structure of society as described by Aldous Huxley in his daring fantastic Utopia, *Brave New World*. The history of civilization teaches us that eventually the fantastic becomes real.

But this is a question of the distant future and for the present one ought hardly to dwell on it. For the present it is interesting to pose the question: what are the chances that in the next few years a method will be found for 'controlling' the direction of mutations? One has to say that we are still very far from solving this difficult problem. Possibly by studying the physical properties of synthetic polymers one might discover a method for more organized action upon their chain structure. It is hard to say how soon this will be done, but in any case the problem is now actual in character, and it lies within the framework of the laws of physics and chemistry which are known to us.

It is hard not to point out that what has seemed the more complex problem – hereditary and direct action on the mutation of species – has appeared to be understood sooner than the other properties of animate nature, and possibly heredity will be the first to be 'controlled' by man.

In conclusion I want to pose the following other fundamental question associated with living Nature and which is imminent.

The question is the following: is our present-day knowledge of the laws of inanimate nature adequate for describing all the phenomena which are characteristic of the living world?

We know that most phenomena are described by existing laws, but all the same I dare say that one of the main properties of living Nature – to reproduce itself – may be the manifestation of various forces in Nature which as yet are still unknown and unexplained by the known laws of interaction between elementary particles. We have no data to affirm that no new property can appear in chains of sufficient length by the arrangement of the atoms according to definite rules which is analogous to that of self-reproduction in living Nature. Of course, in individual atoms and simple molecules such a property may be unnoticeable. That this possibility is not excluded, we can illustrate by the following example.

It is known that only with large accumulations of elementary particles does the force of attraction begin to play its role between them, the force which we call gravitational force. You see, the nature of gravitation is not considered in describing quantum and electrical interaction of atoms and it manifests itself in Nature only in large masses. Likewise other as yet still unknown properties of atomic interaction may be manifest only when their mutual arrangement is ordered.

It is science's task by experimenting to elucidate these laws of self-reproduction and to find those parameters by which it will be possible to describe these laws quantitatively. If this succeeds in being done, new properties of the nature of matter will be discovered which escape us in studying the inanimate world.

But the approach to the solution of this problem is not yet even established, and we can only refer its solution to the distant future.

THE FUTURE AND THE SOCIAL SCIENCES

There is one other field of science on which I shall touch only briefly – this is the science of human society. The social sciences are special in that they have their own theoretical methods of study and generalization which differ from those adopted in the natural sciences for elucidating conformities to law. Owing to this there is a distinct boundary between the natural and social sciences which divides them into two independent and isolated fields of knowledge.

It is generally acknowledged that the basic science of the laws of the development and construction of human society – historical materialism – was founded by Marx. He first elucidated such basic parameters as the class structure, labour productivity and so on, by which one can characterize the structure of society and by which the laws of its development are determined. Marx, Engels, Lenin and their followers created the scientific basis for building a new kind of social structure – communist society.

To the social sciences one ought also to relate the science of man's higher nervous activity. The founders of this basic science are considered to be I. P. Pavlov and Sigmund Freud. They were the first to place the study of thought processes on an experimental basis. By them were found the laws of man's perception of the external medium, the emergence of conditioned reflexes and the influence of the subconscious on man's activity.

The outcome of these investigations and also of those of their pupils is made use of widely in pedagogies, psychiatry, legal practice, etc., and in capitalist countries also for advertisement and propaganda. With the development of the science of higher nervous activity, it is becoming still more closely tied with the social sciences.

It is natural to suppose that just as the successful development of biological sciences must be based on physics and chemistry, so also the science of the laws underlying the organization of society must be based on the science of man's higher nervous activity. Only on this scientific basis can organizations be created for the right education and training of people. Only on this scientific basis can the right forms of organization be sought for man's toil and leisure.

And – this is the main thing – only on a scientific basis can a healthy and effective structure of society be created. We must learn to build the State apparatus on the basis of the science of society, and it is necessary to learn to analyse it just as engineers now analyse electrical machines; it must be simply constructed and operate with a high efficiency. The value of the social sciences for mankind is quite obvious.

It is natural to pose the question: why even in our time, which by many is called the time of the scientific-technical revolution, are the social sciences evolving so feebly? I daresay that the more natural explanation of the hold-up in the development of the social sciences is, as in all such cases, the unfavourable conditions existing right now.

I take the liberty to explain this thought by a somewhat frank analogy which explains the state in which the social sciences find themselves today.

I daresay that the external conditions which have been created today for the social sciences are somewhat similar to those in which the natural sciences were in the Middle Ages. It is well known that the main brake on the development of the natural sciences at that time was the scholastic environment. At that time the Church took upon itself a monopoly of scholastic-dogmatic interpretation of all natural phenomena, decisively sweeping aside whatever in the slightest degree contradicted the canonical writings. This also retarded the development of the natural sciences. Only three hundred years ago did the natural sciences become their own master away from the Church and start to develop rapidly, and the pace of this development goes on building up to this day! Right now in the world there is a great variety of State structures which only acknowledge that what is true in the social sciences is that which demonstrates the efficiency of these structures themselves. Naturally, under such conditions the development of the social sciences is greatly hampered.

Let us take as an illustration how the social sciences are taught to youth in capitalist countries.

There, in secondary schools, as everywhere, the same laws of mechanics of Newton are taught, but there in school they do not teach historical materialism and the laws discovered by Marx on the basis of which capitalist society functions. You see the laws of development of society must also be the same everywhere. The laws of Marx are just as universal as the laws of Newton. Why then, in the social sciences, have not the only laws been acknowledged?

The answer is clear: there is no objective approach to the science of society. Until this is created, the social sciences will develop with great difficulty.

This, it seems to me, explains the striking contrast which exists today in the scale of development of the natural and social sciences.

I daresay that the time is not far distant when people all over the planet will realize the necessity of recognizing the same laws of development of society and on this basis will set up for themselves a more effective and just social system.

It is well known that the technical possibilities which right now are deposited in man's hands by modern science are leading to such sharp contradictions between States that in the face of the possible ruin of the people of all the world they will come to a conclusion regarding the necessity of a scientifically substantiated and effective social system free from war.

It is not hard to predict that the development of the fight for peace and the acknowledgement of the danger of the present deadlock into which mutual relations between different social systems have fallen, inevitably must lead to an intensive development of the social sciences in the near future.

41. GLOBAL SCIENTIFIC PROBLEMS OF THE IMMEDIATE FUTURE

The main reason for the urgency of the problems discussed here is their global nature. In our century, the solution of many problems is beyond the powers of any single country and can only be achieved on a global scale. This realization of the planetary character of relations between man and nature emerged for the first time in the wake of the atomic bomb and the threat of a global nuclear war. It is generally accepted that an outbreak of such a war in any part of the world can poison the entire globe and put an end to human life within a few hours. In view of this threat, mankind must abstain from the use of nuclear weapons.

Some people still believe than man can survive a radioactive disaster by finding refuge in a shelter equipped with suitable filters. But this is a mere illusion, because man can only live in a balanced relationship with nature. And such a balance is unthinkable when a live man is surrounded by an environment devastated by radioactive contamination after a global atomic war. Once out of his shelter, man would be doomed to die of protein starvation in a world where all big fauna have been annihilated and natural equilibrium no longer exists.

Three major aspects of the global problems emerge at present: (1) a technical-economic aspect, linked with the depletion of global natural resources; (2) an ecological aspect, namely the biological equilibrium between man and living nature in the presence of an environment contaminated on a global scale; (3) a social-political aspect, in recognition of the need for solving these problems on a global scale.

Long ago studies of these global problems showed that the usual quantitative parameters which characterize the dynamics of these processes follow a geometric progression, and mathematically form an exponential function with time. A characteristic feature of these processes is that eventually they

---

Address at a meeting of scientists in the editorial office of the journal *Voprosy Filosofii* [Problems of Philosophy], 1972.

[The address was also given at the 4th Session of the 8th Convocation of the Supreme Soviet of the U.S.S.R. in Moscow in September, 1972. A somewhat different translation appears in *Scientific World*, No. 2 (1973).]

speed up and turn into an explosion. A widely known example of such a process is the explosion of the atomic bomb. In the course of a nuclear reaction, each neutron generates more than one neutron; the number of neutrons grows exponentially, and the associated accumulation of energy results in an explosion.

Similar exponential laws apply to processes of demographic growth. Now the world population is estimated at 3.7 billion. Assuming that the present growth rate of about 2% annually will remain unchanged, within 700 years the average density of mankind will be 1 person per square meter! Obviously this is impossible, and the upward trend of the human population will necessarily be reversed much sooner. The main problem of the near future is to determine the time and circumstances of this event, as well as the form of human civilization at the time.

Studies of this complex problem on a large quantitative scale are just beginning with the use of global statistical data. Such studies have become possible owing to the application of electronic computers. Especially interesting and convincing results were published by J. W. Forrester[1] and D. M. and D. L. Meadows[2]. It is shown there that the 'explosive nature' of ecologic processes does not depend solely on the exponential law of population growth. Various other processes – the growth of the consumption of electric power and raw materials, environmental pollution – likewise increase exponentially and may soon lead to a global crisis abrupt enough to have the effect of an explosion.

One of the main global problems is energy, because human consumption of energy resources is the principal factor which determines the level of human civilization and welfare. Coal is the largest energy source in our time; at present consumption levels its reserves will last for about a thousand years. Assuming a stationary population with a per capita energy consumption growth as in the last 100 years, the coal reserves would be exhausted within 100–150 years. With respect to other raw materials the crisis is even more imminent. Thus, silver will last for only 13–40 years, lead 20–60 years, etc. (assuming a fivefold increase of known resources)[3].

Science can show a way out of the impending crisis. Mankind can solve its gravest energy problem by the use of controlled thermonuclear processes. The energy source for these processes is deuterium, a heavy isotope of hydrogen found in the oceans in practically unlimited quantities.

A global crisis due to the depletion of raw materials can likewise be averted with the help of science by the introduction of the so-called 'closed processes' in industry – in much the same way as in nature, where everything is recycled and nothing goes to waste. Closed processes, though much more complex, are quite feasible from a scientific viewpoint. In the introduction of such processes, the main problem will be the need for a greater expenditure of energy. For this reason the adoption of such processes on a global scale must await the development of a practically unlimited energy source, which now can only come from thermonuclear processes.

Exhaustion of raw materials is an actual threat, even to our own generation. Hence the urgent need for solving the technological-economic implication of the 'man-nature' problem. But the social-political aspect emerges immediately too: in view of the global nature of these problems, their solution is beyond the capacity of any single country and requires broad international cooperation on the basis of peaceful co-existence of states with different social orders.

The next problem – the ecological one – arises from the disturbance of natural equilibrium owing to the pollution of the environment on a similarly global scale. Though of lesser gravity than the depletion of raw materials, this problem is more obvious and acutely felt by people, which explains the attention it receives in the individual countries and at the United Nations. The main difficulty in solving this problem lies in the fact that technological processes, carried out on a global scale at the present level of civilization, have begun to affect our environment – air, water, soil – to such an extent that the biologic equilibrium prevailing so far in nature can no longer exist, and fauna and flora essential to mankind are beginning to die.

Technological processes associated with present civilization inevitably disturb ecologic processes; now a different kind of biological equilibrium in nature is required. A major task of ecologists is to find out the necessary conditions for these processes and the ways in which nature can develop in accordance with the needs of human culture.

Ecology has so far studied the equilibrium processes existing in nature; now it must seek new conditions of equilibrium. Lake Baikal provides one example of such a search. Industry needs fresh water. Baikal has lots of it, and this is a valuable resource indeed. But water should not be simply pumped out of Baikal. True, the lake contains vast amounts of pure water, but its real value lies in its function as an enormous biological filter producing pure water. The rivers discharging in the lake supply comparatively polluted water which undergoes purification before leaving Baikal. This purification is due to biological processes in the lake. The penetration of pure, 'distilled' water would have an adverse effect on life there, and Baikal would no longer be able to process the inflowing contaminated water.

To us, the industrial significance of Baikal lies in its function as a large-scale water purifier, and our concern about the lake has to do with finding ways for preserving this function. Those who warn: "Do not touch Baikal!" are therefore wrong. Baikal must be exploited, but in a manner which preserves life in it, and its purification capacity. For this purpose it is necessary to determine the peak pollution load the lake can cope with. Biologists thus face the clear-cut problem of identifying the ecological processes which take place in the lake as a result of the penetration of industrial wastes. Chemists must work out technological processes yielding wastes compatible with the processing capacity of the lake as determined by biologists.

It is known, for example, that the effectiveness of biological processes in water largely depends on the concentration of dissolved oxygen. On these grounds the rate of biological processes in the polluted parts of the lake could be increased by blowing air through the water in order to raise its oxygen content, as is commonly done in aquariums. Modern technology is fully capable not only of destroying life on a global scale, but also of stimulating it. Nature must be cured of diseases just as people are. A correct solution of the problem could even result in an increase in the lake's purifying capacity.

The organization of these works is within the jurisdiction of the U.S.S.R. State Planning Committee [*Gosplan*] and the U.S.S.R. Academy of Sciences. In a Socialist economy the state can create the necessary conditions for concerted work of scientists and industry toward a rational exploitation of the Baikal waters.

A clear example of the fate of lakes as a result of poor management of their waters and neglect of the biological processes which take place in them can be found in the Great Lakes of the U.S.A. and Canada. Wastes of factories using the waters of these lakes have led to a severe pollution which has destroyed all forms of life and has made the water unsuitable for a considerable number of industries. The U.S. government has therefore decided to restore normal life in these lakes. This, however, requires a thorough reorganization of water exploitation procedures in order to establish the ecological background necessary for a reappearance of life in the lakes.

For this purpose the U.S. government will spend 5 billion dollars over the next three years. But this amount is regarded as inadequate for a total rebirth of the Great Lakes. Many experts believe that about 25 billion dollars will be needed.

Ecology must become one of the central biological sciences today. Its main task is not only the study of biological equilibria now existing in nature, but more importantly, the determination of viable equilibria in a nature exploited by modern industrial processes, as well as the equilibrium processes associated with the extensive use of various chemicals in agriculture. All these influences of man on nature are now assuming global proportions; since the processes involved grow exponentially, the lack of control over them can lead to an explosion.

A third aspect of the global problems is the creation of the social conditions that would permit the application of technical and industrial achievements on a scientific basis that would ensure a balanced progress of civilization without the risk of explosive catastrophes.

For solving the first two problems we are in a position to define quite clearly the scientific questions to be answered in the fields of energetics, technology and ecology, in order to avert the impending catastrophe resulting from the depletion of raw materials and the pollution of the environment.

In contrast to the encouraging prospects for a satisfactory solution of the first two problems on the scientific level, planning and implementation of the corresponding measures on a global scale is a social task of unsurmountable magnitude at present.

The fundamental difficulties associated with the solution of these social problems are just beginning to emerge. International measures are clearly needed, but these may conflict with the national interests of individual countries.

Let us consider the following simple example. There are two neighboring countries, one of which produces cellulose causing no water pollution, while the same industry in the other country pollutes the water. Obviously the paper would cost more in the country where the process does not pollute the water. In effect, only one of the two countries would pollute the water with industrial wastes. Thus, water purity is clearly an international problem which requires the cooperation of all maritime countries. The problem boils down to persuading the polluting country to adopt a more expensive process, although such a step may contradict its national interests with a possible loss of markets and an increased expenditure for costlier equipment.

Effective measures for preventing a country from polluting the environment have not yet been devised. Thus, some countries, guided by narrow national interests, carry out nuclear explosions in the atmosphere with little regard for the resulting radioactivity. I believe, therefore, that in the nearest future it will be necessary to set up an authoritative international organization for the control of global matters.

A broad discussion of such world social problems is beginning. One aspect of this discussion is emerging rather clearly. Even in the West, many sociologists and economists are beginning to realize that the global solution of technological-economic problems requires a socialist organization of industry. Among others, the well-known Dutch economist S. Mansholt has expressed such views[4].

Other schools of thought maintain that the capitalist system, having so far found latent possibilities for self-control in terms of prices and taxes, will be able to solve the global problems in a similar way. Such is the opinion, for example, of Carl Kaysen, professor of economics at Harvard University[5]. All these considerations, however, are rather vague.

A socialist organization of the national economy undoubtedly provides a solid foundation for the solution of global problems. Today one can see quite clearly that large-scale ecological problems can be solved in our country. For this reason the case of the utilization of Baikal waters assumes international significance. This example shows that we can exploit the Baikal resources without disturbing the natural equilibrium – something the capitalist countries have been unable to achieve. Here too experience shows the superiority of socialism over capitalism in solving such ecological problems. Not surprisingly, Baikal is now attracting public attention.

I think that our best biological and technological forces should be focused on Baikal in order to analyze the ecological processes of life in the lake with the combined efforts of technologists and biologists.

There is no doubt that the need to solve global problems on an international scale will assist in finding ways for peaceful coexistence and disarmament.

A characteristic aspect of expenditure for armament is the consumption of vast amounts of energy by the fleet, aviation, and armored forces. Military technology consumes enormous amounts of precious materials but contributes nothing to the well-being of mankind. In the manufacture and utilization of arms it is impossible to organize a 'recycling' process as economic equilibrium requires.

When a shortage of energy and raw materials emerges on a global scale and begins to exert an adverse effect on the living standards of mankind there is no choice but to disarm, because the risk of death from aggression becomes less real than the danger associated with a shortage of material resources. And since the solution of global problems requires an atmosphere of close international cooperation all people will feel themselves as neighbors facing a common enemy – the impending global crisis, which makes them forget all quarrels and join forces for a common struggle.

Interest in global problems is rapidly growing, and their discussion inevitably involves many divergencies of opinion with regard to the magnitude of the matter and the measures proposed. But all will agree that these global problems are of vital importance for humanity and must be solved with the participation of the best cultural forces of all countries.

Time is running short for mankind; at any rate, we have less than a century to prevent ecological crisis. To be able to devote its best energies to the solution of these problems, humanity must first realize the magnitude and implications of the global crisis. Explaining such matters to humanity is the task of scientists, who were the first to realize the magnitude of the approaching crisis and who can show ways for averting it.

Therefore scientists, whether specialists in all branches of the natural or the social sciences, must organize the public social consciousness toward a common solution of the ecological problems affecting the world which, as we now realize, is rather small after all.

REFERENCES

[1] J. W. Forrester, *World Dynamics*, Cambridge, 1971.

[2] D. M. Meadows, D. L. Meadows, J. Rangers, and W. W. Behrens III, *The Limits To Growth*, New York, 1972.

[3] R. L. Heilbroner, 'Growth and Survival', *Foreign Affairs* **51**, 139–153 (October 1972).

[4] 'Ecologie et révolution', S. Mansholt *et al., Nouvel observateur* No. 397, Suppl. spéc. 11 (1972).

[5] C. Kaysen, 'The Computer that Printed out W*O*L*F*', *Foreign Affairs*, July 1972, p. 660.

42. GLOBAL PROBLEMS AND ENERGY

Global problems are problems which must be solved on a global scale. Already in ancient times such problems were described in artistic and religious books; they were, for example, The Last Judgement, The Universal Flood.

Now global problems have have become real, and the necessity to solve them is the gravest social and scientific task for humanity as a whole. Scientists must find these solutions and provide the proper foundations for them, while statesman must carry them out on an international scale. Therefore at present the study of such problems becomes more and more a central activity both of scientists and of statesmen.

The reason that such global problems arise is well known: man differs from animals principally by the fact that an animal adapts itself to nature, while man transforms it and adjusts it to his requirements. In our century, owing to the increase in the number of people on the earth's globe and with the growth of material culture, technical and energetic processes have begun to take place which have started to alter the nature of the whole globe. It is now becoming evident that some of these changes are sufficiently significant that they represent a danger for the safe existence of humanity as a whole.

This was first realized by mankind when nuclear war became possible. It is now generally accepted that in the event of nuclear war breaking out the already existing supply of atomic bombs is sufficient not only to destroy a significant part of the population of the globe but in principle to poison the globe with radioactivity to an extent that the remaining part of the population will either perish or will be forced to exist at a level similar to that of prehistoric man.

As long as people directing the affairs of state are conscious of this and are ruled by reason and not by emotion the possibility of nuclear war breaking out will remain under control.

Before proceeding to the analysis of some specific global problems I think

---

Lecture delivered at Stockholm University on 5 May, 1976, and published in *Uspekhi Fiz. Nauk* **122**, 327–337 (1977).

that it is useful to point out a characteristic common to them all to which usually insufficient attention is paid.

This common characteristic consists of the fact that phenomena associated with global problems usually develop according to a law determined by the so-called geometric progression, or, in other words, exponentially. With respect to the growth of population this was first noted already two centuries ago.

The law of geometric growth contains within it a rather unexpected property – it invariably leads to a phenomenon which it is customary to describe as an explosion.

This was noted already in ancient times. In one oriental tale it is related that some wise man had rendered to a certain king a considerable service. Wishing to show his gratitude to the wise man the king offered him to choose his own reward. The wise man asked to be rewarded by grain. He asked that a single grain should be placed on the first square of a chess board, double the amount – two grains – to be placed on the second square, four grains to be placed on the third square, etc. following a geometric progression to the 64th square. The tale relates that the king was surprised by the modesty of the wise man, but when the payment of the reward began it turned out that it exceeded the wealth of the kingdom. This problem is often posed in the upper grades of our schools, and calculations show that the total weight of the grains will exceed one hundred billion tons; this is more than ten times larger than the annual grain harvest of the whole world!

But processes developing according to a geometric progression have yet another property. It is not difficult to show that not only the amount but also the rate of increase of the amount follows a geometric progression. This leads to the fact that after a sufficient lapse of time this rate can also become arbitrarily large. It is customary to describe a rapidly developing process as an explosion.

This explains why at present we speak of a demographic explosion, although the increase in the population on the earth's globe in the present century remains on the average approximately 2% per annum. It is of interest that we also speak of a scientific-technological revolution, as if an unusually vigorous growth of science has at present begun suddenly. In actual fact this is not the case and this can be shown in the following manner.

In Figure 1, years are plotted along the horizontal axis, and along the vertical axis on a logarithmic scale is plotted the number of scientific journals published everywhere[1]. It can be seen that during the last 300 years, i.e., since the time when due to the invention of printing of books and to the

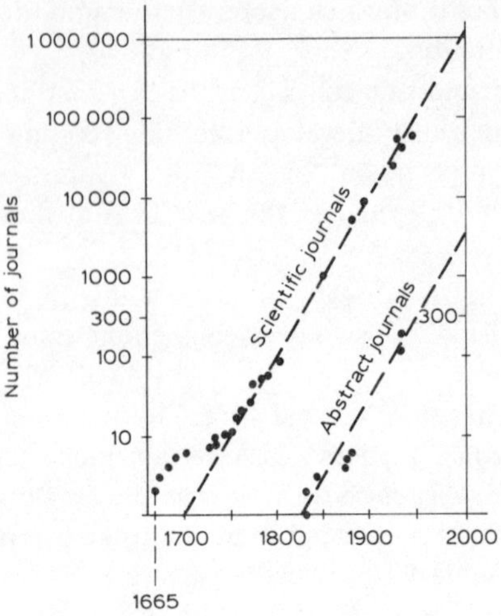

Fig. 1. The growth in the number of scientific journals from the year 1665 onwards.

appearance of postal communication science began to be developed on an international scale, the number of journals has increased according to a geometric progression and their number has invariably doubled every 14 years. Of course, it is natural to consider that the number of published scientific papers is proportional to the activity of scientists; then the growth in the number of journals characterizes the growth of science.

As can be seen from this diagram the law of exponential growth has not been violated until now and mathematically there is no reason to speak of any kind of an explosion. Evidently a process developing in accordance with the geometric progression must reach a limit and stop. The approach to this limit will be in the nature of an explosion.

I will mention another example quite familiar to us of a phenomenon developing according to a geometric progression until it reaches a definite limit, – it consists of infectious diseases. An infection enters a human organism; suppose that this is a bacterium which reproduces by undergoing division every hour. The number of bacteria in the human organism begins to grow just as the number of grains on the squares of a chess board. It is not difficult to calculate that in three days the number of bacteria would reach the astronomical number of 10^{21}, and this is impossible since the weight of bacteria would exceed the weight of the man. Of course, the process must cease before then, and it is well known how this occurs in the course of the

the disease. When the concentration of bacteria attains a certain value a man feels ill and this occurs suddenly like an explosion. Further the process can cease in three different ways: either the increase in the number of bacteria in the man's organism ceases and the man regains his health, or the bacteria continue to multiply and the man succumbs together with the bacteria. Finally, there exists another special solution – this is equilibrium when the man's organism destroys as many bacteria as are produced. Then the illness becomes chronic.

An infectious disease which develops according to a geometric progression is in many respects analogous to those global processes which have begun to occur on the earth's globe. We have suddenly felt ill and in order not to perish it's time to think about a method of treatment. But for this, of course, it is necessary to understand the nature of our illness. Of course, one of the main global problems is the incessant exponential growth of the population of the Earth which is particularly rapid in some countries. This growth must sooner or later cease if for no other reason than because there will not be sufficient food to sustain the life of the people.

The problem which confronts mankind consists of determining how this growth can be stopped painlessly, i.e., not by means of death from starvation as is beginning to occur now.

It is well known that this problem is now being widely discussed, but so far no generally accepted means for solving it have been found. We shall not touch upon this problem, but shall restrict ourselves by assuming that within the next hundred years the number of people on Earth in one way or another will not noticeably vary. We shall concentrate on the problem of how one can guarantee to people a sufficiently high and constant level of civilized life and we shall show that this is attainable only if the energy problem will be solved on a global scale.

The connection between the level of civilized life and the energy supply for mankind is well known. It is clearly illustrated in Figure 2 where along the horizontal axis we have plotted the gross national product for a number of countries calculated per annum per person and expressed in dollars, and along the vertical axis we have plotted the energy utilization per person expressed in terms of coal (in kilograms per annum) [2]. As can be seen from the diagram, within limits of natural fluctuations there exists a simple proportionality between the gross national product per person and the energy resources of a country. This, of course, is quite understandable: in order to manufacture any item it is necessary to perform work and, consequently, to expend energy. Statistics show that in the most developed countries on the

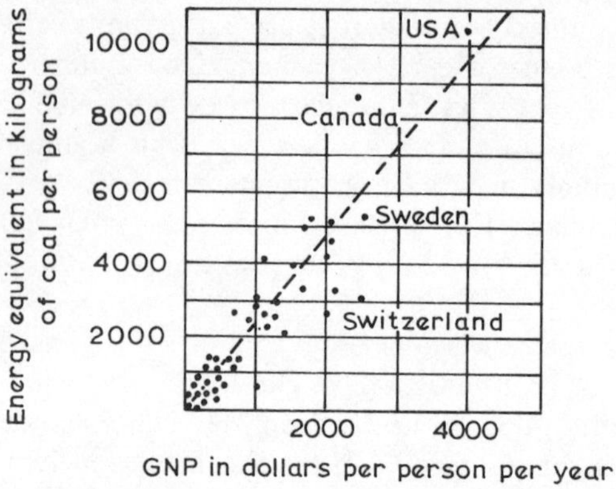

Fig. 2. Utilization of energy and the cross national product (GNP) per person. Data refer to 1968, and are compiled according to the materials of the United Nations and the International Bank for Reconstruction and Development.

average 10 kilowatts are available per person. This is several hundred times more than the energy produced by the muscular work of a person. Thus, the growth in the material well-being of man is most closely connected with the energy produced. At present the utilization of energy is rapidly growing, and not only because the material level of life of people in developed countries is rising, but, principally, because of the necessity to raise the level of life in developing countries. Moreover, an increase in the utilization of energy is associated with the necessity of solving a number of global problems that are arising.

As is well known, at present there is beginning to take place a depletion of reserves of mineral raw materials required for industry, primarily of metals such as silver, tin, copper, and a number of others. This leads to the necessity of extracting them from ores of lower content. Already it turns out to be necessary to extract certain metals (for example, magnesium) dissolved in sea water. This requires large expenditures of energy per unit weight.

The use of energy will also increase in our efforts to combat pollution of the environment which already occurs on a global scale and begins to acquire menacing dimensions. It is well known that technological processes which do not give rise to noxious wastes in manufacturing processes such as, for example, production of paper, lead to an increase in the utilization of energy.

Further increase in the efficiency of agriculture requires the production of mineral fertilizers, in particular, the fixation of nitrogen from the air, and this also leads to an increase in the expenditure of energy.

Finally, one can assume that in the future as a result of the development of chemical technology and the appearance of the possibility of the production from inorganic materials of organic ones, including protein, a possibility will arise to assure the feeding of people by synthetic products making it ever less dependent on products which at present are derived from agriculture. This will also require expenditure of energy.

This role of energy in the development of the material culture of mankind explains why at present the world utilization of energy increases according to a geometric progression and during the last 15 years the growth amounts to 5% per annum. This is the highest index of growth in the world economy, and everywhere the capital expenditures for energy production are the dominant ones.

In view of this dominant role of energy production in world economy the overcoming of the currently imminent energy crisis represents the greatest global problem for mankind.

The cause of the approaching crisis is evident and well understood. Not less than 90% of the sources of energy which are being utilized at present are different types of fuel such as coal, oil, gas, etc. The chemical energy contained in them has been accumulated as a result of biological processes in the course of thousands of years. It turns out that at the present rate of their utilization their reserves on the earth's globe will be used up in the near future. Of course, it is impossible to determine exactly when this will occur, but one can with great certainty assume that this will occur in the course of one–two centuries. Of course, this time can be extended by a more economic use of energy, by means of improvement of technology and by means of stopping the expenditure of energy on armaments, etc. This will only postpone the crisis, but it cannot be prevented, since according to the law of conservation of energy, we cannot realize a *perpetuum mobile* with the aid of which we might produce energy. Therefore, if other sources of obtaining energy are not found to replace the fossil fuel being utilized at present, an inevitable decrease in the utilization of energy will occur, and consequently this will lead to a decrease in the material well-being of people.

The way towards the solution of this problem is quite evident – it is necessary to find sources of energy which would be practically inexhaustible with time. The principal ones among them are well known – this is radiation from the sun, geothermal heat, hydro energy of rivers and of tides. But as analysis has shown [3] they cannot solve the problem on the required scale in a sufficiently economic manner.

The principal difficulty arises because the greatest part of the energy utilized at present in world economy goes to heavy industry (metallurgy, production of machinery, transport, building, etc.). In order to satisfy these requirements cheap energy is needed in amounts of hundreds of millions of kilowatts. This is many times greater than the energy which we called 'consumer' energy and which we utilize for refrigerators, washing machines, television sets, etc. If we would want to satisfy the energy requirements of the whole world economy by utilizing the energy of solar radiation, then in order to obtain a power output of only one million kilowatts it is necessary to absorb it from an area of 10 square kilometers.

Calculations show that in the case of all the proposed methods of transforming radiation into mechanical or electrical energy the capital expenditures are not justified by the energy obtained. This is due to the fact that for effective utilization of energy its flux density must be sufficiently great. Solar energy does not have such a flux density.

Calculations show that geothermal heat due to the poor thermal conductivity of the earth's outer shell also does not have a sufficient energy flux density in order to be profitably exploited. Only in regions where volcanoes are situated does the geothermal heat have a sufficient energy flux density to be profitably utilized. But there are few such regions. For example, in Italy geothermal heat has been successfully utilized for many years, but it amounts to only 2% of the energy balance of the country.

The situation is somewhat better with utilization of hydro energy. Practice shows that electrical energy can be profitably generated only in the case of high heads in mountain regions. This restricts the scale of utilization of hydro energy. In a world energy balance hydro energy at present amounts to not more than 5%, and, apparently, this is its limit.

The situation is still worse with utilization of wind. Its energy flux density is not only small but is extremely inconstant.

Nevertheless, life shows that utilization of solar energy and of the energy of wind and water can be useful for people for the solution of energy problems on the scale of consumer requirements, where one can afford to pay considerably more for energy. As regards the solution of the principal problem of energy requirements at high power these sources cannot aid in overcoming the approaching energy crisis.

If this energy crisis were to have arisen 40 years ago, prior to the discovery of nuclear energy, humanity, doubtlessly, would have been confronted by a catastrophe and human culture would have approached a dead end.

But at the present time one can with complete confidence assert that the sources of nuclear energy provide a scientifically based possibility of solving the approaching energy crisis.

But even along this path there are its own characteristic difficulties which people still have to overcome.

As is well known, at present there exist two real methods of obtaining energy at high power in nuclear processes. The first of these is the evolution of energy in the fission of nuclei of heavy elements, such as uranium, which occurs as a result of neutron bombardment. This process has the nature of a chain reaction and is self-sustaining. It has been well studied, and on its basis an atomic bomb has been realized of the type which was released on Hiroshima and Nagasaki. In energy producing reactors this reaction has been slowed down, has become stable, and the energy obtained is being successfully utilized for profitable production of heat and electrical energy on the scale of millions of kilowatts. The reserves of uranium in nature, if they are utilized economically in reactors of a particular type, the so-called 'breeders,' may be sufficient for thousands of years. Moreover, some scientists assert that extraction of uranium dissolved in sea water in practically unlimited quantities can become profitable even now.

The following generally recognized difficulties[3] lie in the path of transition of the whole energy supply problem to nuclear fuel and they must be overcome.

The principal one of these problems consists of the fact that in the course of utilization of uranium as a nuclear fuel in reactors there occurs an accumulation of a large quantity of radioactive substances, and if an accident should occur and the contents of a reactor should escape into the environment, then poisoning of live organisms and, of course, principally of people will occur. The horrors of poisoning people by radioactive substances are well known as a result of the consequences of atomic bombs dropped by Americans on Hiroshima and Nagasaki. It turns out that due to radioactive poisoning a fraction of the people perishes over a short period of time. But another part of the population perishes over a period of several years, usually from leukemia or other forms of cancer diseases.

The escape outwards of the radioactive content of a reactor can occur if cooling water ceases to be supplied to the reactor. Then the elements of the reactor overheat, the shielding surrounding it is destroyed and the contents escape outwards. In the language of the specialists this is known as 'core meltdown.' In order that this should not occur a number of preventive devices is used. The reliability of these safety measures is estimated numerically: the probability of an accident occurring is calculated. The probability of an accident at nuclear power stations that have been constructed in the U.S.A. until now has been estimated by the constructors of the reactors as being in parts per billion and accidents were assumed to be simply impossible. However, many people have regarded these calculations as being unrealistic.

The reliability of this kind of estimates was radically shaken by an accident which occurred at the Browns Ferry nuclear power station in California* in March 1975. Approximately a year later an official conclusion of the commission[4] appointed to establish the cause of the accident was published. I shall quote from the conclusions only certain characteristic features of this occurrence. The accident occurred as the result of a fire which took place underneath the area where the control center of the electric power station is situated. The cables were set on fire. The cause of the fire was an ordinary candle with the aid of which a worker was trying to discover the leak from a pipe which supplied compressed air.

Only 15 minutes after the conflagration when people were convinced that the available firefighting means were insufficient did they start to call out the fire brigade. Then it turned out that its telephone number was incorrectly recorded. Only an hour after the conflagration did the fire brigade arrive, and then it turned out that there were no instructions as to how one should fight fires in a nuclear power station. Therefore in the course of the next 6 hours people did not know what to do, and then started to put the fire out using water which turned out to be quite successful. The commission established that, since the emergency water supply was disrupted, if the fire had not been put out then, after a short period of time the elements in the reactor would no longer have been cooled by water and a catastrophe would have occurred similar to the one of which I spoke above. It should be noted that the Browns Ferry nuclear power station is situated in a densely populated area. Since the measures provided for the evacuation of population were not put into operation, then in terms of the number of perished and poisoned people the potential catastrophe would have been comparable with the Hiroshima catastrophe.

Of course, this accident has shown that mathematical methods of calculating probabilities of this type of occurrence are inapplicable, since, as has happened in the present case, probabilities have not taken into account what happens as a result of human error: a workman with a candle, a telephone number incorrectly recorded by someone, the fact that no one had provided instructions for fire-fighting in nuclear power stations, etc.

Now, when all the details of this accident have become known, the problem has arisen of the future fate of atomic energy production in the U.S.A.

---

*[Browns Ferry nuclear power station is in Alabama – R.S.C.]

Of all the nuclear power stations in the U.S.A., 35% are situated in California. Therefore the population of that state (20 million persons) is much worried by the fact that the greater number of these stations are constructed in a manner similar to the Browns Ferry station and, moreover, it is proposed to build a number of analogous stations. Half a million persons have signed a petition in which they demanded that the government of the state should not only cease the construction of such stations but should also shut down those already existing in California. According to the laws of the state, a referendum had to be held on this problem, and in order to obtain permission for subsequent construction of atomic stations a two-thirds majority is required.[5] The referendum took place during the summer of 1976 and approximately 70% of those polled voted for the development of nuclear energy, but subject to a tightening up of the measures guaranteeing the reliability of exploitation of nuclear power stations.*

In the U.S.A. at present there are already 59 nuclear reactors in operation, and they produce 4% of all utilized electrical energy. A further widespread development of nuclear energy production is planned, and therefore the Senate reacted very seriously to the accident at Browns Ferry and the question of the reliability of nuclear power stations was investigated in a special committee on atomic energy. At present, extensive testimony to this committee has been published [6] of three leading construction engineers according to whose plans the majority of nuclear reactors has been built, including the one at Browns Ferry. The testimony provides data on a large number of accidents and a completely unsatisfactory guarantee of the safe exploitation of reactors is brought out. They consider that under the conditions existing in the U.S.A. at present for the development of nuclear energy production a catastrophe on the scale of Hiroshima will occur sooner or later. This is further confirmed by the fact that at present in the U.S.A. not a single insurance company undertakes to insure nuclear power stations. These leading reactor engineers have now left the employ of the General Electric Company motivated by the fact that morally they do not feel able to take the responsibility for the safe operation of existing nuclear power stations,

* In order to make possible the further development of atomic energy in the State of California, the State Assembly adopted, just prior to the referendum of June 3, 1976, a number of laws which establish strict control to guarantee the safety of nuclear power stations (both planning and exploitation) including their possible construction underground. Also by legal means the necessity has been established to guarantee safety in handling nuclear fuel and its wastes. These measures influenced the outcome of the referendum which was carried out on June 8, 1976.

and for the consequences of disasters to which possible accidents might lead.

But, of course, under present conditions the global problems of the production of energy at high power will not be able to be solved without nuclear energy, and doubtlessly, a way out of the situation that has been created will be found. It will have to be based on the premise that in case of any accident that might occur in a reactor at an nuclear power station it must under no circumstances acquire the character of the Hiroshima catastrophe. Solutions are already being proposed now, and they consist, for example, of placing the nuclear reactors of such power stations at a sufficient depth underground as is now done in the case of underground tests of atomic explosions. Apparently, although this will make the construction of nuclear power stations more expensive, it can make them completely safe.

Another possible solution is to place the nuclear power stations in regions where there is no population, for example, on uninhabited islands, and to convert the energy generated by them into chemical energy: for example to decompose water and to utilize hydrogen in liquid form as fuel. This fuel will be much better than oil since on burning it will not pollute the air.

Before solving the problem of the widespread development of nuclear energy it should be noted that nuclear energy utilizing uranium fuel requires the solution of several other problems.

The most difficult of them is the storage of radioactive wastes. The difficulty consists of the fact that their radioactivity is quite great and, moreover, decays only very slowly. Therefore any containers in which they might be stored can with time be destroyed under the action of radiation, and the radioactivity could then spread. Calculations show that in going over to nuclear power production as the principal method of power production the resulting amount of radioactive wastes becomes so great that their safe storage becomes a problem which is difficult to solve, and until now there exists no definitive and generally accepted solution.

Finally, in capitalistic countries there exists yet another problem, not of a technical but of an internationally political character which also requires solution. It is well known that for a more complete utilization of uranium, reactors based on fast neutrons are used; they are called 'breeders'. In them uranium is almost completely transformed into plutonium which is the best atomic fuel. In the course of time this will lead to the situation that plutonium will become very widespread and accessible. But plutonium is the basic element from which an atomic bomb is made. In order to make a bomb it is necessary to have only a few kilograms of plutonium and, moreover, at present the construction of an atomic bomb is no longer a secret. Under these conditions the possibility is not excluded that an enterprising group of gangsters can without much effort utilize an atomic bomb for blackmail.

Doubtless people confronted with the approaching energy crisis will find a solution and will be able to overcome the difficulties enumerated above associated with the utilization of nuclear energy generated by the fission of uranium. But it is now already becoming clear that the solution of this problem will have to be effectively realized on an international scale.

I shall also mention, but only briefly, other methods of utilizing nuclear energy since so far they have not yet been realized in practice.

First of all this is the method of obtaining energy not by means of fission of a heavy atom such as uranium, but conversely by the process of fusion of light atoms which, as is well known, can occur with an energy yield.

This is the so-called thermonuclear process. At present it is realized in the hydrogen bomb in which helium and neutrons are produced as a result of fusion of isotopes of hydrogen. This process is accompanied by a large yield of energy and occurs only at a very high temperature of hundreds of millions of degrees. At this temperature all substances are in a gaseous state in the form of plasma when electrons in atoms are completely removed from their nuclei.

It turns out that it is quite possible to produce such a plasma for a short period of time during which an explosion of an atomic bomb takes place. This process occurs in the so-called hydrogen bomb which at present is hundreds of times more powerful than a uranium or a plutonium bomb.

In order to utilize the thermonuclear reaction for obtaining energy one must find a method of carrying it on continuously and, by comparison with the hydrogen bomb, on a small scale. This has turned out in practice to be a problem which is very difficult to realize and which to date has not yet been solved technically.[7]

Scientists and engineers all over the world are at present diligently working on this problem. Although the problem has not yet been solved, nevertheless until now there have not been discovered any scientific or technical obstacles which are in principle insuperable lying on the path to its solution, and I personally think that with time controlled thermonuclear fusion will be realized.

It is very important to find the solution, since the energy obtained by this method is not associated with the difficulties arising in the use of uranium fuel which I have mentioned. In the case of a thermonuclear reaction the amount of accumulated radioactivity is so small that its presence does not create danger. The thermonuclear process cannot be utilized for the production of atomic bombs.

And, finally, the fuel which is utilized in thermonuclear reactors is the hydrogen isotope deuterium and amounts of it stored in the ocean are quite sufficient to supply mankind with energy for many thousands of years. During that time, of course, still other methods of solving the energy problem will be found.

Finally, there exists still another method of obtaining energy on a large scale. In practice it is apparently not realizable, but scientifically it is well founded. On the basis of our present concepts of cosmogony it is considered that when our universe was being created another one could have been created of a similar size but consisting of antimatter. The existence of antimatter has been demonstrated experimentally, it is obtained in accelerators, but only in amounts of a few nuclei. One of the properties of antimatter consists of the fact that on coming into contact with matter they are both annihilated and are converted into energy. It is not difficult to calculate that one gram in such a reaction yields an amount of energy equivalent to that obtained by burning 10,000 tons of coal. Thus, one ton of antimatter would be quite sufficient to supply the whole earth with energy for a year.

But how would we obtain this antimatter from the anti-world? It has been supposed that a small amount of antimatter could penetrate in the form of meterorites into our cosmic space since it is highly rarified. Their collisions with atoms would be so rare that on penetrating into our cosmic space they would not be totally annihilated. By capturing this material from the cosmic space with the aid of satellites and bringing it to earth we could have the most perfect source of energy. It is known that attempts to find antimatter in cosmic space so far have not met with success.

But even if antimatter should exist in our cosmic space in the form of antimeteorites how would we extract it and deliver it to earth in such a manner that it would not be coming in contact with matter? This problem appears to be very difficult and possibly even in general insoluble. But life teaches us that a number of processes which were regarded as improbable have nevertheless been realized. One should not forget this.

In concluding the scientific aspects of discussing the problem of supplying mankind with energy in large quantities I shall touch briefly upon one social and political aspect of this problem associated with its global nature.

It is quite evident that all global problems will have to be solved on an international scale. The principal difficulty in realizing the required solutions will consist of the fact that their demands will often conflict with interests of individual countries. The principal social-political problem reduces to the question of how the interests of individual states can be subordinated to the interest of mankind as a whole.

Here a number of opinions have been expressed. Some people consider that this is in general unrealizable, that complete freedom must be provided for the development of human culture. Until now in the course of millions of years mankind by means of trial and error has itself found the path towards the development of civilization. And it will find such a path again.

Another opinion, a more constructive one, as many justly consider, is that the necessity of solving global problems will lead mankind to build a society with a socialistic structure, and that only with such an organization of society will it be possible to make compatible the interests of individual states with the interests of mankind as a whole [8] . . .

For the solution of global problems it is necessary that a number of sectors of world economy associated with ecological problems should come under international control. There is a tendency towards this already. For example, with ever greater frequency appeals are heard that the exploitation of the world's oceans and, in particular, the extraction of raw materials from their depth, should be controlled by the United Nations.

It is also becoming necessary to solve the problem of the supply of energy and the utilization of energy resources on an international scale. This has already begun to be realized in the creation of the International Atomic Energy Agency whose principal function is the control of resources and of the safety of utilization of atomic energy on a global scale.

An effective solution of global problems will become possible only if their significance for the fate of humanity will be wisely understood by people, and this is possible only if these problems receive wide discussion. Therefore scientists must take care that the discussion should be carried out on a strictly scientific basis. Of course, the solution of global problems must be based on the ethical obligations of man toward society.

REFERENCES

[1] Price, D., *Little Science, Big Science*, N.Y., 1963 (there exists a Russian translation in the collection of articles *Nauka o nauke* (The Science of Science), Moscow, Progress, 1966).

[2] Meadows, D. H., D. L. Meadows, J. Rangers, and W. W. Behrens III, *The Limits to Growth*, N.Y., 1972.

[3] Kapitza, P. L., *Usp. Fiz. Nauk* **118**, 307 (1976) [Sov. Phys. USP **19**, 169 (1976)].

[4] 'Fire at a Nuclear Plant', *U.S. News and World Report*, February 16, 1976.

[5] 'Les déserteurs de l'atome', *Le Nouvel observateur*, 1–7 March 1976.
[6] Testimony of Dale G. Bridenbaugh, Richard B. Hubbard, and Gregory C. Minor before the Joint Committee on Atomic Energy (February 18, 1976), Washington, 1976.
[7] Kapitza, P. L., *Pis'ma Zh. Eksp. Teor. Fiz.* **22**, 20 (1975) [*JETP Lett.* **22**, 9 (1975)].
[8] Kapitza, P. L., *Voprosy filosofii* (Problems of Philosophy) No. 2, 37 (1973).

43. SCIENTIFIC AND SOCIAL APPROACHES FOR THE SOLUTION OF GLOBAL PROBLEMS
THE BERNAL LECTURE, 1976

INTRODUCTION

One of the basic differences between man and animal is that in his evolutionary development man transforms his environment, adapting it to his requirements while evolution of the animal world is based on its adaptability to nature. In order to achieve a more effective transformation of nature, people started acting in an organized fashion, collectively, and this produced a general structure which has now become a state. We call this process civilization. The growth of people's prosperity has led to an exponential increase in world population with increased index. Such a law of growth causes not only the continuous increase in population but also necessarily leads to an increase in speed of such growth. In the final instance, it assumes the characteristics of an explosion, which may not be under the control of man.

It is now universally recognized that the size of the world's population is approaching a critical stage; it is almost 4,000 million and it increases every year by about 100 million. Under present-day civilization, the activity of such a large number of people may lead in the environment to changes of an irreversible character and of such proportions that in the following century they may produce a detrimental effect on the welfare of man; this may not only stop the growth of civilization, but also destroy it. The possibility of such a catastropic process has now been fully appreciated and the necessity for its prevention is beginning to be considered justifiably as the number one problem, by the whole of mankind.

An analysis of this process reveals that in order to prevent an ever deepening crisis, it is essential to solve four basic problems on a global scale:

(1) Control of the growth of population (quantity and quality).

(2) The creation of new energy resources, in view of the depletion of natural reserves of gas, oil, coal, etc.

(3) Depletion of mineral resources necessary for industry.

(4) Global pollution of the environment.

Lecture to the Royal Society of London, delivered October 7, 1976. [First published in *Proc. Roy. Soc. Lond. A* **357**, 1–14 (1977)]

All these problems are interconnected, but such a division appears to be the most convenient in the study of methods for their solution.

It is recognized that solutions have a scientific as well as social side and also touch on various other areas of knowledge. It is impossible, of course, to encompass them in one lecture. Hence, in this address I shall limit myself to the following approach to these problems.

It is not difficult to see that the solution of global problems should be on a solid, scientific basis, i.e. by the use of laws established by physics, chemistry, biology and other natural sciences. But it should also be based on laws which govern the structure of society. These laws too should be scientifically based on the study of the psychic functions of man; for the present they have not been sufficiently clearly expressed to permit their reliable use for practical purposes. There is a widely held view that an objective scientific approach to social problems lies outside the boundaries of human possibilities. Indeed, a number of scientists even consider that the solution of global problems should be left to the 'wisdom of nature' which has governed the evolution of mankind for millions of years. However, I think that we should still try to solve these problems, and not leave the development of civilization completely to chance.

In my lecture, I shall limit myself to an attempt to demonstrate that it is possible to find an objective scientific approach to the solution of a number of sides of global problems even with the limited scientific knowledge of the present day. I shall begin by posing the question of what constitutes a scientific approach to the solution of a problem.

Evolution, as guided by the 'wisdom of nature' is based on 'trial and error' and all the characteristics which meet the demands of evolution consequently are developed; that is how human beings came to be created. Many millions of years have been spent on this process. Man started to transform his environment also by the same method of trial and error. But this process of nature's conquest by man was consolidated by his starting to generalize the experience of successful tests, storing them and transmitting them to other people, thus creating the mechanism for social inheritance and making the repetition of the method of trial and error unnecessary.

This method of trial and error is still the basis of the study of nature and is used to transform it. The process which limits the number of trials and errors being necessary for the solution of a given problem can be regarded as the beginning of a scientific approach. The logical generalization of experience gained from trials and errors leads to the scientific laws occurring in natural processes. The value of a scientific approach for the development of civilization lies in the fact that the experience gained is distributed among people and is preserved throughout time; thus the influence of science on the development of civilization increased with the growth of literacy and printing.

Religion also accumulated experience and served the development of civilization, but it moved away from the scientific approach when people laid down the laws of causality as a basis for the study of nature. The law of causality is taken to state that a given cause leads to a unique effect. Thus a scientifically posed problem has only one solution, just as there is only one multiplication table, and so true scientific generalizations are necessarily objective. That is why science has become international and it accounts for those contradictions which have invariably arisen between religion and science.

The process of scientific thought began to be studied independently in the seventeenth century. Francis Bacon maintained that the inference from experiment to explanation by means of deduction and induction was at the basis of a scientific approach to the solution of a given problem. Bacon described the practical meaning of science in a very vivid way: 'A lame cripple going along the right road can overtake a trotter if the latter is running along the wrong road. Moreover,the faster the trotter runs, once having lost the path, the further he lags behind the cripple'.

The importance of science for the solution of problems concerning the civilization of mankind was prophetically described by Bacon in his *New Atlantis*.

The scientific generalization of experience, gained by trial and error, spread its influence over ever increasing areas of culture. During Bacon's time the basic scientific problems lay in mathematics, mechanics and physics, and then extended to chemistry and biology. An intense development of the applied sciences began in the nineteenth century. The construction of bridges, steam and electrical machines, etc., necessitated the development of such applied sciences as the strength of materials, applied mathematics, mechanics, etc. At the present time the scope of technical services is very wide and serves as the basis of contemporary technology. The last century saw the foundation of a number of technical colleges for teaching applied sciences. In our own twentieth century the scientific approach has extended to the organization of the production process, the so-called 'science of management'. It arose in the U.S.A. by generalization of experience gained from the study of industrial productivity, and is now beginning to have wide application.

A scientific approach to the organization of society also began in the nineteenth century. The path to scientific generalization in this field was shown by Marx and led to some laws which govern economic and social processes. In order to include other aspects of social processes, it was necessary to establish the principles of the emotional interactions between man and society. Only recently have discoveries in these fields begun to appear. For instance, a scientific approach to psychic phenomena was initiated by the researches of Sigmund Freud and I. P. Pavlov.

On coming to the main theme of my lecture, i.e. the scientific approach to the solution of the global problems mentioned above, I must be guided by the already well-known results from various scientific disciplines, and these may sometimes lead to controversial deductions. For this reason I ask that my further analysis be considered tentative and open to discussion.

POPULATION

The first global problem which has to be solved is how we can scientifically determine the policies we must adopt in order to control the size and quality of the world's population.

Of all global problems this would appear to be the most important and difficult. Let us begin with the size of the world population. Statistical information on this problem is now sufficiently comprehensive and reliable. The present size of world population is approaching 4,000 million and increasing exponentially at the rate of 2.4% per annum. It therefore doubles, roughly, every 30 years. Such a rate of growth will necessarily cease in the very near future, since there will be insufficient food. In some countries in Asia and Africa, mass mortality from starvation has already been observed. Some authors estimate that up to 12,000 people a day die on these continents, and the average life-span does not exceed 30–40 years. Such a process of regulating the size of population is not only inhuman, but as is shown in a number of studies, has an effect on the quality of population. Children born of starving parents and growing up on a starvation diet grow both physically and mentally into irreparably defective people. And mankind of course must not allow this to take place. Government and society are thus compelled to organize control of the birth rate to keep the population within the economic possibilities of a country.

Present-day science offers a number of methods of birth control, which are widely and successfully applied even in economically prosperous countries. However, the Universal Declaration of the Rights of Man (U.N.O.), now in force, does not provide for the necessity of controlling the size of the world's population. Article 16 does not include any limits on people's rights in creating a family. This article was last year reviewed by a group of scientific experts from various countries invited to Geneva by the U.N.O. Committee on Human Rights in order to propose that the Declaration of Human Rights be reviewed in the light of the scientific-technological development of contemporary civilization. Here is the conclusion of the experts:

> It is recommended that a better definition be given of the duties of the individual to the community and of the rights of future generations. For example, it seems to us that the crisis in growth of the world's population must lead to some constraint on the individual right to reproduce, and that the right of the child to be born physically and mentally sound takes precedence over the rights of parents to reproduce.

The whole question is one of finding possible social measures for limiting the rights of parents to procreate, an instinct which is deeply inherent in each human being. There are two ways open. First, the establishment of laws limiting the size of a family, or secondly, limiting the size of a family by means of propaganda and incentives. The second method is more acceptable since it corresponds better to the universally accepted principles of ethics.

Statistics show that in developed countries the size of the population tends to stabilize by itself. Thus, in Europe the present yearly growth of population has decreased to 0.4–0.9% and is still diminishing. In developing countries where increase is the highest (often more than 3%) starvation is already setting in and the governments now admit the necessity of state interference. Recently a number of Asiatic countries have started to use various methods of birth control the effect of which is checked by statistical data. There is reason to consider that in the near future, with the growth of culture in developing countries, the size of population will be kept under increasing control.

Of all global problems, the control of the quality of the population is one of the most important inasmuch as its solution to a considerable extent determines the fate of mankind.

At present there is no universally recognized scientific approach to this problem because of the difficulty of objective, non-emotional study of social questions. Nonetheless I shall attempt to approach these problems scientifically.

The qualities of people, as of all living beings, may change from one generation to the next. It is now universally accepted that the direction of this development is determined by the fact that certain genotypes multiply more intensively than others and this makes selection possible. Selection will undoubtedly exist in any fairly close social structure and the quality of its human members will change with time. For instance, statistics show that in most developed countries, qualified people engaged predominantly in intellectual pursuits have the smallest families. The children of two mentally gifted parents may turn out to be even more gifted than their parents. Of course this selection will only make itself apparent in the average statistical data. If such a process is, in fact, taking place it will lead to a reduction in the number of people available for highly qualified mental occupations, but it will raise their average quality and create a mental élite. The question arises – is this good or bad? The answer is that if such a change in the qualitative composition of society brings about its better functioning, then for that given social structure, (i.e. as in Huxley's *Brave New World*), such selection improves the quality of its human composition.

Let us examine the problem of how to study quantitatively the connection between natural selection taking place in a social structure and the effectiveness of its functioning. We shall begin by assuming that in present-day civilization the quality of people's activities is determined by their spiritual and mental abilities. Physical strength is rarely needed in industry. Contemporary medicine, thanks to vaccines, hormones and antibiotics, ensures a good state of health in people of different physical structure and constitution: even diabetics can now lead a happy and useful existence.

The mental and spiritual qualities of an adult are determined by two factors: (1) the quality of his inherent capabilities, and (2) the way he is brought up and educated by society, to make him a useful member of it.

If a person is to develop harmoniously, his upbringing and education must correspond to his spiritual and mental abilities. Thus, the successful formation of a person's individuality depends on the relation between his personal and mental endowments and the activity imposed on him by society.

Such a process has already begun to be subject to scientific analysis. Schematically, a person is born with a number of genetically acquired natural instincts. In order to become a valuable member of society his activities must be directed in accordance with the ethics and traditions of the society in which he lives and works. For this reason some of these natural instincts, mainly those of an egoistic nature, must be suppressed. The process of education and the succeeding compliance with society are a strain on a person's psyche, and if his spiritual and mental qualities are insufficient his nervous system may not stand such a load. According to Freud, if the natural instincts which are genetically inherited by man are suppressed, they will remain unsatisfied in the subconscious and the resultant frustration is often the cause of mental illness. These processes are studied by psychoanalysis, which diagnoses nervous illnesses and tries to alleviate them.

On the other hand, if society places a person in conditions where his spiritual and mental activities exert a greater strain on his psyche than it can cope with, this can also cause mental illness. Such psychic processes, in their more elementary forms, were studied by I. P. Pavlov. Pavlov experimentally studied how conditioned reflexes are acquired by dogs. He observed that in cases where the signals were complex – for example, a definite combination of light and sound – even if the dog finally responded to them, he usually became nervously ill: he suffered from insomnia, loss of appetite, etc. The education of man involves also the adopting of conditioned reflexes. Similarly the stamina of people under the influence of a given social structure will be determined by their ability to adapt to its upbringing and stand up to the psychic load which is imposed on them.

But what is happening to those individuals of 'poor quality' in a given social structure? According to contemporary psychology these individuals tend to develop nervous disorders which may even lead to suicide. Certainly in such cases we can resort to therapy. One of the methods is the use of medicines, i.e. narcotics which help to balance the nervous system. This explains why even in the most primitive societies narcotics in some form or other have always been used. In contemporary society there is a wide use of such stimulants as tea, coffee, alcohol, tobacco, etc.; these undoubtedly help people to cope with their psychic load. The individuals who cannot stand the mental strain of civilization become neurasthenics, drug addicts, alcoholics.

There is yet another symptom of unsuccessful adaptation, i.e. the inability to find satisfaction in life under the conditions inherent in the given social structure. These people cannot suppress their egotistical desires and may even become criminals.

From a critical analysis of the statistical data on the occurrence of such abnormal phenomena we may evidently find a criterion which quantitatively determines the change which takes place in a social structure. Therefore there are three quantitative criteria by which we can evaluate the development of a given social structure:

(1) Improvement in health measured by life expectation.

(2) Growth of material welfare – by the increase in the gross national product per head of population.

(3) Growth of 'spiritual' and mental qualities of the population – by the decrease in criminality, nervous disease, narcotism, etc.

These three factors should not be considered independently. The first and second criteria are well known and are commonly taken as sufficient. The last one should be regarded as the most important, as it determines the fate of a given social structure.

Now we are approaching the most important problem. What quality should a population have if the evolution of mankind is to continue successfully? It is not enough that a given social structure should function effectively. It is also necessary that the structure of society itself should be progressive and conducive to the development of the whole of mankind. Thus the quality of the given social structure should be measured by its participation in solving problems of international importance. Therefore it cannot be based on narrow national interest. One of the criteria is the extent to which a country participates in the progressive development of world civilization and thereby in the solution of global problems. A love of peace and refusal to use force can also be a criterion in the assessment of the quality of the social structure.

The international policy of a state is to a considerable degree determined by the ethics and inner culture on which the structure is founded. The inner culture of a given social structure encompasses the sciences, the arts, literature, religion, the use of leisure, entertainment, education, class relations, distribution of material assets and all other things which go to form people's ideology. Hence, critical study of these factors will show how a present-day social structure may influence the development of the culture of the whole of mankind.

The future of humanity will be determined by the natural selection of social structures themselves and as time goes on this selection will determine the types of structures which will grow in numbers. It is this selection which will in the end determine the fate of mankind. Probably, for the time being, it is not possible for man to influence this selection and we must here still rely on being governed by the 'wisdom of nature'.

ENERGY RESOURCES

The second global problem is the provision of an adequate amount of energy. The use of energy is the basic factor which assures the growth of civilization. It began by man inventing tools so that physical forces could be effectively used. Eventually man harnessed animal power, water energy, wind and fine. A great change took place when man learnt how to transform heat energy into mechanical power and the steam engine was invented. At the beginning of the eighteenth century it was put to practical use. In the nineteenth century man began to use a fundamental new form of energy, i.e. electricity. This led to considerable mechanization in various fields of industry and life. Nuclear energy appeared in the twentieth century and it will undoubtedly play a leading role in future energetics.

The use of energy resources on a global scale is increasing exponentially. Thus, during the last 15 years world usage of energy has increased by 5% each year. Statistical data show that the gross national product of a country is proportional to its use of energy. In the U.S.A. the present output of various energy sources is up to 10 kW per head, i.e. at least 100 times greater than the average muscular power of one person. Hence, the basic function of man now consists in the management of energetics; in other words he presses buttons. For this purpose man only needs to know when and which button he has to press – a mental task, involving no appreciable physical work. It follows that the basic qualities of man required for present civilization are mental and spiritual.

If we start to deprive people of energy resources, civilization will be reversed and we shall return to the level of civilization prevalent in the Middle Ages. Hence, if the possibility were realized, that energy resources will decrease during the centuries to come, this would not only bring a stop to the growth of civilization but also bring about its degradation. The reason for such a crisis is very simple. At present, 90% of the energy used by people is created in the form of heat, derived from burning fuel, i.e. petrol, coal and gas. This fuel has accumulated in our planet thanks to biological processes over the course of many thousands of years.

However economically used, this fuel will eventually be exhausted and a crisis will then ensue. Since it is impossible to determine precisely the extent of world fuel resources, and to foresee the growth of energy consumption, the date of the crisis cannot be estimated reliably, but it is generally recognized that this will happen during the next century or two. The possibility of solving this crisis by exploiting inexhaustible resources of energy such as the sun, geothermal heat, water flow and many others has been studied in detail and there is extensive literature on the subject. But an objective assessment of the situation leads one to the realization that the only way to overcome this crisis and to ensure a supply of energy on a sufficiently large scale is to use nuclear energy.

The existence of nuclear energy was discovered only in this century; 40 years ago this possibility of solving the energy crisis was still unknown. The present situation would be catastrophic were it not for the discovery of nuclear energy. In the past 40 years nuclear physics has been studied so well that it can be scientifically established that the global energy crisis can be averted by its use.

Here, I shall limit myself by concentrating on those scientific and social problems which are still to be solved in order that nuclear energy can be used on a scale large enough to replace the great amount of power now obtained from organically derived fuel.

Two ways are now known by which it is possible to obtain sufficient nuclear energy. One is that when the atoms of one of the heaviest elements, uranium, disintegrate when bombarded by neutrons, under certain conditions a chain reaction can be started and energy will be released. These processes have been well studied and are used in the atom bomb. By slowing down the process it can be used for continuous production of energy and is now successfully employed by electric power stations with a power of about a million kilowatts. It appears that natural resources of uranium will suffice for thousands of years if it can be fully used in reactors (called breeders) which operate on fast neutrons. However, some difficulties which are not yet surmounted still hamper the safe exploitation of uranium as a source of energy. These are basically related to the fact that nuclear processes take place at a very high energy level, corresponding to a temperature of hundreds of millions of degrees centigrade, an order of magnitude many times greater than that which occurs in processes commonly taking place on Earth.

Nuclear processes give rise to various types of radiation which cause materials to deteriorate and are a very great danger to people. The action of such radiation on people is well known from the explosions of atom bombs in Hiroshima and Nagasaki, where tens of thousands of people perished immediately and as many again died in the course of several years from diseases caused by radiation. Modern reactors of a large electric power station contain a concentration of radioactivity far greater than that which existed in the bombs dropped on Hiroshima and Nagasaki. Hence, should this radioactivity escape accidentally it could cause catastrophic consequences of an even greater order. This is why the design of reactors in which nuclear processes are carried out requires a carefully worked out technique of safety precautions. Even in ordinary reactors this proves to be a difficult technical problem, but to ensure the safety of breeder reactors, essential for the full usage of uranium, it appears to be even more difficult.

After uranium has been used there remain the highly radioactive slag residues which after removal from the reactor retain their radioactivity for several hundreds of years. To date no universally accepted method for the safe burial of these slag residues has been found. This is due to the fact that containers made of any material whatever may lose their hermetic properties after long-term exposure to radioactivity.

The problem of a safe exploitation of nuclear electric power stations has now become more acute in connection with the recent major accident at

Browns Ferry, the nuclear power station in California. An analysis of the causes of the accident has shown that when uranium fuel is used present safety measures are not adequate. The difficulty of guaranteeing safe work with uranium reactors is also demonstrated and confirmed by the fact that in spite of comprehensive scientific studies made on all nuclear processes, no universally accepted design of reactors has yet been devised that is recognized as fully ensuring their safe usage. The need for solving the problem of the safe use of uranium, if it is to be the basic source of energy in the future, is now fully appreciated. The issue is being widely discussed at the present time and even legislative decisions are being adopted: there are no scientifically based reasons for assuming that, with time, a solution will not be found to these problems.

There is another difficulty connected with the use of uranium as a nuclear fuel, namely that the plutonium which is formed in reactors may be used for atomic bombs. Wide distribution of nuclear power stations would lead to wide distribution of plutonium and it would be very difficult to enforce a ban on its use in the making of atomic bombs. Indeed, there are already countries which have made atom bombs from plutonium derived from reactors which were built for peaceful purposes. This is another reason why the present method of controlling reactors which use uranium fuel cannot be considered adequate, and this question is also now being widely discussed.

These problems all point to the importance of realizing the alternative methods of deriving nuclear energy by means of thermonuclear reactions where the nuclei of light atoms combine and large quantities of energy are liberated. The heavier elements which thereby emerge are not radioactive so that the thermonuclear reactor does not directly constitute a danger to people; neither can it be used for an atomic bomb. The isotope of hydrogen, deuterium, is an effective fuel and there is enough of it in water so that there would be no shortage of fuel. The possibility of realizing such a reaction and its high energy yield is demonstrated experimentally in the hydrogen bomb, where its inception is achieved by the explosion of a uranium bomb.

However, to achieve this thermonuclear reaction independently so as to ensure a useful source of energy appears to be a difficult scientific task which has not yet been solved, in spite of the constant scientific research efforts in this direction conducted in a number of countries for over 20 years. I personally think that since from a scientific point of view no insurmountable hurdles exist, in time a solution will be found. The necessary conditions for a controllable thermonuclear synthesis are already well established. In outline the solution is as follows: to obtain a continuous thermonuclear reaction we must heat the plasma to a temperature of several hundred million degrees C. During this process there is a great loss of heat which will impede the temperature increase. There are two ways of preventing the loss of heat and of achieving the necessary high temperature. The first is to place the plasma in a strong magnetic field and then we can presume that the heat losses are reduced as the fourth power of the intensity of the necessary magnetic field. The magnetic field needed is 50–60 kG, a field attainable with present-day superconducting solenoids. The second way is the reduction of relative heat loss which takes place in large plasma masses, the produced heat being proportional to the volume, i.e. the cube of the size, and the heat losses to the surface area, i.e. the square of the size. This leads to the necessity of conducting experiments on a big and expensive scale. For instance, the cost of an essential superconducting solenoid of a toroid type is estimated at over a hundred million dollars.

The risk coupled with these expenses is at the moment acting as a deterrent in developing these scientific projects on the required scale.

MINERAL RESOURCES

The third global problem is the exhaustion of mineral resources. Unlike the first two – the problem of controlling world population and the limitation of energy resources – this problem is not in the nature of a crisis. It is simply the depletion of richer sources of minerals and the need to use the poorer ones. In a number of cases, e.g. copper mining, this is already taking place. It necessitates the input of larger amounts of labour and energy, resulting in rising costs of production. This usually leads to the development of a cyclic technology which is more economical as it leaves no wastage. These cyclic processes exist in nature, and this is why life has existed on Earth for millions of years without signs of the exhaustion of minerals.

Agriculture is in a precarious position at the moment. The rich harvests now achieved, 'the green revolution', can be realized only with the wide use of mineral fertilizers. Only 30–40% of mineral fertilizers used in present-day agricultural technology are assimilated by plants, and the rest is often carried away to rivers and lakes where they cause unwelcome pollution. Thus, the natural cyclic process of soil elements in plants is absent, which inevitably leads to the irreplaceable depletion of certain kinds of minerals, e.g. phosphorus, which are an essential part of mineral fertilizers. We are faced with the scientific problem of developing present agricultural technology to achieve a fuller use of mineral fertilizers and the building up of agriculture based on cyclic processes. The transition to cyclic processes in food production for people could be achieved if the food, instead of being produced by nature, were artificially synthesized. Especially in the case of protein it would solve a number of important social and economic problems. For instance, there would be no famine caused by poor harvests, there would be a reduction of gastric diseases, etc. Synthesis of food is a major scientific problem which has scarcely been considered yet and to which the greatest attention should be paid.

To summarize, we see that a scientific approach to the solution of global problems, which is linked with the problem of the depletion of mineral resources, is also linked with the problem of ensuring adequate energy resources, since the use of poorer sources of minerals and the development of cyclic processes and synthetic foods will necessitate an increase of consumed energy. Hence, if energy resources do not grow, there can be no solution to these other problems.

POLLUTION OF THE ENVIRONMENT

The fourth global problem is the pollution of the environment. Of all global problems it is now the most widely discussed, since people are immediately concerned with the pollution of rivers, seas, oceans and air with waste materials of present-day civilization, and the struggle against pollution has now begun everywhere. In a number of cases this struggle has led to the development of the technology of cyclic processes and has presented science with many other problems. The first of these new problems, which we shall now discuss, is related to ecology. Some kinds of industrial wastage, once they enter water, the air or the soil, disturb the biological equilibrium and do harm causing the destruction of certain species of fauna. Would it not be possible to create a new kind of equilibrium in which the wastage would be useful and could be recycled? On a smaller scale biofilters solve this problem but it would be more profitable to solve them on a large scale and to create scientifically biological processes in lakes and rivers, by using mineral fertilizers, and to prevent them from perniciously destroying the existing equilibrium. It is interesting to note that in certain cases new biological equilibriums have already arisen through natural selection. This shows us that ecologists must not only study existing processes but must find new and useful ones which will eventually arise from the existing pollution of our environment.

Another global problem of great importance, now under general consideration, is the pollution of the atmosphere, which can change the climate of the whole world. It is not yet widely realized that the average temperature of the atmosphere of the whole globe is rather unstable. The physical reason for this is well known; the heat which ensures the climatic conditions on Earth is radiated from the Sun. The same amount of energy is released into outer space from the Earth's surface in the form of heat radiation. There must exist an equilibrium between the energy which comes from the solar radiation and the heat radiation from the Earth's surface. With the radiative properties of the Earth the surface must have an absolute average temperature of about 300 K. This means that if either the energy produced from the Sun or the radiative capacity of the Earth's surface changes by only a small percentage, the average temperature of the Earth may be substantially changed. If it cools, an ice age can develop; if it warms, Antarctic and Arctic glaciers will melt and the ocean levels will rise and flood a major part of the European continent.

Climate fluctuations have already occurred on Earth and led to ice ages, confirming the existence of this precarious balance. Contemporary research has shown that ice ages can be traced back millions of years and they have occurred not less than seven times. It is quite possible that the climate of the Earth is changing now, because of present global pollution of the atmosphere, chiefly by automobiles and air transport. The ozone content in the upper layers of the atmosphere may be changing and affecting the transparency of the air, thereby altering radiation from the Earth's surface, which depends upon an infinitesimal admixture in the atmosphere content.The temperature of the Earth's surface will then change. This process is known as the 'greenhouse phenomenon', and although the possibility of its existence is well founded, no quantitative value has been agreed upon, i.e. it is even not known whether the average temperature of the Earth is rising or falling.

The study of radiation from the whole surface of the Earth is a complex scientific problem, which requires considerable organization and can only be solved on a multinational scale. This has not yet been done but there is no special urgency. The inherent heat of a melting glacier is so great that it takes a very long time for changes in climatic processes to take place, so that we have plenty of time to study these processes scientifically.

It is often pointed out that the transition to a nuclear system of energy production could cause a global rise of temperature and produce a so-called 'heat pollution'. Calculations show that this process, when compared to the 'greenhouse phenomenon' caused by pollution of the atmosphere, would be far less significant and would for the moment not threaten to change the global climate of our Earth.

SUMMARY

To summarize our analysis, global problems create new problems for science. The most important of these are:

(1) The study of the influence of the structure of society on the change in the quality of world population.

(2) The safe use of nuclear processes as a basic energy source of the future and, mainly, the creation of a controlled thermonuclear synthesis.

(3) The creation of closed cycles, especially in agricultural technology.

(4) The study of the heat balance of the Earth in connection with pollution of the environment.

We must, however, remember that the time for solving all these problems is limited, possibly to one century, and great intellectual ability and material means are required to solve them in time. To achieve this it is evident that the research projects must be developed on an international scale. An authoritative international organization must be set up which would ensure the effective application of any results achieved. The setting up of such an organization is a major social problem in itself, and beyond the scope of my present communication, but, nevertheless I should like to make brief mention of some of the most important difficulties which arise.

It is evident that the successful solution on a global scale of almost all of the problems mentioned above will necessitate certain responsibilities being placed on governments, which in many instances will be in contradiction to the selfish interests of their state. They will therefore by fulfilled only under external pressure.

The future of civilization depends on whether existing governments are able to provide solutions to global problems. We have already remarked that the part played by a state in solving global problems is determined by the social structure and ideology which lie at its basis. For instance, since this would require a firm control on the economy of the country there are now prominent economists who insist that the problem can be solved only by states with a socialist structure.

What else could ensure the solution of global problems? One can only say that modern science is sufficiently advanced to help find a speedier solution. To organize the implementation of solutions is a major international problem in itself and can only be realized by reason and by the ability of the leading powers to assess correctly the catastrophic consequences to the whole of mankind which will ensue if these problems are not solved in time.

One of the most pressing problems at this time is to carry out essential scientific work which, as has been indicated, must be completed in time, before the crisis. Will there be enough time? This will depend on how projects are organized, and to ensure their effectiveness the best intellectual resources, on an international scale, should be directed to them. They must also be assured of adequate material means; this is not the case at present.

With present productivity, not more than one third of the population in developed countries is engaged in the production of the vital essentials of the country. A great part of the surplus production forces are spent on so-called defence costs and the maintenance of armies which do not take part in productive work. In fact in all countries these expenses by far exceed the amount spent on scientific work. It would seem wiser to divert at least a part of the money spent on armaments and, in particular, a part of those intellectual resources which are put into the development of armaments to the organization of scientific research for the solution of global problems.

This would not affect national budgets or lead to unemployment, since jobs would not be taken away but just directed towards other tasks.

At the present time, when most of the international differences dominate international cooperation, there is no sign of such measures being put into practice. The only hope is that in the end, reason will triumph. But, for this, problems must be expressed clearly and convincingly and widely discussed. This can be done mainly by scientists, since they can talk with sufficient authority on the possible solution of global problems for the benefit of the whole of mankind. Thus, we should not stand aside from the solution of such problems but realize their connection with our scientific work.

An example of the adoption of such behaviour by a scientist was given by Bernal in his day. He was an outstanding scientist and in his scientific investigations and public appearances he invariably underlined the connection between science and social problems. For this, Bernal received universal acknowledgement and gratitude.

44. THE IMPACT OF MODERN SCIENTIFIC IDEAS ON SOCIETY

The theme of our discussion is interesting in that it concerns the scientific basis of the structure and working of our society. Today, I am attracted most of all by global problems whose practical solution is directly relevant to the social structure of society, and I propose to speak about the role of science in this connection.

Science's leading role in our civilization is now, of course, universally recognized. Science has even been described – and probably rightly so – as a productive force. History has invariably shown that practically every major scientific discovery or theory has an effect on the development of our civilization.

The following examples illustrate this particularly clearly. Although seemingly limited in scope and not leading initially to any major results, the discoveries concerning electricity made over the past two centuries by Franklin, Galvani, Oersted and Faraday, and the theoretical description of those discoveries made by Maxwell, have led to today's electrical technology, on which the everyday operation and industrial production of our modern society are largely based.

The role of science can be seen no less clearly when we consider radioactivity, discovered by Becquerel in 1896. His discovery was at first regarded as a curious but fairly insignificant phenomenon of nature. Research carried out by Curie and Rutherford showed that this phenomenon was of fundamental importance and was connected with the processes that take place in the nuclei of atoms. Less than a hundred years have passed since these discoveries, yet they have already given man his mightiest source of energy – to which we now look for a solution to the universal crisis caused by the depletion of energy resources. Nuclear energy, moreover, has also put into man's hands a weapon of such destructive force that fear of its possible use has compelled states to revise radically their attitude to armed conflicts.

The link between scientific discoveries and their practical applications is rather unpredictable and unexpected, as is well illustrated by a single,

---

A lecture to the UNESCO symposium on the Centenary of Albert Einstein, held at Munich and Ulm, Federal Republic of Germany, September 18–20, 1978.

remarkable example of Einstein's endeavours. I am referring to his work on induced radiation, published in 1916.[1] I think I am right in saying that of all Einstein's major works this publication attracted least notice, and yet its practical value is now unquestionable.

The modern laser, which today plays an important part in both science and various practical fields, is based – as we all know – on the phenomenon of stimulated emission; its basic theory was provided by Einstein as early as 1916 in the above-mentioned work. Scientific experimental technique was sufficiently advanced at that time for the laser to have been built then, and yet it was not developed until the 1960's. The examples I have given thus show that science advances practice only when there exists a close interrelationship between theory and experiment. The separation of theory from experience is what causes the time lag in the introduction of a scientific discovery into practical life.

Speaking about the role of science, it occurs to me that perhaps I should clarify more precisely what science really is, for it seems to me that these days matters that in no way constitute science are often called science to give them the proper image.

The concept of science goes back to ancient Greece, but its modern meaning emerged only in the 16th century. I believe that, in broad outline, the meaning of the term 'science' can be defined as follows:

It is well-known that human beings, unlike animals, shape their own well-being, remoulding nature rather than adapting to it as the rest of the animal world does. As this has always been done collectively it has given rise to the state. At the basis of evolution, guided by 'the wisdom of nature', lies the method of trial-and-error, in which experiences that prove to be in keeping with the requirements of the species continue to develop. This is what is meant by the law of natural selection. It is how our natural environment was formed and how man was formed – but it took millions of years to form man.

Man also began to use the trial-and-error method in reshaping nature. But the essential factor to ensure the effectiveness of this process consists in not repeating mistakes and in forming generalized theories from empirically discovered experience.

Thus the process of social succession for man came into being – a process that was able to operate effectively given the opportunity for wide

---

[1] 'Strahlungs-Emission und -Absorption nach der Quantentheorie', *Verhandl. Dtsch. Phys. Ges.*, **18**, 318–323 (1916).

dissemination and preservation of the experience transmitted from generation to generation. At first, this was accomplished through the establishment of traditions; it was assisted to a significant degree by the rites that were developed by religion. The written word certainly contributed greatly to the preservation of accumulated experience and the wider dissemination of profitable experience. The process of social succession that had been formed by experimental trial-and-error began to exert a stronger influence on the development of civilization once it had acquired the form we now call *science*. Maybe it is no accident that we trace the origin of modern science to the time when printing was widely introduced.

Religion at first played a progressive role, summing up elements of acquired experience, but – unlike science – it lacked objectivity in its analysis of positive tests and did not evolve because of its dogmatic approach .

With the trial-and-error method, acquired experience becomes scientific when it is interpreted in accordance with the law of causality, i.e. that a given cause always produces a given effect and that every problem therefore admits of only one solution. This is what constitutes the fundamental characteristic of scientific analysis: its objectivity renders it universal, and therein lies the essential difference between it and religion.

Hence the only interpretation of empirical facts that may be considered scientific is one that is objective and universally recognized.

It is a well-known fact that religion is capable of a blithe disregard for the laws of causality and consequently provides answers to problems that can have no scientific solution, such as the creation of the world, free will, the existence of a divine force and so on. This is why it is possible for many different religions to exist, whereas science – like the multiplication table – is unique.

Science began to acquire an influence as an independent field in the organization of society at the time of the Renaissance. The clearest statement of the nature of scientific analysis and of its significance at that time was provided by Francis Bacon, who held that empirical data obtained by observation for use in science are analysed by means of the logical methods of induction and deduction. The role of dialectics in the development of science was demonstrated later, beginning with Hegel and Kant. Bacon gave a most picturesque illustration of the importance of scientific cognition of nature as the most effective way of solving practical problems: "A lame cripple going along the right road can overtake a trotter if the latter is running along the wrong road. Moreover, the faster the trotter runs, once having lost the path, the further he lags behind the cripple". Bacon prophetically described the social significance of science in his *New Atlantis*, in which he painted a Utopian picture of a state structure organized on a scientific basis.

It was then that differences between religion and science became acutely apparent in the clash between the teaching of the Church and the scientific concepts of Copernicus and Galileo on the question of the structure of the universe.

The reason for these conflicting descriptions is now perfectly clear: it lay in the fact that on one and the same question – for example, the description of the world – science came up with an answer that differed from the mythological picture accepted by religion. The scientific explanation was based exclusively on the objective laws of mechanics as established by Galileo and subjected to theoretical analysis by Newton. The universe as described by Copernicus did not correspond to the picture presented in the Bible and accepted by the Catholic Church. Such defiance undermined the authority of the church, on which the social structure of the time was based and which ensured the stability of the foundations on which the ruling power rested. This opposition between science and religion not only retarded the development of science but frequently took a tragic turn that cost a scientist his life – as in the case of Giordano Bruno, who perished at the stake.

The opposition between science and religion has continued to this day. Naturally, it does not take such an acute form as it did with Galileo and Copernicus but, as recently as the last century, it reached a heated pitch when Darwin proclaimed his theory of the origins of animal species as having evolved by way of natural selection. Nor did he hesitate to extend this theory to include the origin of man, despite the fact that religion held man to have been created by God. The arguments between science and religion on this matter took on no less a proportion than they had on the question of the universe; the backward influence of religion cost some scientists their jobs, although this time there were no human sacrifices. In time, these differences began to assume a more peaceful form, leading to the division of world outlooks into the materialist and the idealist.

Attempts are now being made to resolve these differences, chiefly on the grounds that the social function of religion is today no longer based on the questions around which such arguments arise. The hindrance that religion has imposed on science for over 300 years is now coming to an end.

In addition to the teaching of theology in universities, leading scientists have been able to pass on to youth their experience in the realm of the natural sciences, too. The number of universities has started to increase rapidly, and in almost all the countries of Europe academies of sciences have sprung up and have engaged in scientific co-operation. Postal links have developed, and printing has provided for international collaboration among scientists. The first scientific journal appeared in 1650 and, according to research carried out by the scientific historian, D. J. de Solla Price,[2] from that date to the present day the number of scientific journals has been growing exponentially – doubling every 10–15 years and now nearing 100,000.

In the development of scientific disciplines a certain progression has been discernible: in Bacon's time development was mainly in physics, mathematics, mechanics, chemistry and other natural sciences; biology began to develop somewhat later.

In the last century, the development of technology and industry gave rise to new directions in science, which today we call its applied aspects. These were particularly necessary for the introduction of electricity to industry and the application of electricity to everyday life. It was also the beginning of the development of applied disciplines such as construction engineering, strength of materials, hydraulics, and many others. The applied sciences, although firmly based on fundamental sciences such as mathematics, physics, chemistry and mechanics, exist in their own right since their content is determined by the branch of industry or technology they serve.

Whereas up to the 18th century, the higher educational establishments – universities in most cases – developed the fundamental sciences (or pure sciences, as they were then called), at the beginning of the 19th century a new type of educational establishment, known as the polytechnic, was set up to train engineers in the applied sciences.

The Germans were the first to introduce, on a wide scale, specialized higher educational establishments for engineering, which of course explains the high level of technology – especially electrical technology – achieved in Germany towards the end of the last century and the beginning of this. It was an age that produced world-famous scientific engineers such as Siemens, Arnold, Walker, Steinmetz, Stodola, Tesla and Loewi. It is interesting to note that these technical educational establishments were of such a high calibre that many of their graduates went on to become great scientists. Einstein, for instance, graduated from the Federal Polytechnic Institute in Zürich (ETH), while such eminent scientists as Dirac, Langevin, Debye, Ioffe, Lebedev, Poincaré, Cockcroft and many others also studied at technical engineering institutes.

In this century, the scientific method has spread to a new field – that of organizing industrial production and management. In the United States, it has

---

[2] D. J. de Solla Price, *Little Science, Big Science.*

been developed to the highest degree, mainly through the introduction of standardization and through the mass production methods devised by Henry Ford. It was also in the United States that Taylor developed his scientific method for studying the actual process of production, thus giving rise to the field of applied science known as management theory. It now relies heavily on the use of computers, which serve to establish a functional relationship between a multitude of factors governing the efficiency of production processes. The scientific approach to production processes that has been developed in the United States no doubt explains the high and as yet unsurpassed level of productivity reached there. This new field of applied sciences is now being widely used in both capitalist and socialist countries.

There is one important sphere, however, in which science has still great difficulty in influencing development – that of the social sciences, which study the laws governing the functioning of the state. The practical role of these sciences is to accomplish the effective organization of a country's national economy. One would think that, if it is possible to establish a science for organizing production at factory level, it should also be possible to do so on a national scale. This area of social science is generally termed political economy. Although it has long been in existence, for a long time it could not, from our viewpoint, be regarded as a science because it did not possess the necessary objectivity. Economists were like doctors telling a patient, on the strength of their empirical experience, what treatment to follow – but often not understanding the mechanism that had caused the disease. Economists, in like manner, would give advice on how to overcome difficulties usually without any knowledge of the natural scientific laws creating those difficulties.

The first to achieve a scientific approach to economics was Karl Marx. His role may be compared to that of Newton, who, as everyone knows, broadened the concept of force in mechanics by introducing the force of inertia and thus discovered, from the condition of equilibrium, the fundamental law governing the motion of material bodies. At the basis of economics, Marx postulated, are processes of the movement of capital and the manifestation of the social processes that produce such movement. In this connection Marx extended the concept of 'capital', defining its size not in terms of accumulated cash but in terms of everything that constitutes the true wealth of a country or a person.

The rate of growth of capital is determined by laws that are constant under all social structures. These laws, discovered by Marx, are – like Newton's mechanical law – entirely objective and therefore scientific. Marx investigated these natural laws by studying the economics of capitalism. The fundamental law he discovered leads to the conclusion that when production is based on hired labour, capital growth is determined by the profit the owner of production receives. Marx demonstrated that in such a case the dynamics of the capital growth process was unstable because of the spontaneous nature of capitalist economy. One of the principal reasons for instability of the capitalist economy lies in the fact that the profit goes to the capitalists; there is therefore nothing to prevent capital from accumulating in the hands of the employers, which inevitably leads to impoverishment of the workers. In the final analysis Marx believed that in the industrialized countries this would result in impoverishment of the masses and thus in a situation that would be resolved by revolution. The spontaneous economy of capitalism would then be replaced by a planned national economy like those due to come into being under socialism. As history has shown, this did not happen. The reason is that, although Marx's scientific construction was right, it was based on the rate of development in his time – in the last century. As a result of the scientific and technological revolution, this rate began to increase rapidly at the beginning of the present century. It is a known fact that the rate of capital growth is determined by the productivity of labour, which is in turn almost wholly a function of the amount of energy available to the worker. In Marx's time this energy supply was small and consisted almost entirely in the worker's own muscular strength; today the position has altered substantially. In the developed countries, physical labour accounts for less than 1 per cent of the total energy expended on production. The result of this has been that the growth of the total capital of countries – the 'gross national product', as it is now called – has become so great that the impoverishment of the proletariat is not taking place. The 'affluent society' – as it is now referred to – has come into being.

Marx pointed out yet another factor that was bound to produce unsteady economic growth and was also connected with the unplanned nature of capitalism. He demonstrated that, under capitalism, the capital in a country increases unevenly and that from time to time there must be crises that cause industrial depression and unemployment. The mechanism of such crises is explained by the fact that the growth of any branch of industry is determined by capital investment. Under capitalism, the volume of such investments is governed by capital returns, which are in turn determined by demand. When the market becomes saturated in a given branch of industry, profits fall and there has to be a reduction in capital growth in that branch in order to restore the balance. Hence, there exists a reciprocal relationship between profits and capital investment. A time lag naturally occurs before equilibrium is re-established, and the technical term for this is 'relaxation'. The process leads to oscillatory fluctuations that may be used to described periodic economic crises. Similar processes in mechanics have been thoroughly studied, and the way in which they can lead to auto-oscillation has been demonstrated. The duration and intensity of such oscillation are determined essentially by the relaxation time, which is in its turn determined by the effectiveness of the feed-back. It is possible to dampen these oscillations by increasing the effectiveness of the feed-back; this is being done in planned economies and is practised in socialist countries. This is confirmed by the fact that, in the existing socialist countries, where there is planned control of capital investment, counteraction has been so greatly improved that, although a certain degree of instability still remains, it is many times smaller than in the capitalist countries.

Under the capitalist system, the growth in production brought about at the beginning of this century by the scientific and technological revolution was accompanied by a rapid increase in the amplitude in the processes of oscillation that took place in capital investment; finally, the crisis of 1929 reached such proportions that it turned into economic catastrophe.

It would seem logical that, in order to overcome the crisis, measures should have been taken, in accordance with Marx' scientific analyses, to institute a planned economy – but this was not done.

It is instructive to recall the way in which attempts were made to deal with the advent of crises of the dimensions of the one that occurred in 1929. A method of tackling them was proposed, as we know, by the British economist, John Maynard Keynes. An exceptionally talented and widely educated scientist, Keynes began his scientific career as a mathematician, working in the field of probability theory. Later he worked as a consultant to insurance companies; after World War I he demonstrated the unsoundness of the Versailles peace treaty and then began to concern himself with economics on a national scale. Keynes knew and valued the work of Marx but, being a realist, he came to the conclusion that, so long as capital was in private hands, a direct transition to socialism as a means of fighting the crisis was not a practical proposition. According to Marx, of course, the effort should be directed against the uncontrolled growth of capital investment. It was therefore necessary, as far as was practically possible, to limit freedom of investment and step up state control over capital. This could be achieved in one of two ways. The first was to increase taxes, thereby placing part of capital growth under the control of the state. Since taxation had always existed in every country, what had to be done was to increase it unobtrusively. The second way proposed by Keynes was bolder and more original. It consisted in running the national budget, contrary to all the accepted rules of financial management, actually at a deficit, which would naturally lead to inflation. The reasoning behind this proposal was that capital depreciates when it is lying idle – which, it goes without saying, is unprofitable. Inflation, therefore, would tend to stimulate capital investment; the counter-effect would be improved; and that would lead to the development of new lines of industry. And it is indeed a fact that moderate permanent inflation, as suggested by Keynes, has had the effect of mitigating the crises.

This recipe of Keynes was readily applied, and the capitalist economy became considerably more stable for a period of 40 years. Crises still occurred, but they were of acceptable dimensions.

Three years ago that recipe suddenly stopped working. In all capitalist countries inflation began to shoot upwards until it reached such a pitch that it became almost impossible for the national economy to develop and function normally. Growth of the national product dwindled; unemployment became permanent, the major currency exchange rates became unstable. A series of attempts to battle with the crisis during the past three years have yielded no effective method, and the crisis is beginning to take on a chronic character.

And yet it can be asserted with reasonable accuracy that world crises originate according to the laws discovered by Marx as applying to the capitalist economy of a single country. It is certainly not hard to see that great changes have taken place in the world economy over the past 40 years. Certain branches of industry – together with the capital invested in them – began to be shared by a few highly developed countries. That community of interests led to a specialization of industry in certain states necessitating the use of the raw materials and manpower resources of other countries. The pooling of capital investments resulted in the formation of multinational companies.

Although within the framework of a single country it has proved possible over the past 40 years to control the dynamics of capital sufficiently to stabilize the economy by means of taxation and the creation of artificial inflation, on an international scale this method of stabilization becomes useless. The reason is that each state resists interference in its affairs and acts in accordance with its own national interests, which may be contrary to the interests of other countries. Even if the governments of individual states were able to reach agreement on measures for economic development, the free rein given to private capital would normally make it impossible to implement such measures.

The need to coordinate the economies of individual countries on a global scale is now so keenly felt that a certain amount of integration is taking place in developed countries: instances of this are the Common Market and CMEA. Mention should also be made of the serious research work undertaken by the Club of Rome, which has also begun to study the stability of existing national economic processes on a world scale. Its published studies[3] on these problems bear witness (despite the constant criticism to which they are subjected) to the fact that scientific investigations are undoubtedly being conducted on the right lines and are furnishing valuable material for scientific understanding of the present crises.

In time, of course, a way will be found to achieve economic stability on a global scale. According to the natural laws discovered by Marx, all that is needed is to find a way within each country of placing capital – and the returns on capital – under effective state control. Only then will the governments of individual countries be able to reach agreement and begin to implement a concerted economic policy. That international economic stabilization is a practical possibility where there is full control of capital, as there is under socialism, is demonstrated by the existence of CMEA.

Most leading economists already admit that control over the dynamics of capital investment has become essential in view of the global proportions attained by the present-day economy as a result of the high labour productivity brought about by the scientific and technological revolution. Like Keynes, the economists are seeking appropriate measures, but at the same time they are trying as far as possible to preserve the principle of '*laisser faire, laisser passer*' that forms the basis of the capitalist system. Attempts to find a compromise solution are being started; these have given rise, for example, to the theory of convergence – which states that as the present social systems evolve and develop they will in time merge into a single system that will preserve the best features of each and constitute a compromise between socialism and capitalism.

It is interesting to note the biased reaction, inspired by a dread of socialism, that was provoked by Marx' research demonstrating the need to set up a planned economy, despite the fact that his work was purely scientific and therefore objective – so that there was no justification for an emotional

[3] D. Meadows *et al.*, *The Limits to Growth.*
M. Mesarovic, E. Pestel, *Mankind at the Turning Point.*

attitude to it. The natural laws discovered by Marx would seem to have given rise to situations similar to that which arose in the last century, when development of the natural sciences was inhibited because they were undermining the authority of the church. Development of the social sciences is now being similarly inhibited in so far as they frequently lead to conclusions that undermine the authority of the state.

Besides the problem of a stable world economy, it is certain that in the next century, mankind will have to solve a number of global problems connected with the world-wide depletion of energy and raw materials resources and pollution of the environment. All these problems can be solved only by a strictly scientific approach and on an international scale.

Human beings, like gregarious animals, live in closed communities, i.e. states. Although today the social structures of the developed states are similar in many respects – each one having organizations such as a police force, an army, monetary system, a national education system, and so on, that come under the direct management of the state – they nonetheless differ in the nature of their social structure and their degree of cultural development. And it is these last which are the indicators that reflect man's evolutionary development.

The question arises of how to define the progressiveness or backwardness of any of the existing social structures. Which of them is following the road leading to the evolutionary progress of mankind? Can this question be answered on a strictly scientific basis, in such a way that the answer will be objective and unambiguous?

Let us attempt to analyse this question.

Human culture can be divided into the material and the spiritual, each of which is capable of evolving on its own. The material culture of a state is made up essentially of the capital at the disposal of the state. It is now generally considered that economically developed countries are those in which the annual gross national product per capita exceeds approximately $2,000. The total number of people in such countries comprises approximately one third of the population of the world, and it is in those countries, by and large, that the scientific activity of mankind in concentrated. The practical role of science, as can be seen from the foregoing analysis and as put forward long ago by Bacon, consists in organizing human labour in such a way as to insure the most effective growth of material culture.

Spiritual culture, on the other hand, corresponds to human mental activity. Although it cannot be measured quantitatively, the evolutionary growth of man is in fact determined by his culture inasmuch as it gives him mastery over nature. Between the material and spiritial cultures there is a link: spiritual culture points out the direction in which the wealth of the state should grow in order to meet the material needs of society. It is therefore customary to regard material culture as the foundation on which spiritual culture develops – like a kind of superstructure.

Such a construction, however, can scarcely be considered accurate. Material wealth is situated outside man, is progressively expended and has to be renewed. Spiritual wealth, such as science and art, is handed on from generation to generation; it is preserved and can therefore evolve independently. There is no limitation to its growth, whereas the growth of material culture is limited by the number of food calories a person needs, how many items of clothing there are in his wardrobe, and the amount of living space he requires, etc. In the developed countries these limits are now fully within reach.

Unlimited growth of material consumption by one person inhibits his spiritual development and soon becomes harmful (an excess of food, for instance, leads to overweight). People who are engaged in intensive creative work therefore do not usually waste time consuming large quantities of material resources.

Endeavours to push growth of material resources beyond a certain limit produces a situation in which people have to work under greater pressure than their nervous system can normally stand. The organization of labour under conditions of high productivity calls for the tightly regulated management of production, and, so far as the daily routine of the worker is concerned, it also leads to a lack of freedom of choice in the type of work he does. All this puts pressure on a person's morale, and the price he has to pay for a high level of material well-being is the loss of his spiritual contentment. That this is already happening is borne out by statistical indicators. In the richest countries the number of suicides is increasing – a sure sign that there is a growing number of unhappy people. A similar indicator is the fact that, in those countries, the numbers of drug addicts and alcoholics are going up. It is a known fact that the physiological action of narcotics consists in shielding a person's psyche temporarily from the constraining pressures imposed on him by social life. What all this indicates, in the final analysis, is that the social structure of society, in its progressive evolution towards the goal of providing people with the happiest possible existence, should monitor both the size and the quality of the gross national product. This is diametrically opposed to the yardstick applied in the developed countries to a country's level of development, which they gauge by the material prosperity of the population – calculated merely from the size of the per capita gross national product.

In so far as spiritual culture guarantees people the happiest existence, it should serve as the basis, with regard to the evolutionary development of mankind, for evaluating the progressiveness of the social structure of a country.

A person's spiritual life is composed of three elements: his private life, his relations with society (principally the people with whom he lives and works), and his activity as a citizen of the state.

Meeting the needs of a person's private life is a matter that varies with the individual person, depending on his natural abilities and the demands made on him by society through its ethics and traditions. From a scientific viewpoint these processes can be studied by means of psychology, as was done most successfully by Pavlov and Freud – although the practical significance of such research is restricted to its application to curative psychiatry. At the present time there are still no generally accepted, objective laws governing the development of the spiritual culture of man and society. What is important and well-known, however, is that a person's spiritual happiness is associated with the feeling of freedom. People desire freedom in the choice of a husband or wife, friends, religion and occupation, and they want to be free to live peacefully.

The attainment of a high level of material culture in a modern developed society demands such high labour productivity that – as has already been indicated – it creates conditions in the work and life of the individual that are hardly distinguishable from forced labour. This is most clearly evident in the case of assembly-line production, in which freedom of work is totally absent. With the highly organized daily life of people these days, many aspects of life are strictly regimented, and people find themselves constantly under control. They are obliged to obtain an education, to get to work on time and to wear certain clothes – and they cannot even cross the street where they want to.

The state compels a person to abide by its laws, and society compels him to live in accordance with its traditions – and even, in wartime, to kill people. As has been shown, after a certain degree of well-being has been reached people begin to lose their liberty as their prosperity increases, and already the freedom of the individual in the developed countries is now extremely limited. Modern society strives to organize people's lives in such a way that they will have an *impression* of freedom despite the lack of it. This is achieved by means of propaganda and by turning people's interest towards sport, sex and entertainment so as to distract them from reality. Such were the principles on which society was organized in Aldous Huxley's anti-Utopian novel, *Brave New World*.

The civilization of a state is determined not only by the social and economic processes that take place within its frontiers but also by the international relations that influence the development of culture. We have already spoken about economic relations, but no less a role is played by culture. These latter relations are of an emotional nature and may take different forms. They may, for example, create ideological, national, racial or religious disagreements between countries, leading to conflicts and wars. However, in the fields of science, art, sport and so on such relations can promote cultural growth.

The processes associated with the development of human culture are studied by history. From our point of view, history cannot be considered a science, because it is not based on the discovery of objective natural laws. Historical processes such as the establishment of power, class conflicts, expansion and so forth are always associated with emotional activity on the part not only of individual men but of entire communities. As J. Piaget rightly showed, objective study of such processes cannot be achieved, because they do not lend themselves to scientific investigation.

It is possible, nevertheless, to exert an influence on man's emotional activity and organize that activity. It can be done through art. Art has played a major organizing role among all peoples and in every age. It has developed and has been transmitted from generation to generation and, like science, it constitutes a national cultural heritage. To this day music is played at funerals; every people has its national anthem; religion makes wide use of music.

Literature and the figurative arts have a particularly strong impact on the development of spiritual culture, because they exert an influence on the emotions that are involved in the formation of morals and ethics both in personal interrelationships between individuals and in social interrelationships.

There is a certain similarity between the impact of science and the impact of art on the organization of social life. As indicated earlier, these scientific laws that are a generalization of experience are fixed for all time and are internationally accepted.

Similarly, art is a generalization of processes which take place in the lives of people and of society. A truly great work of art that has an effective impact on the growth of spiritual culture cannot be affected by any specific political situation, for it becomes international and everlasting. In literature, the works of Cervantes, Shakespeare and Tolstoy have for hundreds of years exerted a world-wide impact on man. The same applies to figurative art. We have only to think of the paintings and sculptures of Titian, Michelangelo. Goya, Rembrandt . . . Or, in music, the works of Beethoven, Mozart, Chopin, Moussorgsky . . .

Thus the impact of art on the cultural growth of the whole of mankind is at least as great as the impact of science. It was not without reason that Jean Cocteau said: "Poetry is indispensible, although why – I know not".

To try now to answer the question we formulated earlier – which state social structure in the evolutionary development of mankind is the most advanced – I think there is full justification for believing that such an evaluation should be based on the quality of a country's spiritual culture. Since the process of human evolution unfolds by means of competition between different social structures, on the final count the states that survive will be those in which the spiritual culture meets the requirements of mankind's evolutionary development. We may expect, however, that in the process of evolution the law of natural selection will also extend its effects to individual men. The evolutionary development of mankind operates in the selection not only of social structures but also of man himself.

Discovering the natural laws that govern these two processes will probably remain forever beyond the reach of human capability. In this matter, we must place our trust in the 'wisdom of nature', which has unfailingly guided the development of the whole of nature over the course of hundreds of millions of years.

24/X/78.

P.L. KAPITZA – BIBLIOGRAPHY

SCIENTIFIC PAPERS

1. 'Electron inertia in molecular Ampère currents', *J. Russ. Phys. Chem. Soc., Phys. Sect.* **48**, 297 (1916).
2. 'The preparation of Wollaston fibres', *J. Russ. Phys. Chem. Soc., Phys. Sect.* **48**, 324 (1916).
3. 'A method of reflection from crystals', *Bull. Roentgenol. Radiol.* **1**, 33 (1919).
4. 'The Koch recording microphotometer', *Bull. Roentgenol. Radiol.* **1**, 54 (1919).
5. 'The dependence of the emission edge of the continuous X-ray spectrum on the azimuth of the emission and the influence of the anticathode metal', *Usp. Fiz. Nauk.* **2**, 322 (1921).
6. 'On the possibility of an experimental determination of the magnetic moment of an atom', *J. Russ. Phys. Chem. Soc., Phys. Sect.* **50**, 159 (1922); with N.N. Semenov.
7. 'The loss of energy of an α-ray beam in its passage through matter. I. Passage through air and CO_2', *Proc. Roy. Soc.* **A 102**, 48 (1922).
8. 'Note on the curved tracks of β-particles', *Proc. Camb. Phil. Soc.* **21**, 129 (1922).
9. 'On the theory of δ-radiation', *Phil. Mag.* **45**, 989 (1923).
10. 'Some observations on α-particle tracks in a magnetic field', *Proc. Camb. Phil. Soc.* **21**, 511 (1923).
11. 'A method of producing strong magnetic fields', *Proc. Roy. Soc.* **A 105**, 691 (1924).
12. 'α-ray tracks in a strong magnetic field', *Proc. Roy. Soc.* **A 106**, 602 (1924).
13. 'The Zeeman effect in strong magnetic fields', *Nature* **114**, 273 (1924); with H.W.B. Skinner.
14. 'The Zeeman effect in strong magnetic fields', *Proc. Roy. Soc.* **A 109**, 224 (1925); with H.W.B. Skinner.
15. 'Over-tension in a condenser battery during a sudden discharge', *Proc. Camb. Phil. Soc.* **23**, 144 (1926).
16. 'Further developments of the method of obtaining strong magnetic fields', *Proc. Roy. Soc.* **A 115**, 658 (1927).
17. 'The study of the specific resistance of bismuth crystals and its change in strong magnetic fields and some allied problems', *Proc. Roy. Soc.* **A 119**, 358 (1928).
18. 'The change of electrical conductivity in strong magnetic fields', *Proc. Roy. Soc.* **A 123**, 292 (1929).
19. 'Magnetostriction and the phenomena of the Curie point', *Proc. Roy. Soc.* **A 124**, 1 (1929); with R.H. Fowler.
20. 'Metallic conductivity and its change in a magnetic field', *Metallwirtsch.* **19**, 443 (1929).
21. 'A property of superconducting metals', *Nature* **123**, 870 (1929).
22. 'Magnetostriction of diamagnetic substances in strong magnetic fields', *Nature* **124**, 53 (1929).

23. 'The change of resistance of gold crystals at very low temperatures in a magnetic field and supra-conductivity', *Proc. Roy. Soc.* A **126**, 683 (1930); *Physik. Zs.* **31**, 713 (1930).
24. 'Methods of experimenting in strong magnetic fields', *Proc. Phys. Soc.* **42**, 425 (1930).
25. 'Experimental studies in strong magnetic fields', *Usp. Fiz. Nauk* **11**, 533 (1931); Proceedings of the 6th Solvay Congress, Le Magnetisme, Gauthier-Villars, Paris, p. 458.
26. 'The study of the magnetic properties of matter in strong magnetic fields' I and II, *Proc. Roy. Soc.* A **131**, 224, (1931); III-V; *Proc. Roy. Soc.* A **135**, 537 (1932).
27. 'A method of measuring magnetic susceptibilities', *Proc. Roy. Soc.* A **132**, 442 (1931); with W.L. Webster.
28. 'Reply to some remarks by O. Stierstadt about an error in principle in my measurements of the change in resistivity in strong magnetic fields', -*Zs. f. Physik* **69**, 421 (1931).
29. 'Hydrogen liquefaction plant at the Royal society Mond Laboratory', *Nature* **129**, 224 (1932); with J.D. Cockcroft.
30. 'The change of resistance of metals in magnetic fields', Leipziger Vorträge 1933, S. Hirzel, Leipzig, p. 1.
31. 'The reflection of electrons from standing light waves', *Proc. Camb. Phil. Soc.* **29**, 297 (1933); with P.A.M. Dirac.
32. 'Liquefaction of Helium by an adiabatic method without precooling with liquid hydrogen', *Nature* **133**, 708 (1934).
33. 'The liquefaction of helium by an adiabatic method', *Proc. Roy. Soc.* A **147**, 189 (1934); *Usp. Fiz. Nauk* **16**, 145 (1936).
34. 'A modified potentiometer for measuring very small resistance', *J. Sci. Instrum.* **14**, 165 (1937); with C.J. Milner.
35. 'Note on the use of liquid nitrogen in magnetic experiments', *J. Sci. Instrum.* **14**, 201 (1937); with C.J. Milner.
36. 'Viscosity of liquid helium below the λ-point', *Dokl. Akad. Nauk* **18**, 21 (1938); *C.R. Acad. Sci. USSR* **18**, 21 (1938); *Nature* **141**, 74 (1938).
37. 'The Zeeman and Paschen-Back effects in strong magnetic fields', *Proc. Roy. Soc.* A **167**, 1 (1938); *Zh. Eksp. Teor. Fiz.* 8, 276 (1938); with P.G. Strelkov and E. Laurman.
38. 'Expansion turbine producing low temperatures applied to air liquefaction', *J. Phys. USSR* **1**, 7 (1939); *Zh. Tekh. Fiz.* 9, 99 (1939).
39. 'Influence of friction forces on the stability of high-speed rotors', *J. Phys. USSR* **1**, 29 (1939); *Zh. Tekh. Fiz.* 9, 124 (1939).
40. 'Problems of liquid Helium', *Sov. Nauk.* **1**, 33 (1941).
41. 'The study of heat transfer in Helium II', *J. Phys. USSR* **4**, 181 (1941); *Zh. Eksp. Teor. Fiz.* **11**, 1 (1941).
42. 'Heat transfer and superfluidity of Helium II', *J. Phys. USSR* **5**, 59 (1941); *Zh. Eksp. Teor. Fiz.* **11**, 581(1941).
43. 'On the superfluidity of liquid Helium II', *Usp. Fiz. Nauk.* **26** 133 (1944).
44. 'On Professor S.Ya. Gersh's paper "Low and high pressures in deep-cooling systems" ', *Bull. Mech. Engn.* **7/8**, 43 (1944).

45. 'Theoretical and empirical expressions for the heat transfer in a two-dimensional turbulent flow', *Dokl. Akad. Nauk.* 55, 595 (1947).
46. 'Wave flow of thin layers of a viscous fluid' I and II, *Zh. Eksp. Teor. Fiz.* **18**, 319 (1948); III (with S.P. Kapitza), *Zh. Eksp. Teor. Fiz.* **19**, 105 (1949).
47. 'On the problem of the formation of sea waves by the wind', *Dokl. Akad. Nauk* **64**, 513 (1949).
48. 'Dynamical stability of a pendulum when its point of suspension vibrates', *Zh. Eksp. Teor. Fiz.* **21**, 588 (1951).
49. 'Pendulum with a vibrating suspension', *Usp. Fiz. Nauk* **44**, 7 (1951).
50. 'Evaluations of the sum of negative even powers of roots of Bessel functions', *Dokl. Akad. Nauk* 77, 561 (1951).
51. 'Heat conductivity and diffusion in a liquid medium under periodic flow conditions', *Zh. Eksp. Teor. Fiz.* **21**, 964 (1951).
52. 'The hydrodynamic theory of lubrication in the presence of rolling' , *Zh. Tekh. Fiz.* **25**, 747 (1955).
53. 'On the nature of ball-lightning', *Dokl. Akad. Nauk* **101**, 245 (1955).
54. 'Design of a helium-liquefying cycle with expansion engines connected in cascade', *Zh. Teor. Fiz.* 29, 427 (1959); *Soviet Phys.-Techn. Phys.* 4, 377 (1959).
55. 'Static boundary problem for a hollow cylinder of finite length', *Zh. Teor. Fiz.* **29**, 1177 (1959); *Soviet Phys.-Techn. Phys.* **4**, 1077 (1960); with V.A. Fok and L.A. Vainshtein.
56. 'Symmetric electric oscillations of a perfectly conducting hollow cylinder of finite length', *Zh. Teor. Fiz.* **29**, 1188 (1959); *Soviet Phys.-Techn. Phys.* **4**, 1088 (1960); with V.A. Fok and L.A. Vainshtein.
57. 'Expansion engine for liquefaction of helium', *Zh. Teor. Fiz.* **31**, 486 (1961); *Soviet Phys.-Techn. Phys.* **6**, 349 (1961); with I.B. Danilov.
58. 'Cascade compression Helium liquefier with no external refrigeration', *Zh. Teor. Fiz.* **32**, 457 (1962); *Soviet Phys.-Techn. Phys.* **7**, 333 (1962); with I.B. Danilov.
59. 'High power electronics', from *High-Power Microwave Electronics*. Vol. 1, *Izd. Akad. Nauk SSSR*, 1962, p. 9; English edition; Pergamon Press, Oxford, 1964, p. 3; *Usp. Fiz. Nauk* 78, (1962); *Soviet Phys.-Uspekhi* **5**, 777 (1963).
60. 'Characteristic oscillations of grigged resonators', from *High-Power Microwave Electronics*, Vol. I, *Izd. Akad. Nauk SSSR*, 1962, p. 158; English edition; Pergamon Press, Oxford, 1964, p. 117.
61. 'The theory of electronic processes in a continuous high-power generator of the magnetron type', from *High-Power Microwave Electronics*, Vol. III, *Izd. Akad. Nauk SSSR*, 1964, p. 7; English edition; Pergamon Press, Oxford, 1967, p. 3; with S.I. Filimonov and S.P. Kapitza.
62. 'Transducers of H-waves into E-waves', *High-Power Electronics*, No. 4, Nauka, Moscow, 7-52 (1965).
63. 'Experimental study of a transducer', *High-Power Electronics*, No. 4, 53-65 (1965); with L.A. Prozorova.
64. 'Absolute measurements of a high-frequency field in a resonator', *High-Power Electronics*, No. 4, 206-211 (1965).
65. 'New H-wave transducers', *High-Power Electronics*, No. 5, Nauka, Moscow, 209-214 (1968); with L.A. Prozorova.

66. 'Ball lightning and radioactive emanation of linear lightnings', *Zh. Teor. Fiz.* **38** (II), 1829-1830 (1968).
67. 'Cascade helium liquefiers with piston type engines', The proceedings of the first Cryogenic Engineering Conference held in Tokyo and Kyoto, Japan, 9-13 April 1967, Heywood Temple Industrial Publication Ltd., London 1968, p.p. 228-231 (with I.B. Danilov).
68. 'Solenoid producing a magnetic field up to 30 kOe in a volume of 5 liters', *Usp. Fiz. Nauk* **95**, 35 (1968); *Soviet Phys.-Uspekhi* **11**, 299 (1968); with S.I. Filimonov.
69. 'Two-level nigotron with high continuous power', *High-Power Electronics*, No. 6, Nauka, Moscow, 7-36 (1969); with S.I. Filimonov and S.P. Kapitza.
70. 'Free plasma filament in high frequency field at high pressure', *Zh. Eksp. Teor. Fiz.* **57**, 1801 (1969); *Soviet Phys. – JETP* **30**, 973 (1970).
71. 'A thermonuclear reactor with a plasma filament freely floating in a high frequency field', *Zh. Eksp. Teor. Fiz.* 58, 377 (1970); *Soviet Phys. – JETP* **31**, 199 (1970).
72. 'Apparatus for production of a free plasma filament. Determination of the current and resistance of the filament', *Zh. Eksp. Teor. Fiz.* **61**, 1016 (1971); *Soviet Phys. – JETP* **34**, 542 (1972); with S.I. Filimonov.
73. 'Plasma heating by magnetoacoustic oscillations', *Zh. Eksp. Teor. Fiz.* **67**, 1411 (1974); *Soviet Phys. – JETP* **40**, 701 (1975); with L.P. Pitaevskii.
74. 'Modern developments in studies of magnetism', *Priroda*, N2, 50 (1974).
75. 'Useful energy from thermonuclear reactors', *Pisma v ZETF* **22**, 20 (1975); *JETP Letters* **22**, 9 (1975).
76. 'Energy and physics', *Usp. Fiz. Nauk* 118, 307 (1976); *Soviet Phys. – Uspekhi* **19**, 169 (1976).
77. 'Plasma and the controlled thermonuclear reactions' (Nobel lecture), *Rev. Mod. Phys.* **51**, 417 (1979); *Science* **205**, 959 (1979).

PUBLIC TALKS AND POPULAR PAPERS

1. 'Cod-liver oil', *Argus* **10**, 76 (1913).
2. 'The future of magnetism', *Nature* **119**, 809 (1927).
3. 'The production of and experiments in strong magnetic fields', *Trans. Oxford Univ. Jun. Sc. Club* (5) 4, 129 (1931).
4. 'A new method for the liquefaction of Helium', *Sotsialisticheskaya Rekonstruktsiya i Nauka*, No. 8, 104 (1934).
5. 'A Great Man' (in memoriam of I.P. Pavlov), *Pravda*, 27 February, 1936.
6. The construction and work of the Institute of Physical Problems of the USSR Academy of Sciences', *Izv. Akad. Nauk SSSR, Ser. fiz.* **3**, 278 (1937).
7. 'Lord Rutherford', *Nature* **140**, 1053 (1937).
8. 'Recollections of Professor E. Rutherford', *Usp. Fiz. Nauk* **19**, 1 (1938).
9. 'Liquid air', *Tekhnika Molodezhi* **3**, 28 (1939); *Planovoe Khozyaïstvo* **2**, 73 (1939).
10. 'A new method of producing low temperatures for air liquefaction', *Sov. Nauka* **5**, 150 (1939).
11. 'Science for the people', *Komsomolskaya Pravda*, 1 January, 1940. [Also titled 'For the good of the people']
12. 'On stereoscopic films', *Mashinostroenie*, 12 March, 1940.
13. 'The unity of science and engineering', *Pravda*, 17 June, 1941.

14. 'E. Rutherford', *Bolshaya Sovetskaya Entsiklopediya* **48**, 503 (1941).
15. 'Science and war', *Krasnaya Zvezda*, 26 November, 1942; *Science* **95**, 396 (1942).
16. 'On the 25th anniversary of the establishment of the Soviet State', *J. Phys. USSR.* **6**, 233 (1942).
17. 'We fight for freedom', *Trud*, 28 October, 1942.
18. 'Report on the organization of scientific work at the Institute for Physical Problems of the Academy of Sciences of the USSR', *Vestnik Akad. Nauk* **6**, 75 (1943).
19. 'Problems of intensification of technological progress by oxygen', *Kislorod* **1**, 1 (1944).
20. 'Superfluidity', *Bolshaya Sovetskaya Entsiklopediya* **50**, 423 (1944).
21. 'Isaac Newton', *Izv. Akad. Nauk SSSR, Ser. fiz.* 8, 169 (1944).
22. 'The scientific activity of Benjamin Franklin', *Usp. Fiz. Nauk* **58**, 169 (1956).
23. 'How is atomic war to be prevented?', *Novoe Vremya* **14**, 8 (1956).
24. 'Some aspects of the organization of scientific work', *Pravda*, 4 May, 1957; *World of Science* **1**, 14 (1959).
25. 'The future of science', *World of Science* **4**, No. 3-4, 2, (1960); *Nauka i Zhizn'* **3**, 18 (1962).
26. 'Lomonosov and world science', *Izvestia*, 18 November, 1961; *Zhizn' dlya Nauki*, Znanie, Moscow, 1965, p. 3: *Usp. Fiz. Nauk* 87, 155 (1965).
27. 'Theory, experiment, practice', *Econom. Gazeta*, 26 March, 1962, p. 10.
28. 'In commemoration of Niels Bohr', *Priroda* **1**, 67 (1963).
29. 'The physicist and public figure Paul Langevin', *Zhizn' dlya Nauki*, Znanie, Moscow, 1965, p. 54.
30. 'Effectiveness of Scientific Work', *Vestnik Akad. Nauk SSSR*, **2**, 55 (1965).
31. 'Adopting Achievements of Science and Engineering', *Komsomolskaya Pravda*, 20 January, 1965.
32. 'Recollections of Lord Rutherford', *Proc. Roy. Soc.* **A, 297**, 123 (1966).
33. 'Aleksandr Aleksandrovitch Fridman', from *Selected Works of A.A. Fridman*, Nauka, Moscow, 1966, p. 397.
34. 'Philosophy and the ideological struggle', *Voprosy filosofii* No. 5, 147 (1969).
35. 'Lev Davydovitch Landau', *Biog. Memoirs of the Fellows Roy. Soc.* **15**, 141 (1969).
36. 'Science teaching and scientific method', *Scientific World* **15**, N 1, 12, 197 (1971); *Voprosy filosofii*, N 7, 16 (1971); *The Physics Teacher*, 9, 429 (1971).
37. 'The role of an eminent scientist in the development of science', *Tekhnika molodezhi*, N 1, 14 (1972); *New Scientist* **51**, N 769, 639 (1971).
38. 'Remarks on the anniversary of the Physico-Technical Institute', from *Physics Problems*, Znanie, Moscow, 1972, p. 42.
39. 'The scientific activity of Rutherford', from *Ernest Rutherford Selected Scientific Papers Atomic Structure and Artificial Disintegration of Elements*, Nauka, Moscow, 1972.
40. 'Three aspects of the global problem of relation between man and nature', *Voprosy filosofii*, N 2, 37 (1973); *Scientific World* N 2, 17 (1973).
41. 'Basic factors in the organization of science and how they are handled in the USSR', *Daedalus* **102**, Spring 1973, p. 167-176.
42. 'The centenary of the Journal of experimental and theoretical physics and the role of journals in the development of science', *Usp. Fiz. Nauk.* **111**, 535 (1973); *Soviet Phys. – Uspekhi* **16**, 928 (1974).

43. 'The Professor and the student', *Khimiya i Zhis'n*, N 7, 21 (1974).
44. 'Reading M.M. Prishvin's diary', *Sever*, N 6, 69, 1975; from 'Prishvin and Present Times', Sovremennik, Moscow, 1978, p. 147.
45. 'Our home, the planet Earth', in *Society and Environment*, Progress, Moscow 1976; English edition 1977, p. 9-14.
46. 'Scientific and social approaches for the solution of global problems' (Bernal lecture), *Proc. Roy. Soc.* **A 357**, 1 (1977); *Voprosy filosofii* N 1, 46 (1977).
47. 'Global problems and energy', *Usp. Fiz. Nauk* **122**, 327 (1977); *Soviet Phys. – Uspekhi* **20**, 547 (1977).
48. 'The impact of modern scientific ideas on society', *Voprosy filosofii*, N 1, 61 (1979).

BOOKS

1. *Collected Papers of P.L. Kapitza*, edited by D. ter Haar; Pergamon Press, Oxford; Vol. 1, 1964; Vol. 2, 1965; Vol. 3, 1967.
2. *High-Power Microwave Electronics*, Nauka, Moscow, 1962, 194 pp.; English edition; Pergamon Press, Oxford, 1964, 148 pp.
3. *Life Devoted to Science (Lomonosov, Franklin, Rutherford, Langevin)* Znanie, Moscow, 1965, 62 pp.
4. *Theory, Experiment, Practice*, Znanie, Moscow, 1966, 47 pp.
5. *Physics Problems*, Znanie, Moscow, 1972, 46 pp.
6. *Experiment, Theory, Practice*, Nauka, Moscow, 1974, 288 pp.; 2nd edition, 1977, 350 pp; English translation (edited by R. S. Cohen) with additions; Dordrecht, Boston and London, D. Reidel Publ. Co., 1980.
7. *Science, Mankind, Organization*; in Japanese; Misuzu Shobo, Tokyo, 1974, 273 pp.
8. *Le Livre du Problème de Physique*, Edition CEDIC, Paris, 1977, 70 pp.

INDEX OF NAMES

BOSTON STUDIES IN THE PHILOSOPHY OF SCIENCE

1. Marx W. Wartofsky (ed.), *Proceedings of the Boston Colloquium for the Philosophy of Science 1961-1962.* 1963.
2. Robert S. Cohen and Marx W. Wartofsky (eds.), *In Honor of Philipp Frank.* 1965.
3. Robert S. Cohen and Marx W. Wartofsky (eds.), *Proceedings of the Boston Colloquium for the Philosophy of Science 1964-1966. In Memory of Norwood Russell Hanson.* 1967.
4. Robert S. Cohen and Marx W. Wartofsky (eds.), *Proceedings of the Boston Colloquium for the Philosophy of Science 1966-1968.* 1969.
5. Robert S. Cohen and Marx W. Wartofsky (eds.), *Proceedings of the Boston Colloquium for the Philosophy of Science 1966-1968.* 1969.
6. Robert S. Cohen and Raymond J. Seeger (eds.), *Ernst Mach: Physicist and Philosopher.* 1970.
7. Milic Capek, *Bergson and Modern Physics.* 1971.
8. Roger C. Buck and Robert S. Cohen (eds.), *PSA 1970. In Memory of Rudolf Carnap.* 1971.
9. A. A. Zinov'ev, *Foundations of the Logical Theory of Scientific Knowledge (Complex Logic).* (Revised and enlarged English edition with an appendix by G. A. Smirnov, E. A. Sidorenka, A. M. Fedina, and L. A. Bobrova.) 1973.
10. Ladislav Tondl, *Scientific Procedures.* 1973.
11. R. J. Seeger and Robert S. Cohen (eds.), *Philosophical Foundations of Science.* 1974.
12. Adolf Grünbaum, *Philosophical Problems of Space and Time.* (Second, enlarged edition.) 1973.
13. Robert S. Cohen and Marx W. Wartofsky (eds.), *Logical and Epistemological Studies in Contemporary Physics.* 1973.
14. Robert S. Cohen and Marx W. Wartofsky (eds.), *Methodological and Historical Essays in the Natural and Social Sciences. Proceedings of the Boston Colloquium for the Philosophy of Science 1969-1972.* 1974.
15. Robert S. Cohen, J. J. Stachel and Marx W. Wartofsky (eds.), *For Dirk Struik. Scientific, Historical and Political Essays in Honor of Dirk Struik.* 1974.
16. Norman Geschwind, *Selected Papers on Language and the Brain.* 1974.
18. Peter Mittelstaedt, *Philosophical Problems of Modern Physics.* 1976.
19. Henry Mehlberg, *Time, Causality, and the Quantum Theory*: Vol. 1, *Essays on the Causal Theory of Time.* 1980.
20. Kenneth F. Schaffner and Robert S. Cohen (eds.), *Proceedings of the 1972 Biennial Meeting, Philosophy of Science Association.* 1974.
21. R. S. Cohen and J. J. Stachel (eds.), *Selected Papers of Léon Rosenfeld.* 1978.
22. Milic Capek (ed.), *The Concepts of Space and Time. Their Structure and Their Development.* 1976.

23. Marjorie Grene, *The Understanding of Nature. Essays in the Philosophy of Biology.* 1974.
24. Don Ihde, *Technics and Praxis. A Philosophy of Technology.* 1978.
25. Jaakko Hintikka and Unto Remes, *The Method of Analysis. Its Geometrical Origin and Its General Significance.* 1974.
26. John Emery Murdoch and Edith Dudley Sylla, *The Cultural Context of Medieval Learning.* 1975.
27. Marjorie Grene and Everett Mendelsohn (eds.), *Topics in the Philosophy of Biology.* 1976.
28. Joseph Agassi, *Science in Flux.* 1975.
29. Jerzy J. Wiatr (ed.), *Polish Essays in the Methodology of the Social Sciences.* 1979.
32. R. S. Cohen, C. A. Hooker, A. C. Michalos, and J. W. van Evra (eds.), *PSA 1974: Proceedings of the 1974 Biennial Meeting of the Philosophy of Science Association.* 1976.
33. Gerald Holton and William Blanpied (eds.), *Science and Its Public: The Changing Relationship.* 1976.
34. Mirko Drazen Grmek, Robert S. Cohen and Guido Cimino (eds.), *On Scientific Discovery.* 1980.
35. Stefan Amsterdamski, *Between Experience and Metaphysics. Philosophical Problems of the Evolution of Science.* 1975.
36. Mihailo Marković and Gajo Petrović (eds.), *Praxis. Yugoslav Essays in the Philosophy and Methodology of the Social Sciences.* 1979.
37. Hermann von Helmholtz: *Epistemological Writings. The Paul Hertz/Moritz Schlick Centenary Edition of 1921 with Notes and Commentary by the Editors.* (Newly translated by Malcolm F. Lowe. Edited, with an Introduction and Bibliography, by Robert S. Cohen and Yehuda Elkana.) 1977.
38. R. M. Martin, *Pragmatics, Truth, and Language.* 1979.
39. R. S. Cohen, P. K. Feyerabend, and M. W. Wartofsky (eds.), *Essays in Memory of Imre Lakatos.* 1976.
42. Humberto R. Maturana and Francisco J. Varela, *Autopoiesis and Cognition. The Realization of the Living.* 1979.
43. A. Kasher (ed.), *Language in Focus: Foundations, Methods and Systems. Essays Dedicated to Yehoshua Bar-Hillel.* 1976.
46. Peter Leonidovich Kapitza, *Experiment, Theory, Practice.* 1980.
47. Maria Luisa Dalla Chiara, *Italian Studies in the Philosophy of Science.* 1980.
48. Marx W. Wartofsky, *Models: Representation and the Scientific Understanding.* 1979.
50. Yehuda Fried and Joseph Agassi, *Paranoia: A Study in Diagnosis.* 1976.
51. Kurt H. Wolff, *Surrender and Catch: Experience and Inquiry Today.* 1976.
52. Karel Kosík, *Dialectics of the Concrete.* 1976.
53. Nelson Goodman, *The Structure of Appearance.* (Third edition.) 1977.
54. Herbert A. Simon, *Models of Discovery and Other Topics in the Methods of Science.* 1977.
55. Morris Lazerowitz, *The Language of Philosophy. Freud and Wittgenstein.* 1977.
56. Thomas Nickles, *Scientific Discovery, Logic, and Rationality.* 1980.
57. Joseph Margolis, *Persons and Minds. The Prospects of Nonreductive Materialism.* 1977.

58. Gerard Radnitzky and Gunnar Andersson (eds.), *Progress and Rationality in Science*. 1978.
59. Gerard Radnitzky and Gunnar Andersson (eds.), *The Structure and Development of Science*. 1979.
60. Thomas Nickles, *Scientific Discovery: Case Studies*. 1980.
61. Maurice A. Finocchiaro, *Galileo and the Art of Reasoning*. 1980.